## OCEAN WAVES AND OSCILLATING SYSTEMS

Second Edition

Understand the interaction between ocean waves and oscillating systems with this useful new edition. With a focus on linear analysis of low-amplitude waves, you are provided with a thorough understanding of wave interactions, presented to be easily accessible to non-specialist readers. Topics covered include the background mathematics of oscillations, gravity waves on water, the dynamics of wave–body interactions, and the absorption of wave energy by oscillating bodies and oscillating water columns. Featuring new content throughout, including three new chapters on oscillating-body wave energy converters, oscillating water columns and other types of wave energy converters, and wave energy converter arrays, this book is an excellent resource for students, researchers, and engineers who are new to the subject of wave energy conversion, as well as those with more experience.

JOHANNES FALNES is a professor emeritus of experimental physics at the Department of Physics of the Norwegian University of Science and Technology noted for his contributions to wave-energy research. He is one of the pioneers of modern wave-energy research.

ADI KURNIAWAN is a research fellow at the Oceans Graduate School of the University of Western Australia where he conducts research on wave energy. When the book manuscript was drafted and completed, he was an assistant professor at the Department of Civil Engineering of Aalborg University. His research has appeared in scholarly journals such as the *Journal of Fluid Mechanics* and *Proceedings of the Royal Society*.

**Cambridge Ocean Technology Series**

# OCEAN WAVES AND

# OSCILLATING SYSTEMS

## LINEAR INTERACTIONS INCLUDING WAVE-ENERGY EXTRACTION

Second Edition

**JOHANNES FALNES**

Norwegian University of Science and Technology

**ADI KURNIAWAN**

The University of Western Australia

CAMBRIDGE
UNIVERSITY PRESS

# CAMBRIDGE
## UNIVERSITY PRESS

University Printing House, Cambridge CB2 8BS, United Kingdom

One Liberty Plaza, 20th Floor, New York, NY 10006, USA

477 Williamstown Road, Port Melbourne, VIC 3207, Australia

314–321, 3rd Floor, Plot 3, Splendor Forum, Jasola District Centre, New Delhi – 110025, India

79 Anson Road, #06–04/06, Singapore 079906

Cambridge University Press is part of the University of Cambridge.

It furthers the University's mission by disseminating knowledge in the pursuit of
education, learning, and research at the highest international levels of excellence.

www.cambridge.org
Information on this title: www.cambridge.org/9781108481663
DOI: 10.1017/9781108674812

First published 2020

Printed in the United Kingdom by TJ International, Padstow Cornwall

*A catalogue record for this publication is available from the British Library.*

*Library of Congress Cataloging-in-Publication Data*
Names: Falnes, Johannes, 1931– author. | Kurniawan, Adi, author.
Title: Ocean waves and oscillating systems : linear interactions including
    wave-energy extraction / Johannes Falnes and Adi Kurniawan.
Description: Second edition. | New York : Cambridge University Press, 2020. |
    Series: Cambridge ocean technology series | Includes bibliographical
    references and index.
Identifiers: LCCN 2019038349 (print) | LCCN 2019038350 (ebook) |
    ISBN 9781108481663 (hardback) | ISBN 9781108674812 (epub)
Subjects: LCSH: Ocean waves. | Ocean wave power.
Classification: LCC GC211.2 .F35 2020 (print) | LCC GC211.2 (ebook) |
    DDC 551.46/3–dc23
LC record available at https://lccn.loc.gov/2019038349
LC ebook record available at https://lccn.loc.gov/2019038350

ISBN 978-1-108-48166-3 Hardback

# Contents

# Prefaces

This book is intended to provide a thorough consideration of the interaction between waves and oscillating systems under conditions where amplitudes are sufficiently small that linear theory is applicable. In practice, this small-wave assumption is reasonably valid for most of the time, during which, for example, a wave-energy converter is generating most of its income. During the rather rare extreme-wave situations, however, nonlinear effects may be significant, and such situations influence design loads, and hence the costs, for ships and other installations deployed at sea. This matter is treated in several other books.

The present book is mainly based on lecture notes from a postgraduate university course on water waves and extraction of energy from ocean waves, which I taught many times since 1979 until my retirement at the end of 2001. Except in 1983, my course was taught every second year, mainly for doctorate students at the university in Trondheim, but other interested students have also attended. Moreover, a similar two-week course was given in 1986 with participants from the Norwegian industry. In addition, the materials have also been used in two-week courses with international participation in 1998 and 2005 at Chalmers University of Technology, Gothenburg, Sweden, and in 2010 at the Norwegian University of Science and Technology (NTNU), Trondheim. The latter was when I served as a visiting scientist at NTNU's Centre for Ships and Ocean Structures (CeSOS) during 2005–2010.

In February 1980, the lecture notes were issued in a bound volume entitled *Hydrodynamisk teori for bølgjekraftverk* (*Hydrodynamic Theory for Wave Power Plants*) by L. C. Iversen and me. One hundred copies were published by the University of Trondheim, Division of Experimental Physics. Later I revised the lecture notes and translated them into English. In 1993, this process resulted in a two-volume work entitled *Theory for Extraction of Ocean-Wave Energy*.

I wish to thank the course participant Knut Bønke for his inspiring encouragement to have the lecture notes typed in 1979 and issued in a bound volume, and for his continued encouragement over many years to write a textbook based on the notes. Moreover, I would like to thank Jørgen Hals, a course participant from 1997, for working out the subject index of the first-edition book. I am also in debt to my other students for their comments and proofreading. In this connection, I wish to mention, in particular, the following graduate students (the years they completed their doctorate degrees are given): L. C. Iversen (1980), Å. Kyllingstad (1982), O. Malmo (1984), G. Oltedal (1985), A. Brendmo (1995), H. Eidsmoen (1996), and Torkel Bjarte-Larsson (2005). Also my collaborator over many years, P. M. Lillebekken, who attended the course in 1981, has made many valuable comments. I also wish to acknowledge important discussions and cooperation with two students who completed their PhD at CeSOS, Department of Marine Technology, NTNU, namely Jørgen Hals (2010) and Adi Kurniawan (2013). Without Adi's acceptance to be my coauthor, I could not have, at an age of 85, in 2017, accepted Cambridge University Press's proposal to author a revised edition of my 2002 textbook. I also wish to thank Dr. Jens Peter Kofoed and Aalborg University, Denmark, for allowing Adi to be my coauthor. While Adi was a student at my two-week course in 2010, Jens Peter was one of my two-week-course students in 1998.

Most of all, I am in debt to my late colleague Kjell Budal (1933–1989), whose initiative inspired my interest in wave-power utilisation at an early stage. During the oil crisis at the end of 1973, we started a new research project aimed at utilising ocean-wave energy. At that time, we did not have a research background in hydrodynamics, but Budal had carried out research in acoustics, developing a particular microphone, whereas I had studied waves in electromagnetics and plasma physics. In 1974, we jointly published a (Norwegian) textbook *Bølgjelære* (*Wave Science*) for the second-year undergraduate students in physics. This is an interdisciplinary text on waves, with particular emphasis on acoustics and optics.

With this background, our approach and attitude toward hydrodynamic waves have perhaps been more interdisciplinary than traditional. In my view, this has influenced our way of thinking and stimulated our contributions to the science of hydrodynamics. This background is also reflected in the present book, notably in Chapter 3, where interaction between oscillations and waves is considered in general; water waves, in particular, are treated in the subsequent chapters.

I wish to thank Professor J. N. Newman and Dr. Alain Clément for suggesting the use of computer codes WAMIT and AQUADYN, which have been used to compute the numerical results presented in Sections 5.2.4 and 5.7.3, respectively. I am also grateful to Dr. Stephen Barstow for permission to use Problem 4.4, which he formulated.

Finally, I wish to acknowledge the support given, to my achievements, from my parents, Brita, b. Nygaard (1900–1977) and Peder Magne Falnes (1902–1983), and from my wife, Dagny Elisabeth, b. Agle (1934–2006).

Johannes Falnes
Trondheim, Norway, July 23, 2019

The book by Professor Johannes Falnes has greatly benefited me ever since I was a PhD student at CeSOS, NTNU, and has been a main reference for me in the years that have followed. So upon receiving Professor Falnes's invitation to coauthor a possible second edition, I was delighted with the prospect but, at the same time, filled with the trepidation of carrying out such a task alongside someone who is a renowned figure in the field of wave energy. Professor Falnes's consistent encouragement has kept me in this course.

This edition features new materials, some of which are the result of our recent collaboration. We have maintained the style of the first edition, where theoretical results are derived step by step so that the reader can follow the derivations closely and clearly. In various places, equations and text have been revised to ease understanding and avoid redundancies. We hope that with these improvements, the book will continue to be a useful reference for students as well as more experienced practitioners.

My involvement in writing this book would not have been possible without the support and permission of Dr. Jens Peter Kofoed from the Department of Civil Engineering at Aalborg University and the partial support from the Danish Energy Agency. In this connection, I would also like to thank Dr. Kim Nielsen, Chairman of the Danish Partnership for Wave Power. I am also grateful to my colleagues at the Oceans Graduate School of the University of Western Australia for their support while final work on the book was carried out. Finally, I owe my gratitude to my wife, Graciana, and children, Sethia, Hana, Musa, and Betha, for their unconditional love.

Adi Kurniawan
Albany, Australia, January 30, 2020

# CHAPTER ONE

# Introduction

In this book, gravity waves on water and their interaction with oscillating systems (having zero forward speed) are approached from a somewhat interdisciplinary point of view. Before the matter is explored in depth, a comparison is briefly made between different types of waves, including acoustic waves and electromagnetic waves, drawing the reader's attention to some analogies and dissimilarities. Oscillating systems for generating or absorbing waves on water are analogous to loudspeakers or microphones in acoustics, respectively. In electromagnetics, the analogues are transmitting or receiving antennae in radio engineering, and light-emitting or light-absorbing atoms in optics.

The discussion of waves in this book is almost exclusively limited to waves of sufficiently low amplitudes for linear analysis to be applicable. Several other books (see, e.g., the monographs by Mei et al. [1], Faltinsen [2], Sarpkaya [3] or Chakrabarti [4]) treat the subject of large ocean waves and extreme wave loads—which are important for determining the survivability of ships, harbours and other ocean structures, including devices for conversion of ocean-wave energy. In contrast, the purpose of the present book is to convey a thorough understanding of the interaction between waves and oscillations when the amplitudes are low, which is true most of the time. For example, for a wave-power plant, the income is determined by the annual energy production, which is essentially accrued during most times of the year, when amplitudes are low—that is, when linear interaction is applicable. On the other hand, as with many other types of ocean installations, wave-power plants have their expenses, to a large extent, determined by the extreme-load design.

There are several steps in the process of converting wave energy to some useful form, such as electrical energy, by means of an immersed oscillating wave-energy-conversion (WEC) unit. In the first conversion step, energy may be converted to some form of kinetic and/or potential energy, to possibly be stored in flywheels and or gas accumulators of a hydraulic machinery. Such energy storage may, for certain applications, be desirable in order to even out fluctuations of the incident wave energy on a minute-to-minute scale [5]. In a second conversion step, mechanical energy may be delivered, for example, through the shaft of a turbine or some other type of hydraulic motor, to, for example, an electrical generator, which may serve a tertiary conversion to electric energy. The second conversion step may possibly be circumvented by the application of linear electric generators. The present book deals with the first conversion step, the *primary wave-energy conversion*—that is, how to reduce the energy in an incident wave while, correspondingly, increasing the energy input to an oscillating WEC system. The technological/engineering aspects related to conversion and useful application of wave energy are not covered in the present book. Readers interested in such subjects are referred to other literature [5–11].

The content of the subsequent chapters is outlined in the following paragraphs. At the end of each chapter, except the first, there is a collection of problems.

Chapter 2 gives a mathematical description of free and forced oscillations in the time domain as well as in the frequency domain. An important purpose is to introduce students to the very useful mathematical tool of the complex representation of sinusoidal oscillations. The mathematical connection between complex amplitudes and Fourier transforms is treated. Linear systems are discussed in a rather general way, and for a causal linear system in particular, the Kramers–Kronig relations are derived. A simple mechanical oscillating system is analysed to some extent. Then the concept of mechanical impedance is introduced. Moreover, a discussion of energy accounting in the system is included to serve as a tool for physical explanation, in subsequent chapters, of the so-called hydrodynamic added mass and other analogue quantities.

In Chapter 3, a brief comparison is made of waves on water with other types of waves, in particular with acoustic waves. The concepts of wave dispersion, phase velocity and group velocity are introduced. In addition, the transport of energy associated with propagating waves is considered, and the radiated power from a radiation source (wave generator) is mathematically expressed in terms of a phenomenologically defined radiation resistance. The radiation impedance, which is a complex parameter, is introduced in a phenomenological way. For mechanical waves (such as acoustic waves and waves on water), its imaginary part may be represented by an added mass. Finally, in Chapter 3, an analysis is given of the absorption of energy from a mechanical wave by means of a mechanical oscillation system of the simple type considered in Chapter 2. The optimum parameters of this system for maximising the absorbed energy are discussed. The maximum is, for such a simple oscillating system, obtained at resonance. Chapter 3 is rather elementary and intended as a general introduction for readers not familiar with hydrodynamics. It may be skipped by readers already familiar with hydrodynamics.

From Chapter 4 and onward, a deeper hydrodynamic discussion of water waves is the main subject. With the assumption of inviscid and incompressible fluid and irrotational fluid motion, the hydrodynamic potential theory is developed. With the linearisation of fluid equations and boundary conditions, the basic equations for low-amplitude waves are derived. In most of the following discussions, either infinite water depth or finite, but constant, water depth is assumed. Dispersion and wave-propagation velocities are studied, and plane and circular waves are discussed in some detail. Also, non-propagating, evanescent, plane waves are considered. Another studied subject is wave-transported energy and momentum. The spectrum of real sea waves is treated only briefly in the present book. The rather theoretical Sections 4.7 and 4.8, which make extensive use of Green's theorem, may be omitted at the first reading, and then be referred to as needed during the study of the remaining chapters of the book. Whereas most of Chapter 4 is concerned with discussions in the frequency domain, the last section contains discussions in the time domain. As 'Linear Interactions' is part of the present book's title, nonlinear wave energy converters, such as the Norwegian 'Tapchan' or the Danish 'Wave Dragon', are outside the scope of the present book, except for a simple exercise, though, related to Chapter 4 (Problem 4.11).

The subject of Chapter 5 is the interactions between waves and oscillating bodies, including wave generation by oscillating bodies as well as forces induced by waves on the bodies. Initially, six-dimensional generalised vectors are introduced which correspond to the six degrees of freedom for the motion of an immersed (three-dimensional) rigid body. The radiation impedance, known from the phenomenological introduction in Chapter 3, is now defined in a hydrodynamic formulation, and, for a three-dimensional body, extended to a $6 \times 6$ matrix. In a later part of the chapter, the radiation impedance matrix is extended to the case of a finite number of interacting, radiating, immersed bodies. For this case, the generalised excitation force vector is decomposed into two parts, the Froude–Krylov part and the diffraction part, which are particularly discussed in the 'small-body' (or 'long-wavelength') approximation. From Green's theorem (as mentioned in the summary of Chapter 4), several useful reciprocity theorems are derived, which relate excitation force and radiation resistance to each other or to 'far-field coefficients' (or 'Kochin functions'). Subsequently, these theorems are applied to oscillating systems consisting of concentric axisymmetric bodies or of two-dimensional bodies. The occurrence of singular radiation-resistance matrices is discussed in this connection. While most of Chapter 5 is concerned with discussions in the frequency domain, two sections, Sections 5.3 and 5.9, contain discussions in the time domain. In the latter section, motion response is the main subject. In the former section, two hydrodynamic impulse-response functions are considered, where one of them is causal and, hence, has to obey the Kramers–Kronig relations.

The extraction of wave energy by means of an oscillating immersed WEC body is the subject of Chapter 6, which starts by explaining wave absorption as a wave-interference phenomenon: to absorb a wave, the oscillating WEC body needs to radiate a wave which interferes destructively with the incident wave. Thus, energy is removed from the incident wave and converted to some form of useful energy. The WEC body thus needs to displace water, push water during half part of a wave period and suck water during the remaining part. Unfortunately, some proposed devices, claimed to be wave energy absorbers, seem to essentially just follow the water motion of the incident wave.

Moreover, in Chapter 6, WEC units are classified according to their horizontal extension relative to the wavelength. A formula is derived for the power that is absorbed from an incident plane wave by means of a WEC body, if it is oscillating in only one degree of freedom. This absorbed power is positive, provided the body's oscillation has an amplitude and a phase within a certain region relative to the amplitude and phase of the incident wave. This region may be illustrated by the wave-power 'island', where the highest point corresponds to the maximum power which may be absorbed from the incident wave. Moreover, a section of Chapter 6 is devoted to the subject of optimum control aiming at maximising absorbed wave power. The Budal upper bound (BUB), as well as the Keulegan–Carpenter number, are additional important matter in a section of Chapter 6. Here, in contrast to the previous sections, the WEC body's hull volume size and shape are matters of concern. The final section of Chapter 6 concerns wave-power absorption by means of a WEC body oscillating in more than one mode.

Toward the end of the chapter (Section 6.5), a study is made of the absorption of wave energy by means of an immersed body oscillating in several (up to six) degrees of freedom. This discussion provides a physical explanation of the quite frequently encountered cases of singular radiation-resistance matrices, as mentioned earlier (also see Sections 5.7 and 5.8). However, the central part of Chapter 6 is concerned with wave-energy conversion which utilises only a single body oscillating in just one degree of freedom. With the assumption that an external force is applied to the oscillating system, for the purpose of power takeoff and optimum control of the oscillation, this discussion has a different starting point than that given in the last part of Chapter 3. The conditions for maximising the converted power are also studied for the case in which the body oscillation has to be restricted as a result of its designed amplitude limit or because of the installed capacity of the energy-conversion machinery.

Oscillating water columns (OWCs) are mentioned briefly in Chapter 4 and considered in greater detail in Chapter 7, where their interaction with incident waves and radiated waves is the main subject of study. Two kinds of interaction are considered: the radiation problem and the excitation problem. The radiation problem concerns the radiation of waves resulting from an oscillating dynamic air pressure above the internal air–water interfaces of the OWCs. The excitation problem concerns the oscillation which is due to an incident wave when the dynamic air pressure is zero within the OWC's air chamber. Comparisons are made with corresponding wave–body interactions. Also, wave-energy extraction by an OWC WEC is discussed. The last part of Chapter 7 concerns WECs which oscillate in unconventional modes of motion—that is, modes other than the six rigid-body modes considered in Chapter 5. The analysis of such WECs is facilitated by the use of generalised modes, which are described by shape functions.

The main subject of Chapter 8, the last chapter of the book, is related to possible WEC systems for a more distant future, arrays of WEC bodies and WEC arrays based on oscillating bodies as well as OWCs. Also, two-dimensional WEC devices, verbally described in Section 6.1, are considered in some mathematical detail in Chapter 8.

# CHAPTER TWO

# Mathematical Description of Oscillations

In this chapter, which is a brief introduction to the theory of oscillations, a simple mechanical oscillation system is used to introduce concepts such as free and forced oscillations, state-space analysis and representation of sinusoidally varying physical quantities by their complex amplitudes. In order to be somewhat more general, causal and noncausal linear systems are also looked at and Fourier transform is used to relate the system's transfer function to its impulse response function. With an assumption of sinusoidal (or 'harmonic') oscillations, some important relations are derived which involve power and stored energy on one hand and the parameters of the oscillating system on the other hand. The concepts of resonance and bandwidth are also introduced.

## 2.1 Free and Forced Oscillations of a Simple Oscillator

Let us consider a simple mechanical oscillator in the form of a mass–spring–damper system. A mass $m$ is suspended through a spring and a mechanical damper, as indicated in Figure 2.1. Because of the application of an external force $F$, the mass has a position displacement $x$ from its equilibrium position.

Newton's law gives

$$m\ddot{x} = F + F_R + F_S, \tag{2.1}$$

where $F_S$ is the spring force and $F_R$ is the damper force.

If we assume that the spring and the damper have linear characteristics, then we can write $F_S = -Sx$ and $F_R = -R\dot{x}$, where the 'stiffness' $S$ and the 'mechanical resistance' $R$ are coefficients of proportionality, independent of the displacement $x$ and the velocity $u = \dot{x}$. Thus, we have the following linear differential equation with constant coefficients:

$$m\ddot{x} + R\dot{x} + Sx = F, \tag{2.2}$$

where an overdot is used to denote differentiation with respect to time $t$.

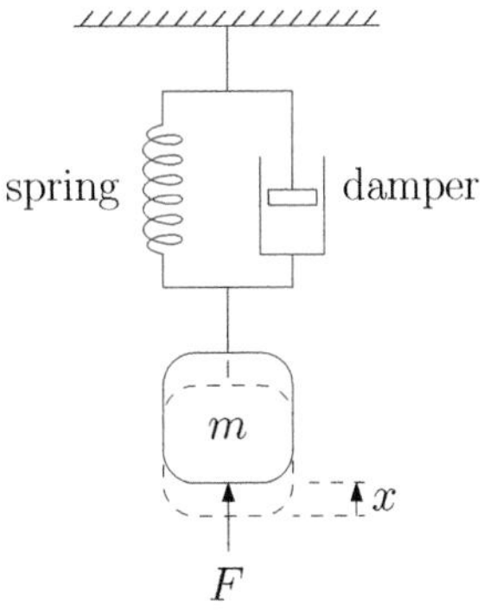

Figure 2.1: Mechanical oscillator in the
form of a mass–spring–damper system.

### 2.1.1 Free Oscillation

If the external force is absent—that is, $F = 0$—we may have the so-called free
oscillation if the system is released at a certain instant $t = 0$, with some initial
energy

$$W_0 = W_{p0} + W_{k0} = Sx_0^2/2 + mu_0^2/2, \tag{2.3}$$

written here as a sum of potential and kinetic energy, where $x_0$ is the initial
displacement and $u_0$ the initial velocity. It is easy to show (see Problem 2.1) that
the general solution to Eq. (2.2), when $F = 0$, is

$$x = (C_1 \cos \omega_d t + C_2 \sin \omega_d t) e^{-\delta t}, \tag{2.4}$$

where

$$\delta = R/2m, \quad \omega_0 = \sqrt{S/m}, \quad \omega_d = \sqrt{\omega_0^2 - \delta^2} \tag{2.5}$$

are the damping coefficient, the undamped natural angular frequency and the
damped angular frequency, respectively. The integration constants $C_1$ and $C_2$
may be determined from the initial conditions as (see Problem 2.1)

$$C_1 = x_0, \quad C_2 = (u_0 + x_0\delta)/\omega_d. \tag{2.6}$$

For the particular case of zero damping force, the oscillation is purely
sinusoidal with a period $T_0 = 2\pi/\omega_0$, which is the so-called natural period of
the oscillator. The free oscillation as given by Eq. (2.4) is an exponentially
damped sinusoidal oscillation with 'period' $T_d = 2\pi/\omega_d$, during which a fraction
$1 - \exp(-4\pi\delta/\omega_d)$ of the energy in the system is lost, as a result of power
consumption in the damping resistance $R$.

The time histories of the normalised displacement $x/x_0$ and energy $W/W_0 =$
$(Sx^2 + mu^2)/2W_0$ of an oscillator with an initial displacement $x_0$ when $F = 0$
and $u_0 = 0$ are shown in Figure 2.2. The amplitude of the oscillation decays

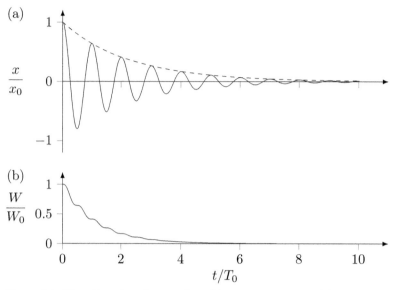

Figure 2.2: Time histories of normalised (a) displacement and (b) energy of a mechanical oscillator for a given initial displacement $x = x_0$, in the absence of external force, $F = 0$, and initial velocity, $u_0 = 0$.

exponentially. If the ratio between two successive peaks is known—for example, from physical measurements—the damping coefficient can be obtained as

$$\delta = \log\left(x(t)/x(t + T_d)\right)/T_d \tag{2.7}$$

according to Eq. (2.4). As expected, the rate of loss of energy in the system is largest when the instantaneous velocity is largest. The energy curve has a horizontal tangent and an inflection point at instants when the velocity is zero and changes sign.

We define the oscillator's *quality factor* $Q$ as the ratio between the stored energy and the average energy loss during a time interval of length $1/\omega_d$:

$$Q = \left(1 - e^{-2\delta/\omega_d}\right)^{-1}. \tag{2.8}$$

If the damping coefficient $\delta$ is small, then $Q$ is large. When $\delta/\omega_0 \ll 1$, the following expansions (see Problem 2.2) may be useful:

$$Q = \frac{\omega_0}{2\delta}\left(1 + \frac{\delta}{\omega_0} - \frac{1}{6}\frac{\delta^2}{\omega_0^2} + \mathcal{O}\left\{\frac{\delta^3}{\omega_0^3}\right\}\right)$$

$$\approx \frac{\omega_0}{2\delta} = \frac{\omega_0 m}{R} = \frac{S}{\omega_0 R} = \frac{(Sm)^{1/2}}{R}, \tag{2.9}$$

$$\frac{\delta}{\omega_0} = \frac{1}{2Q}\left(1 + \frac{1}{2Q} + \frac{5}{24Q^2} + \mathcal{O}\left\{Q^{-3}\right\}\right) \approx \frac{1}{2Q}. \tag{2.10}$$

As a result of the energy loss, the freely oscillating system comes eventually to rest. The free oscillation is 'overdamped' if $\omega_d$ is imaginary—that is, if

$\delta > \omega_0$ or $R > 2(Sm)^{1/2}$. [The quality factor $Q$, as defined by Eq. (2.8), is then complex, and it loses its physical significance.] Then the general solution of the differential equation (2.2) is a linear combination of two real, decaying exponential functions. The case of 'critical damping'—that is, when $R = 2(Sm)^{1/2}$ or $\omega_d = 0$—needs special consideration, which we omit here. (See, however, Problem 2.11.)

### 2.1.2 Forced Oscillation

When the differential equation (2.2) is inhomogeneous—that is, if $F = F(t) \neq 0$—the general solution may be written as a particular solution plus the general solution (2.4) of the corresponding homogeneous equation (corresponding to $F = 0$).

Let us now consider the case in which the driving external force $F(t)$ has a sinusoidal time variation with angular frequency $\omega = 2\pi/T$, where $T$ is the period. Let

$$F(t) = F_0 \cos(\omega t + \varphi_F), \tag{2.11}$$

where $F_0$ is the amplitude and $\varphi_F$ the phase constant for the force. It is convenient to choose a particular solution of the form where

$$x(t) = x_0 \cos(\omega t + \varphi_x) \tag{2.12}$$

is the position and

$$u(t) = \dot{x}(t) = u_0 \cos(\omega t + \varphi_u) \tag{2.13}$$

is the corresponding velocity of the mass $m$. Here the amplitudes are related by $u_0 = \omega x_0$ and the phase constants by $\varphi_u - \varphi_x = \pi/2$.

For Eq. (2.12) to be a particular solution of the differential equation (2.2), it is necessary (see Problem 2.3) that the excursion amplitude is

$$x_0 = \frac{u_0}{\omega} = \frac{F_0}{|Z|\omega} \tag{2.14}$$

and that the phase difference

$$\varphi = \varphi_F - \varphi_u = \varphi_F - \varphi_x - \pi/2 \tag{2.15}$$

is an angle which is in quadrant no. 1 or no. 4 and which satisfies

$$\tan \varphi = (\omega m - S/\omega)/R. \tag{2.16}$$

Here

$$|Z| = \sqrt{R^2 + (\omega m - S/\omega)^2} \tag{2.17}$$

is the absolute value (modulus) of the complex mechanical impedance, which is discussed later.

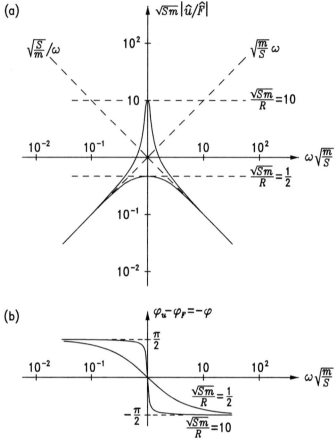

Figure 2.3: Frequency response of relation between velocity $u$ and applied force $F$ in normalised units, for two different values of the damping coefficient. (a) Amplitude (modulus) response with both scales logarithmic. (b) Phase response with linear scale for the phase difference.

The 'forced oscillation', Eq. (2.12) or (2.13), is a response to the driving force, Eq. (2.11). Let us now assume that $F_0$ is independent of $\omega$ and then discuss the responses $x_0(\omega)$ and $u_0(\omega)$, starting with $u_0(\omega) = F_0/|Z(\omega)|$. Noting that $|Z|_{\min} = R$ for $\omega = \omega_0 = \sqrt{S/m}$ and that $|Z| \to \infty$ for $\omega = 0$ as well as for $\omega \to \infty$, we see that $(u_0/F_0)_{\max} = 1/R$ for $\omega = \omega_0$ and that $u_0(0) = u_0(\infty) = 0$. We have resonance at $\omega = \omega_0$, where the 'reactive' contribution $\omega m - S/\omega$ to the mechanical impedance vanishes.

Graphs of the non-dimensionalised velocity response $\sqrt{Sm}\, u_0/F_0$ versus $\omega/\omega_0$ are shown in Figure 2.3 for $\sqrt{Sm}/R$ equal to 10 and 0.5. Note that the graphs are symmetric with respect to $\omega = \omega_0$ when the frequency scale is logarithmic. The phase difference $\varphi$ as given by Eq. (2.16) is also shown in Figure 2.3. The graphs of Figure 2.3, where the amplitude response is presented in a double logarithmic diagram and the phase response in a semilogarithmic diagram, are usually called Bode plots or Bode diagrams [12].

Next, we consider the resonance bandwidth, the frequency interval $(\Delta\omega)_{\mathrm{res}}$ where

$$\frac{u_0(\omega)}{F_0} > \frac{1}{\sqrt{2}}\left(\frac{u_0}{F_0}\right)_{\max} = \frac{1}{R\sqrt{2}}, \tag{2.18}$$

that is, where the kinetic energy exceeds half of the maximum value. At the upper and lower edges of the interval, $\omega_u$ and $\omega_l$, the two terms of the radicand in Eq. (2.17) are equally large. Thus, we have

$$\omega_u m - S/\omega_u = R = S/\omega_l - \omega_l m. \tag{2.19}$$

Instead of solving these two equations, we note from the aforementioned symmetry that $\omega_u \omega_l = \omega_0^2 = S/m$—that is, $S/\omega_u = m\omega_l$ and $S/\omega_l = m\omega_u$. Evidently,

$$(\Delta\omega)_{\mathrm{res}} = \omega_u - \omega_l = R/m = 2\delta. \tag{2.20}$$

The relative bandwidth is

$$\frac{(\Delta\omega)_{\mathrm{res}}}{\omega_0} = \frac{2\delta}{\omega_0} = \frac{R}{\sqrt{Sm}}, \tag{2.21}$$

and it is seen that this is inverse to the maximum non-dimensionalised velocity response $\sqrt{Sm}u_0(\omega_0)/F_0 = \sqrt{Sm}/R$, as indicated on the graph in Figure 2.3. From Eq. (2.9), we see that this is approximately equal to the quality factor $Q$, when this is large ($Q \gg 1$). In the same case,

$$(\Delta\omega)_{\mathrm{res}}/\omega_0 \approx 1/Q. \tag{2.22}$$

Next we consider the excursion response—which, in non-dimensionalised form, may be written as $Sx_0(\omega)/F_0$. It equals unity for $\omega = 0$ and zero for $\omega = \infty$. At resonance, its value is

$$S x_0(\omega_0)/F_0 = S/\omega_0 R = \sqrt{Sm}/R = \omega_0/2\delta, \tag{2.23}$$

as obtained by using Eqs. (2.14) and (2.17). Note that $x_0(\omega)$ has its maximum at a frequency which is lower than the resonance frequency. It can be shown (see Problem 2.4) that, if $R < \sqrt{2Sm}$ or $\delta < \omega_0/\sqrt{2}$, then

$$\frac{Sx_{0,\max}}{F_0} = \frac{\omega_0}{2\delta}\left(1 - \frac{\delta^2}{\omega_0^2}\right)^{-1/2} \quad \text{at } \omega = \omega_0\left(1 - \frac{2\delta^2}{\omega_0^2}\right)^{1/2}, \tag{2.24}$$

and if $R > \sqrt{2Sm}$, then

$$\frac{Sx_{0,\max}}{F_0} = \frac{S x_0(0)}{F_0} = 1. \tag{2.25}$$

For large values of $Q$, there is only a small difference between $x_0(\omega_0)/F_0$ and $x_{0,\text{max}}/F_0$. Using Eq. (2.10), we find

$$\frac{Sx_0(\omega_0)}{F_0} = Q - \frac{1}{2} + \frac{1}{24Q} + \mathcal{O}\{Q^{-2}\} \tag{2.26}$$

and

$$\frac{Sx_{0,\text{max}}}{F_0} = Q - \frac{1}{2} + \frac{1}{6Q} + \mathcal{O}\{Q^{-2}\} \tag{2.27}$$

for

$$\omega = \omega_0\left(1 - \frac{1}{4Q^2} + \mathcal{O}\{Q^{-3}\}\right). \tag{2.28}$$

### 2.1.3 Electric Analogue: Remarks on the Quality Factor

For readers with a background in electric circuits, it may be of interest to note that the mechanical system of Figure 2.1 is analogous to the electric circuit shown in Figure 2.4, where an inductance $m$, a capacitance $1/S$, and an electric resistance $R$ are connected in series.

The force $F$ is analogous to the driving voltage, the position $x$ is analogous to the electric charge on the capacitance, and the velocity $u$ is analogous to the electric current. If Kirchhoff's law is applied to the circuit, Eq. (2.2) results. The 'capacitive reactance' $S/\omega$ is related to the capacitance's ability to store electric energy (analogous to potential energy in the spring of Figure 2.1), and the 'inductive reactance' $\omega m$ is related to the inductance's ability to store magnetic energy (analogous to kinetic energy in the mass of Figure 2.1).

The electric (or potential) energy is zero when $x(t) = 0$, and the magnetic (or kinetic) energy is zero when $u(t) = \dot{x}(t) = 0$. The instants for $x(t) = 0$ and those for $u(t) = 0$ are displaced by a quarter of a period $\pi/2\omega_0$, and at resonance the maximum values for the electric and magnetic (or potential and kinetic) energies are equal, $mu_0^2/2 = m\omega_0^2 x_0^2/2 = Sx_0^2/2$, because $m\omega_0 = S/\omega_0$. Thus, at resonance, the stored energy is swinging back and forth between the two energy stores, twice every period of the system's forced oscillation.

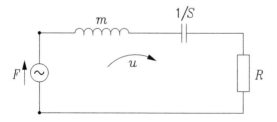

Figure 2.4: Electric analogue of the mechanical system shown in Figure 2.1.

By Eq. (2.8), we have defined the quality factor $Q$ as the ratio between the stored energy and the average energy loss during a time interval $1/\omega_d$ of the free oscillation. An alternative definition would have resulted if instead the forced oscillation at resonance had been considered, for a time interval $1/\omega_0$ (and not $1/\omega_d$). The stored energy is $mu_0^2/2$ and the average lost energy is $Ru_0^2/2\omega_0$ during a time $1/\omega_0$ (as is shown in more detail later, in Section 2.3).

Such an alternative quality factor equals the right-hand side of the approximation (2.9) and would have been equal to the inverse of the relative bandwidth (2.21), to the non-dimensionalised excursion amplitude (2.23) at resonance (that is, the ratio between the excursion at resonance and the excursion at zero frequency), and to the ratio of the reactance parts $\omega_0 m$ and $S/\omega_0$ to the damping resistance $R$. The term quality factor is usually used only when it is large, $Q \gg 1$. In that case, the relative difference between the two definitions is of little importance.

## 2.2 Complex Representation of Harmonic Oscillations

### 2.2.1 Complex Amplitudes and Phasors

When dealing with sinusoidal oscillations, it is mathematically convenient to apply the method of complex representation, involving complex amplitudes and phasors. A great advantage with the method is that differentiation with respect to time is simply represented by multiplying with $i\omega$, where $i$ is the imaginary unit ($i = \sqrt{-1}$). We consider again the forced oscillations represented by the excursion response $x(t)$ or velocity response $u(t)$ due to an applied external sinusoidal force $F(t)$, as given by Eqs. (2.12), (2.13) and (2.11), respectively.

With the use of Euler's formula

$$e^{i\psi} = \cos \psi + i \sin \psi, \tag{2.29}$$

or, equivalently,

$$\cos \psi = (e^{i\psi} + e^{-i\psi})/2, \qquad \sin \psi = (e^{i\psi} - e^{-i\psi})/2i, \tag{2.30}$$

the oscillating quantity $x(t)$ may be rewritten as

$$\begin{aligned}
x(t) &= x_0 \cos (\omega t + \varphi_x) \\
&= \tfrac{1}{2}x_0[e^{i(\omega t+\varphi_x)} + e^{-i(\omega t+\varphi_x)}] \\
&= \tfrac{1}{2}x_0[e^{i\varphi_x}e^{i\omega t} + e^{-i\varphi_x}e^{-i\omega t}].
\end{aligned} \tag{2.31}$$

Introducing the *complex amplitude* (see Figure 2.5)

$$\hat{x} = x_0 e^{i\varphi_x} = x_0 \cos \varphi_x + ix_0 \sin \varphi_x \tag{2.32}$$

and the complex conjugate of $\hat{x}$

$$\hat{x}^* = x_0 e^{-i\varphi_x} = x_0 \cos \varphi_x - ix_0 \sin \varphi_x, \tag{2.33}$$

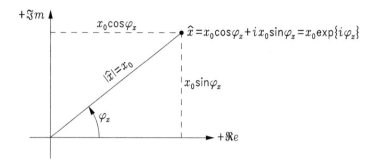

Figure 2.5: Complex-plane decomposition of the complex amplitude $\hat{x}$.

we have

$$2x(t) = \hat{x}e^{i\omega t} + \hat{x}^* e^{-i\omega t}. \tag{2.34}$$

Note that the sum is real, while the two terms are complex (and conjugate to each other). Another formalism is

$$2x(t) = \hat{x}e^{i\omega t} + \text{c. c.}, \tag{2.35}$$

where c. c. denotes complex conjugate of the preceding term.

The complex amplitude (2.32) contains information on two parameters:

1. the (absolute) amplitude $|\hat{x}| = x_0$, which is real and positive, and
2. the phase constant $\varphi_x = \arg \hat{x}$, which is an angle to be given in units of radians (rad) or degrees (°).

Multiplying Eq. (2.32) with $e^{i\omega t}$, we have

$$\hat{x}e^{i\omega t} = x_0 e^{i(\omega t + \varphi_x)} = x_0[\cos(\omega t + \varphi_x) + i\sin(\omega t + \varphi_x)], \tag{2.36}$$

from which we observe that

$$x(t) = x_0 \cos(\omega t + \varphi_x) = \text{Re}\{\hat{x}e^{i\omega t}\} = \text{Re}\{x_0 e^{i(\omega t + \varphi_x)}\}, \tag{2.37}$$

or that the instantaneous value equals the real part of the product of the complex amplitude $\hat{x}$ and $e^{i\omega t}$. On the complex plane, $x(t)$ equals the projection of the rotating vector $\hat{x}e^{i\omega t}$ on the real axis (see Figure 2.6). The value $x(t)$ is negative when $\omega t + \varphi_x$ is an angle in the second or third quadrant of the complex plane.

A summary is given in Table 2.1, where the real function $x(t)$ denotes a physical quantity in harmonic oscillation. [Note: frequently, one may see it written as $x(t) = \hat{x}e^{i\omega t}$. In this case, because $x(t)$ is complex, it must be implicitly understood that it is the real part of $x(t)$ that represents the physical quantity.]

Let us now consider position, velocity and acceleration and their interrelations. If the position of an oscillating mass point is given by Eq. (2.37), then the velocity is

Table 2.1: Various mathematical expressions for a sinusoidal oscillation, where $\hat{x} = x_0 \exp(i\varphi_x)$ is the complex amplitude.

---

$x(t) = x_0 \cos(\omega t + \varphi_x)$

$x(t) = \frac{1}{2}(\hat{x}e^{i\omega t} + \hat{x}^* e^{-i\omega t})$

$x(t) = \frac{1}{2}(\hat{x}e^{i\omega t} + \text{c.c.})$

$x(t) = \text{Re}\{\hat{x}e^{i\omega t}\}$

---

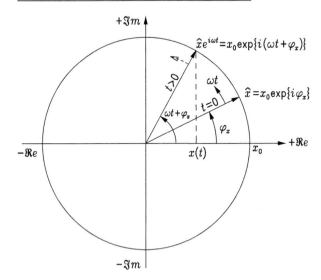

Figure 2.6: Phasor shown at times $t = 0$ and $t > 0$.

$$u = \dot{x} = \frac{dx}{dt} = -\omega x_0 \sin(\omega t + \varphi_x)$$
$$= \omega x_0 \cos(\omega t + \varphi_x + \pi/2) \tag{2.38}$$
$$= \text{Re}\{\omega x_0 \exp[i(\omega t + \varphi_x + \pi/2)]\}.$$

Because $e^{i\pi/2} = \cos(\pi/2) + i\sin(\pi/2) = i$, this gives

$$u(t) = \dot{x} = \text{Re}\{i\omega x_0 e^{i(\omega t + \varphi_x)}\} = \text{Re}\{i\omega \hat{x} e^{i\omega t}\}. \tag{2.39}$$

Differentiation of Eq. (2.34) gives the same result:

$$u = \dot{x} = \frac{1}{2}(i\omega \hat{x} e^{i\omega t} - i\omega \hat{x}^* e^{-i\omega t})$$
$$= \frac{1}{2}(i\omega \hat{x} e^{i\omega t} + \text{c.c.}) \tag{2.40}$$
$$= \text{Re}\{i\omega \hat{x} e^{i\omega t}\}.$$

A more straightforward way to arrive at the same result, however, is by using the fact that $d(\text{Re}\{\hat{x}e^{i\omega t}\})/dt = \text{Re}\{d(\hat{x}e^{i\omega t})/dt\}$, which immediately gives the result. Since $u(t) = \text{Re}\{\hat{u}e^{i\omega t}\}$, the complex velocity amplitude is therefore

$$\hat{u} = i\omega \hat{x}. \tag{2.41}$$

Thus, to differentiate a physical quantity (in harmonic oscillation) with respect to time $t$ means to multiply the corresponding complex amplitude by $i\omega$. Formally, $i\omega$ is a simpler operator than $d/dt$. This is one advantage of using the complex representation of harmonic oscillations.

Similarly, the acceleration is

$$a(t) = \dot{u} = \frac{du}{dt} = \text{Re}\{i\omega\hat{u}e^{i\omega t}\}. \tag{2.42}$$

Thus, the complex acceleration amplitude is

$$\hat{a} = i\omega\hat{u} = i\omega\, i\omega\hat{x} = -\omega^2\hat{x}. \tag{2.43}$$

In analogy with Eq. (2.32), we may use Eqs. (2.11) and (2.13) to write the complex amplitudes for the external force and the velocity as

$$\hat{F} = F_0 e^{i\varphi_F}, \quad \hat{u} = u_0 e^{i\varphi_u}, \tag{2.44}$$

respectively. Because

$$\hat{u} = \dot{\hat{x}} = \omega x_0 \exp[i(\varphi_x + \pi/2)], \tag{2.45}$$

we have

$$u_0 = \omega x_0, \qquad \varphi_u = \varphi_x + \pi/2. \tag{2.46}$$

Similarly,

$$a_0 = \omega u_0 = \omega^2 x_0, \qquad \varphi_a = \varphi_u + \pi/2 = \varphi_x + \pi. \tag{2.47}$$

The complex amplitudes $\hat{x}$, $\hat{u}$ and $\hat{a}$ are illustrated in the phase diagram of Figure 2.7. Instead of letting the complex-plane vectors (the phasors) rotate in

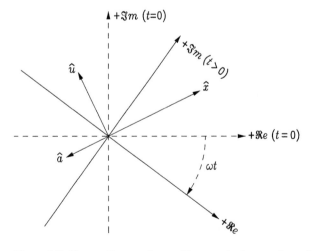

Figure 2.7: Phasor diagram for position $x$, velocity $u$ and acceleration $a$.

the counterclockwise direction, we may envisage the phasors being stationary, while the coordinate system (the real and imaginary axes) rotates in the clockwise direction. Projection of the phasors on the real axis equals the instantaneous values of position $x(t)$, velocity $u(t)$ and acceleration $a(t)$.

### 2.2.2 Mechanical Impedance

We shall now see how the dynamic equation (2.2) will be simplified, when we represent sinusoidally varying quantities by their complex amplitudes. Inserting

$$F(t) = \tfrac{1}{2}(\hat{F}e^{i\omega t} + \hat{F}^* e^{-i\omega t}), \tag{2.48}$$

$$x(t) = \tfrac{1}{2}(\hat{x}e^{i\omega t} + \hat{x}^* e^{-i\omega t}) \tag{2.49}$$

into Eq. (2.2) gives

$$\left[\hat{F} - \left(R + i\omega m + \frac{S}{i\omega}\right)\hat{u}\right]e^{i\omega t} + \left[\hat{F}^* - \left(R - i\omega m - \frac{S}{i\omega}\right)\hat{u}^*\right]e^{-i\omega t} = 0. \tag{2.50}$$

Introducing the complex *mechanical impedance*

$$Z = R + i\omega m + S/(i\omega) = R + i(\omega m - S/\omega) \tag{2.51}$$

gives

$$(\hat{F} - Z\hat{u})e^{i\omega t} + (\hat{F}^* - Z^*\hat{u}^*)e^{-i\omega t} = 0. \tag{2.52}$$

This equation is satisfied for arbitrary $t$ if $\hat{F} - Z\hat{u} = 0$, giving

$$\hat{u} = \hat{F}/Z \tag{2.53}$$

or

$$\hat{x} = \hat{u}/(i\omega) = \hat{F}/(i\omega Z). \tag{2.54}$$

When the absolute value (modulus) is taken on both sides of this equation, Eq. (2.14) is obtained—if Eq. (2.44) is also observed.

An electric impedance is the ratio between complex amplitudes of voltage and current. Analogously, the mechanical impedance is the ratio between the complex force amplitude and the complex velocity amplitude. In SI units, mechanical impedance has the dimension $[Z] = \text{Ns/m} = \text{kg/s}$. The impedance $Z = R + iX$ is a complex function of $\omega$. Here $R$ is the *mechanical resistance*, and $X = \omega m - S/\omega$ is the *mechanical reactance*, which is indicated as a function of $\omega$ in Figure 2.8. Note that $X = 0$ for $\omega = \omega_0 = \sqrt{S/m}$. For this frequency, corresponding to resonance, the velocity-amplitude response is maximum:

$$|\hat{u}/\hat{F}|_{\max} = |u_0/F_0|_{\max} = 1/|Z|_{\min} = 1/R. \tag{2.55}$$

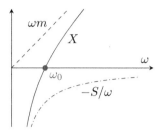

Figure 2.8: Mechanical reactance of the oscillating system versus frequency.

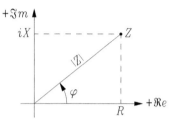

Figure 2.9: Complex-plane decomposition of the mechanical impedance $Z$.

The impedance $Z$, as shown in the complex plane of Figure 2.9, is

$$Z = R + iX = |Z|e^{i\varphi}, \tag{2.56}$$

where

$$|Z| = \sqrt{R^2 + X^2}, \tag{2.57}$$

in accordance with Eq. (2.17). Further, we have

$$\cos \varphi = R/|Z|, \quad \sin \varphi = X/|Z|, \tag{2.58}$$

from which Eq. (2.16) follows. Because $R > 0$, we have $|\varphi| < \pi/2$. At resonance, $\varphi = 0$ because $X = 0$.

Inserting Eq. (2.56) into Eq. (2.53), we have

$$\hat{u} = \frac{\hat{F}}{Z} = \frac{|\hat{F}|e^{i\varphi_F}}{|Z|e^{i\varphi}} = |\hat{F}/Z| \exp\left[i(\varphi_F - \varphi)\right] \tag{2.59}$$

or

$$|\hat{u}| = |\hat{F}|/|Z|, \quad \varphi_u = \varphi_F - \varphi. \tag{2.60}$$

We may choose the time origin $t = 0$ such that $\varphi_F = 0$; that is, $\hat{F}$ is real and positive. Then $\varphi_u = -\varphi$. A case with positive $\varphi$ is illustrated with phasors in Figure 2.10 and with functions of time in Figure 2.11.

When $\varphi > 0$ (as in Figures 2.10 and 2.11), the velocity $u$ is said to lag in phase, or to have a phase lag, relative to the force $F$. In this case, corresponding to $\omega > \omega_0$, the reactance is dominated by the inertia term $\omega m$. In the opposite case, $\omega_0 > \omega$, $\varphi < 0$, the reactance is dominated by the stiffness, and the velocity leads in phase (or has a phase advance) relative to the force.

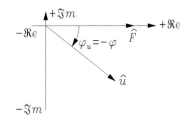

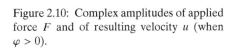
Figure 2.10: Complex amplitudes of applied force $F$ and of resulting velocity $u$ (when $\varphi > 0$).

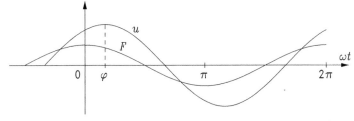

Figure 2.11: Applied force $F$ and resulting velocity $u$ versus time (when $\varphi > 0$). The velocity has a phase lag relative to the force.

Frequency-response diagrams for two different values of damping resistance $R$ are shown in Figure 2.3. If $\omega_0 \ll \omega$, we have $Z \approx i\omega m$; that is, $\hat{u} \approx \hat{F}/(i\omega m)$ or $\hat{a} = i\omega\hat{u} \approx \hat{F}/m$. Thus,

$$ma(t) \approx F(t). \tag{2.61}$$

In contrast, if $\omega_0 \gg \omega$, we have $Z \approx -iS/\omega$; that is, $\hat{u} \approx i\omega\hat{F}/S$ or $\hat{x} = \hat{u}/i\omega \approx \hat{F}/S$. Hence,

$$Sx(t) \approx F(t). \tag{2.62}$$

The differential equation (2.2) expresses how the applied external force is balanced against inertia, damping and stiffness forces. The force balance may also be represented by complex amplitude relations:

$$m\hat{a} + R\hat{u} + S\hat{x} = \hat{F}, \tag{2.63}$$

where $\hat{a} = i\omega\hat{u}$ and $\hat{x} = \hat{u}/i\omega$, or

$$(R + iX)\hat{u} = Z\hat{u} = \hat{F}, \tag{2.64}$$

or, graphically, by a force-phasor diagram. Figure 2.12 shows a phasor diagram for the numerical example $R = 4$ kg/s, $\omega m = 5$ kg/s and $S/\omega = 2$ kg/s. For those values, we obtain $X = 3$ kg/s, $|Z| = 5$ kg/s and $\varphi = 0.64$ rad $= 37°$. The damping force $R\hat{u}$ is in phase with the velocity $\hat{u}$. The stiffness force $S\hat{x}$ and the inertia force $m\hat{a}$ are in anti-phase. The former has a phase lag of $\pi/2$, and the latter a phase lead of $\pi/2$, relative to the velocity.

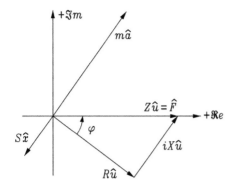

Figure 2.12: Phasor diagram showing balance of forces.

### 2.3 Power and Energy Relations

We shall now consider energy delivered by the external force to the simple mechanical oscillating system, shown in Figure 2.1, and see how this energy is exchanged with energy consumed in the damping resistance $R$ and with stored potential and kinetic energy in the spring $S$ and mass $m$. The mechanical power (rate of work) delivered by supplying the external force $F(t)$ is

$$P(t) = F(t)u(t) = F(t)\dot{x}(t), \tag{2.65}$$

where

$$\begin{aligned} F(t) &= F_m(t) - F_R(t) - F_S(t) \\ &= ma(t) + Ru(t) + Sx(t), \end{aligned} \tag{2.66}$$

in accordance with the dynamic equation (2.2). Thus, we have

$$P(t) = P_R(t) + \left[ P_k(t) + P_p(t) \right], \tag{2.67}$$

where the powers delivered to $R$, $m$ and $S$ are

$$P_R(t) = -F_R(t)u(t) = Ru^2, \tag{2.68}$$

$$P_k(t) = F_m(t)u(t) = m\dot{u}u = \frac{d}{dt}W_k(t), \tag{2.69}$$

$$P_p(t) = -F_S(t)u(t) = S\dot{x}x = \frac{d}{dt}W_p(t), \tag{2.70}$$

respectively, where

$$W_k(t) = m[u(t)]^2/2 \tag{2.71}$$

is the kinetic energy and

$$W_p(t) = S[x(t)]^2/2 \tag{2.72}$$

is the potential energy stored in spring $S$. The energy stored in the oscillating system is

$$W(t) = W_k(t) + W_p(t), \tag{2.73}$$

and the corresponding power or rate of change of energy is

$$P_k(t) + P_p(t) = \frac{d}{dt} W(t). \tag{2.74}$$

The delivered power $P(t)$ has two components:

1. $P_R(t)$, which is consumed in the damping resistance $R$ (cf. instantaneous 'active power'), and
2. $P_k(t) + P_p(t)$ (cf. instantaneous 'reactive power'), which is exchanged with stored kinetic energy (in mass $m$) and potential energy (in spring $S$).

### 2.3.1 Harmonic Oscillation: Active Power and Reactive Power

With harmonic oscillation, we have

$$
\begin{aligned}
P(t) &= F(t)\,u(t) \\
&= \tfrac{1}{2}(\hat{F}e^{i\omega t} + \hat{F}^* e^{-i\omega t})\tfrac{1}{2}(\hat{u}e^{i\omega t} + \hat{u}^* e^{-i\omega t}) \\
&= \tfrac{1}{4}(\hat{F}\hat{u}^* + \hat{F}^*\hat{u} + \hat{F}\hat{u}e^{2i\omega t} + \hat{F}^*\hat{u}^* e^{-2i\omega t}).
\end{aligned} \tag{2.75}
$$

The sum of the two last terms, being complex conjugate to each other, is a harmonic oscillation with angular frequency $2\omega$, and hence, it has a time-average value equal to zero. The delivered time-average power is

$$P \equiv \overline{P(t)} = \frac{1}{4}\hat{F}\hat{u}^* + \text{c. c.} = \frac{1}{2}\text{Re}\{\hat{F}\hat{u}^*\} \tag{2.76}$$

$$= \frac{1}{2}\text{Re}\{Z\hat{u}\hat{u}^*\} = \frac{1}{2}\text{Re}\{Z|\hat{u}|^2\} = \frac{R}{2}|\hat{u}|^2$$

$$= \frac{R}{2}\left|\frac{\hat{F}}{Z}\right|^2 = \frac{R}{2|Z|^2}|\hat{F}|^2 = \frac{R}{2(R^2 + X^2)}|\hat{F}|^2. \tag{2.77}$$

The consumed power is

$$
\begin{aligned}
P_R(t) &= Ru^2 = \frac{R}{4}\left(\hat{u}e^{i\omega t} + \hat{u}^* e^{-i\omega t}\right)^2 \\
&= \frac{R}{4}\left(2\hat{u}\hat{u}^* + \hat{u}^2 e^{i2\omega t} + \hat{u}^{*2}e^{-i2\omega t}\right).
\end{aligned} \tag{2.78}
$$

Here, the two last terms have a zero time-average sum. Thus,

$$P_R = \overline{P_R(t)} = \frac{R}{2}\hat{u}\hat{u}^* = \frac{R}{2}|\hat{u}|^2 = \overline{P(t)}. \tag{2.79}$$

Hence, the consumed power and the delivered power are equal in time average.

The instantaneous values are (if we choose $\varphi_F = 0$)

$$
\begin{aligned}
P(t) &= \overline{P(t)} + (\tfrac{1}{4}\hat{F}\hat{u}e^{2i\omega t} + \text{c. c.}) \\
&= \overline{P(t)} + \tfrac{1}{2}|\hat{F}\hat{u}|\cos(2\omega t - \varphi)
\end{aligned}
\tag{2.80}
$$

for the delivered power and

$$
\begin{aligned}
P_R(t) &= \overline{P(t)} + (\tfrac{1}{4}\hat{R}\hat{u}^2 e^{2i\omega t} + \text{c. c.}) \\
&= \overline{P(t)} + \tfrac{1}{2}\hat{R}|\hat{u}|^2\cos(2\omega t - 2\varphi)
\end{aligned}
\tag{2.81}
$$

for the consumed power. Thus, in general, $P_R(t) \neq P(t)$. The difference $P(t) - P_R(t) = P_k(t) + P_p(t)$ is instantaneous reactive power exchanged with stored energy in the system.

In the special case of resonance ($\omega = \omega_0$, $\varphi = 0$, $Z = R$), we have $P_R(t) = P(t) = (R/2)|\hat{u}|^2(1 + \cos 2\omega t) = R|\hat{u}|^2\cos^2 \omega t$. Hence, at resonance, there is no reactive power delivered. The stored energy is constant. It is alternating between kinetic energy and potential energy, which have equal maximum values.

In general, the stored instantaneous kinetic energy is

$$
\begin{aligned}
W_k(t) &= \frac{m}{2}[u(t)]^2 = \frac{m}{2}|\hat{u}|^2 \cos^2 (\omega t - \varphi) \\
&= \frac{m}{4}|\hat{u}|^2 [1 + \cos(2\omega t - 2\varphi)],
\end{aligned}
\tag{2.82}
$$

where the average kinetic energy is

$$
W_k \equiv \overline{W_k(t)} = \frac{m}{4}|\hat{u}|^2 = \frac{m}{4}\hat{u}\hat{u}^*,
\tag{2.83}
$$

and the instantaneous potential energy is

$$
\begin{aligned}
W_p(t) &= \frac{S}{2}[x(t)]^2 = \frac{S}{2}|\hat{x}|^2 \sin^2 (\omega t - \varphi) \\
&= \frac{S}{4}|\hat{x}|^2 [1 - \cos (2\omega t - 2\varphi)],
\end{aligned}
\tag{2.84}
$$

where the average potential energy is

$$
W_p \equiv \overline{W_p(t)} = \frac{S}{4}|\hat{x}|^2 = \frac{S}{4\omega^2}|\hat{u}|^2.
\tag{2.85}
$$

The instantaneous total stored energy is

$$
\begin{aligned}
W(t) &= W_k(t) + W_p(t) \\
&= W_k + W_p + (W_k - W_p)\cos (2\omega t - 2\varphi),
\end{aligned}
\tag{2.86}
$$

which is of time average

$$
\begin{aligned}
W \equiv \overline{W(t)} &= W_k + W_p = \frac{1}{4}\left(m|\hat{u}|^2 + S|\hat{x}|^2\right) \\
&= \frac{m}{4}|\hat{u}|^2\left[1 + \left(\frac{\omega_0}{\omega}\right)^2\right].
\end{aligned}
\tag{2.87}
$$

The amplitude of the oscillating part of the total stored energy is

$$W_k - W_p = \frac{1}{4\omega}\left(\omega m - \frac{S}{\omega}\right)|\hat{u}|^2 = \frac{X}{4\omega}|\hat{u}|^2 = \frac{X}{4\omega}\hat{u}\hat{u}^*. \tag{2.88}$$

The instantaneous reactive power is

$$P_k(t) + P_p(t) = \frac{d}{dt}W(t) = -2\omega(W_k - W_p)\sin(2\omega t - 2\varphi)$$
$$= -\frac{X}{2}|\hat{u}|^2 \sin(2\omega t - 2\varphi). \tag{2.89}$$

We observe that the reactance and the reactive power may be related to the difference between the kinetic energy and the potential energy stored in the system. At resonance ($\omega = \omega_0$), the maximum kinetic energy $m|\hat{u}|^2/2$ equals the maximum potential energy $S|\hat{x}|^2/2 = S\omega_0^{-2}|\hat{u}|^2/2$, and the reactance vanishes. If $\omega < \omega_0$, the maximum potential energy is larger than the maximum kinetic energy, and the mechanical reactance differs from zero (and is negative) because some of the stored energy has to be exchanged with the external energy, back and forth twice every oscillation period, due to the imbalance between the two types of energy store within the system. If $\omega > \omega_0$, the maximum kinetic energy is larger than the maximum potential energy, and the mechanical reactance is positive. It may be noted that the average values $W_k$ and $W_p$ are just half of the maximum values $m|\hat{u}|^2/2$ and $S|\hat{x}|^2/2$ of the kinetic and potential energies, respectively.

If we define the delivered 'complex power' as

$$\mathcal{P} = \tfrac{1}{2}\hat{F}\hat{u}^* = \tfrac{1}{2}Z\hat{u}\hat{u}^* = \tfrac{1}{2}R|\hat{u}|^2 + i\tfrac{1}{2}X|\hat{u}|^2, \tag{2.90}$$

we see from Eqs. (2.76) and (2.81) that Re$\{\mathcal{P}\} = P$ equals the average delivered power, the average consumed power and the amplitude of the oscillating part of the consumed power. Moreover, from Eq. (2.89), we see that Im$\{\mathcal{P}\}$ is the amplitude of the instantaneous reactive power. Finally, we see from Eq. (2.80) that $|\mathcal{P}|$ is the amplitude of the oscillating part of the delivered power. Note that the time-independent quantities Re$\{\mathcal{P}\}$, Im$\{\mathcal{P}\}$ and $|\mathcal{P}|$ are sometimes called *active power*, *reactive power* and *apparent power*, respectively.

## 2.4 State-Space Analysis

The simple oscillator shown in Figure 2.1 is represented by Eq. (2.2), which is a second-order linear differential equation with constant coefficients. It is convenient to reformulate it as the following set of two simultaneous differential equations of first order:

$$\dot{x}_1 = x_2, \quad \dot{x}_2 = -\frac{R}{m}x_2 - \frac{S}{m}x_1 + \frac{1}{m}u_1, \tag{2.91}$$

where we have introduced the state variables

$$x_1(t) = x(t), \qquad x_2(t) = u(t) = \dot{x}(t) \tag{2.92}$$

and the input variable

$$u_1(t) = F(t). \tag{2.93}$$

Equations (2.91) may be written in matrix notation as

$$\begin{bmatrix} \dot{x}_1 \\ \dot{x}_2 \end{bmatrix} = \begin{bmatrix} 0 & 1 \\ -S/m & -R/m \end{bmatrix} \begin{bmatrix} x_1 \\ x_2 \end{bmatrix} + \begin{bmatrix} 0 \\ 1/m \end{bmatrix} u_1. \tag{2.94}$$

For a more general case of a linear system which is represented by linear differential equations with constant coefficients, these may be represented in the state-variable form [13]:

$$\dot{\mathbf{x}} = \mathbf{A}\mathbf{x} + \mathbf{B}\mathbf{u}, \tag{2.95}$$

$$\mathbf{y} = \mathbf{C}\mathbf{x} + \mathbf{D}\mathbf{u}. \tag{2.96}$$

If there are $n$ state variables $x_1(t), \ldots, x_n(t)$, $r$ input variables $u_1(t), \ldots, u_r(t)$ and $m$ output variables $y_1(t), \ldots, y_m(t)$, then the system matrix (or state matrix) $\mathbf{A}$ is of dimension $n \times n$, the input matrix $\mathbf{B}$ of dimension $n \times r$ and the output matrix $\mathbf{C}$ of dimension $m \times n$. The $m \times r$ matrix $\mathbf{D}$, which is zero in many cases, represents direct coupling between the input and the output.

When the identity matrix is denoted by $\mathbf{I}$, the solution of the vectorial differential equation (2.95) may be written as (see Problem 2.9)

$$\mathbf{x}(t) = e^{\mathbf{A}(t-t_0)}\mathbf{x}(t_0) + \int_{t_0}^{t} e^{\mathbf{A}(t-\tau)}\mathbf{B}\mathbf{u}(\tau)d\tau, \tag{2.97}$$

where the matrix exponential is defined as the series

$$e^{\mathbf{A}t} = \mathbf{I} + \mathbf{A}t + \frac{1}{2!}\mathbf{A}^2 t^2 + \frac{1}{3!}\mathbf{A}^3 t^3 + \cdots , \tag{2.98}$$

which is a matrix of dimension $n \times n$ and which commutes with the matrix $\mathbf{A}$. Hence, we may perform algebraic manipulations (including integration and differentiation with respect to $t$) with $e^{\mathbf{A}t}$ in the same way as we do with $e^{st}$ where $s$ is a scalar constant. Inserting Eq. (2.97) into Eq. (2.96) gives the output vector

$$\mathbf{y}(t) = \mathbf{C}e^{\mathbf{A}(t-t_0)}\mathbf{x}(t_0) + \int_{t_0}^{t} \mathbf{C}e^{\mathbf{A}(t-\tau)}\mathbf{B}\mathbf{u}(\tau)d\tau + \mathbf{D}\mathbf{u}(t) \tag{2.99}$$

in terms of the input vector $\mathbf{u}(t)$ and the initial-value vector $\mathbf{x}(t_0)$. In many cases, we analyse situations where $\mathbf{u}(t) = 0$ for $t < t_0$ and $\mathbf{x}(t_0) = 0$. Then the first right-hand term in Eq. (2.99) vanishes. Note from the integral term in Eq. (2.99) that the output $\mathbf{y}$ at time $t$ is influenced by the input $\mathbf{u}$ at an earlier time $\tau$ ($\tau < t$).

It may be difficult to make an accurate numerical computation of the matrix exponential, particularly if the matrix order $n$ is large. Various methods have been proposed for computation and discussion of the matrix exponential [14, 15]. Which method is best depends on the particular problem.

In some cases (but far from always), the system matrix $\mathbf{A}$ has $n$ linearly independent eigenvectors $\mathbf{m}_i$ ($i = 1, 2, \ldots, n$), where

$$\mathbf{A}\mathbf{m}_i = \Lambda_i \mathbf{m}_i. \tag{2.100}$$

The eigenvalues $\Lambda_i$, which may or may not be distinct, are solutions of (roots in) the $n$th degree equation

$$\det(\mathbf{A} - \Lambda\mathbf{I}) \equiv |\mathbf{A} - \Lambda\mathbf{I}| = 0. \tag{2.101}$$

It may be mathematically convenient to transform the system matrix $\mathbf{A}$ to a particular (usually simpler) matrix $\mathbf{A}'$ by means of a similarity transformation [13, 16]

$$\mathbf{A}' = \mathbf{M}^{-1}\mathbf{A}\mathbf{M}, \tag{2.102}$$

where $\mathbf{M}$ is a non-singular $n \times n$ matrix, which implies that its inverse $\mathbf{M}^{-1}$ exists. It is well-known that $\mathbf{A}'$ has the same eigenvalues as $\mathbf{A}$. If the matrix $\mathbf{A}$ has $n$ linearly independent eigenvectors $\mathbf{m}_i$, then $\mathbf{A}'$ is diagonal,

$$\mathbf{A}' = \text{diag}(\Lambda_1, \Lambda_2, \ldots, \Lambda_n), \tag{2.103}$$

if $\mathbf{M}$ is chosen as

$$\mathbf{M} = (\mathbf{m}_1, \mathbf{m}_2, \ldots, \mathbf{m}_n), \tag{2.104}$$

which is the $n \times n$ matrix whose $i$th column is the eigenvector $\mathbf{m}_i$. Because all the vectors $\mathbf{m}_i$ are linearly independent, this implies that the inverse matrix $\mathbf{M}^{-1}$ exists. In this case the matrix exponential is also simply a diagonal matrix,

$$e^{\mathbf{A}'t} = \text{diag}\left(e^{\Lambda_1 t}, e^{\Lambda_2 t}, \ldots, e^{\Lambda_n t}\right). \tag{2.105}$$

Introducing also the transformed state vector

$$\mathbf{x}' = \mathbf{M}^{-1}\mathbf{x} \quad (\text{i.e., } \mathbf{x} = \mathbf{M}\mathbf{x}') \tag{2.106}$$

and the transformed matrices

$$\mathbf{B}' = \mathbf{M}^{-1}\mathbf{B}, \quad \mathbf{C}' = \mathbf{C}\mathbf{M}, \tag{2.107}$$

we rewrite Eqs. (2.95) and (2.96) as

$$\dot{\mathbf{x}}' = \mathbf{A}'\mathbf{x}' + \mathbf{B}'\mathbf{u}, \tag{2.108}$$

$$\mathbf{y} = \mathbf{C}'\mathbf{x}' + \mathbf{D}\mathbf{u}. \tag{2.109}$$

In cases in which the matrix $\mathbf{A}'$ and hence also its exponential are diagonal, as in Eqs. (2.103) and (2.105), it is easy to compute the output $\mathbf{y}(t)$ from Eq. (2.99) with $\mathbf{A}$, $\mathbf{B}$, $\mathbf{C}$ and $\mathbf{x}_0$ replaced by $\mathbf{A}'$, $\mathbf{B}'$, $\mathbf{C}'$ and $\mathbf{x}_0'$, respectively.

The use of state-space analysis is a well-known mathematical tool, for instance in control engineering [13]. Except when the state-space dimension $n$ is rather low, such as $n = 2$ in the example below, the state-space analysis is performed numerically [14, 15].

As a simple example, we again use the oscillator shown in Figure 2.1, in which we shall consider the force $F(t)$ as input, $u_1(t) = F(t)$, and the displacement $x(t)$ as output, $y_1(t) = x(t) = x_1(t)$. The matrices to use with Eq. (2.96) are

$$\mathbf{C} = \begin{bmatrix} 1 & 0 \end{bmatrix}, \quad \mathbf{D} = \begin{bmatrix} 0 & 0 \end{bmatrix}, \tag{2.110}$$

and with Eq. (2.95),

$$\mathbf{A} = \begin{bmatrix} 0 & 1 \\ -S/m & -R/m \end{bmatrix}, \quad \mathbf{B} = \begin{bmatrix} 0 \\ 1/m \end{bmatrix}. \tag{2.111}$$

See Eq. (2.94). Thus, with $r = 1$ and $m = 1$, this is a SISO (single-input single-output) system. The state matrix is of dimension $2 \times 2$. With this example, the eigenvalue equation (2.101) has the two solutions

$$\Lambda_1 = -\delta + i\omega_d, \quad \Lambda_2 = -\delta - i\omega_d, \tag{2.112}$$

where $\delta$ and $\omega_d$ are given by Eq. (2.5). The corresponding eigenvectors are (see Problem 2.10)

$$\mathbf{m}_1 = c_1 \begin{bmatrix} 1 \\ \Lambda_1 \end{bmatrix}, \quad \mathbf{m}_2 = c_2 \begin{bmatrix} 1 \\ \Lambda_2 \end{bmatrix}, \tag{2.113}$$

where $c_1$ and $c_2$ are arbitrary constants, which may be chosen equal to 1. These eigenvectors are linearly independent if $\Lambda_2 \neq \Lambda_1$ (i.e., $\omega_d \neq 0$). The application of Eqs. (2.99) to (2.109) results in an output which agrees with Eqs. (2.4) and (2.6) for the case of free oscillations—that is, when $F(t) = 0$ (see Problem 2.10). For the case of 'critical damping' ($\omega_d = 0$ and $\Lambda_1 = \Lambda_2 = \Lambda = -\delta$), there are not two linearly independent eigenvectors. Then neither $\mathbf{A}'$ nor $e^{\mathbf{A}'t}$ is diagonal (see Problem 2.11).

## 2.5 Linear Systems

We have discussed in detail the simple mechanical system in Figure 2.1 (or the analogous electric system shown in Figure 2.4). It obeys the constant-coefficient linear differential equation (2.2). Further, we have just touched upon another linear system, represented by a more general set of linear differential equations with constant coefficients, Eq. (2.95). Next, let us discuss linear systems more generally.

A system may be defined as a collection of components, parts or units which influence each other mutually through relations between causes and resulting effects. Some of the interrelated physical variables denoted by $u_1(t), u_2(t), \ldots, u_r(t)$ may be considered as input to the system, whereas other variables, denoted by $y_1(t), y_2(t), \ldots, y_m(t)$, are considered as outputs. We may contract these variables into an $r$-dimensional input vector $\mathbf{u}(t)$ and an $m$-dimensional output vector $\mathbf{y}(t)$.

An example is the mechanical system shown in Figure 2.1, where the external force $F(t) = u_1(t)$ is the input variable, and the output variables are the excursion $x(t) = y_1(t)$, the velocity $u(t) = y_2(t)$ and the acceleration $a(t) = y_3(t)$ of the oscillating mass. In this case, the dimensions of the input and output vectors are $r = 1$ and $m = 3$, respectively. If the system parameters $R$, $m$ and $S$ are constant or, more precisely, independent of the oscillation amplitudes, then this particular system is linear. However, if the power $P_R(t) = R[u(t)]^2 = y_4(t)$ consumed by the resistance $R$ had also been included as a fourth member of our chosen set of output variables, then our mathematically considered system would have been nonlinear. For instance, if the input $u_1(t) = F(t)$ had been doubled, then also $y_1(t)$, $y_2(t)$ and $y_3(t)$ would have been doubled, but $y_4(t)$ would have been increased by a factor of four.

The mathematical relationship between the variables may be expressed as

$$\mathbf{y}(t) = \mathbf{T}\{\mathbf{u}(t)\}, \tag{2.114}$$

where the operator symbol $\mathbf{T}$ designates the law for determining $\mathbf{y}(t)$ from $\mathbf{u}(t)$. If we in Eq. (2.99) set $\mathbf{x}(t_0) = \mathbf{0}$ and $\mathbf{D} = \mathbf{0}$, we have the following example of such a relationship:

$$\mathbf{y}(t) = \int_{t_0}^{t} \mathbf{C} e^{\mathbf{A}(t-\tau)} \mathbf{B} \mathbf{u}(\tau) \, d\tau. \tag{2.115}$$

With this example, the state variables obey a set of differential equations with constant coefficients and all the initial values at $t = t_0$ are zero. The output $\mathbf{y}$ at time $t$ is a result of the input $\mathbf{u}(t)$, only, during the time interval $(t_0, t)$.

A system is linear if, for any two input vectors $\mathbf{u}_a(t)$ and $\mathbf{u}_b(t)$ with corresponding output vectors $\mathbf{y}_a(t) = \mathbf{T}\{\mathbf{u}_a(t)\}$ and $\mathbf{y}_b(t) = \mathbf{T}\{\mathbf{u}_b(t)\}$, we have

$$\alpha_a \mathbf{y}_a(t) + \alpha_b \mathbf{y}_b(t) = \mathbf{T}\{\alpha_a \mathbf{u}_a(t) + \alpha_b \mathbf{u}_b(t)\} \tag{2.116}$$

for arbitrary constants $\alpha_a$ and $\alpha_b$. The output and input in Eq. (2.116) are sums of two terms. It is straightforward to generalise this to a finite number of terms. Extension to infinite sums and integrals is an additional requirement which we shall include in our definition of a linear system. According to this definition, the superposition principle applies for linear systems. Since $\mathbf{A}$, $\mathbf{B}$ and $\mathbf{C}$ are constant matrices, the system represented by Eq. (2.115) is linear.

In some simple cases, there is an instantaneous relation between input and output—for instance, between the voltage across an electric resistance and the resulting electric current or between the acceleration of a body and the net force applied to it. In most cases, systems are dynamic. Then the instantaneous output variables depend both on previous and present values of the input variables. For instance, the velocity of a body depends on the force applied to the body at previous instants.

In some cases, it may be convenient to define a system in which both the input and the output are chosen to be variables caused by some other variable(s). Then it may happen that the present-time output variables may even be influenced by future values of the input variables. Such a system is said to be noncausal. Otherwise, for causal systems the output (the response) cannot exist before the input (the cause).

In the remainder of this chapter, let us consider a linear system with a single input ($r = 1$) and a single output ($m = 1$). The input function $u(t)$ and the output function $y(t)$ are then scalar functions. For this case, we write

$$y(t) = L\{u(t)\} \tag{2.117}$$

instead of Eq. (2.114). (Because the system is linear, we have written $L$ instead of $T$ to designate the law determining the system's output from its input.) In Sections 2.5.2 and 2.6.2, we introduce a mathematical function, the transfer function, which characterises the linear system.

### 2.5.1 The Delta Function and Related Distributions

As a preparation, let us first give a few mathematical definitions. Let $\varphi(\tau)$ be an arbitrary function which is continuous and infinitely many times differentiable at $t = \tau$. Then the impulse function $\delta(t)$, also called the Dirac delta function, delta function, or, more properly, delta distribution, is defined [17] by the property

$$\varphi(0) = \int_{-\infty}^{\infty} \delta(t)\varphi(t)\, dt \tag{2.118}$$

or, more generally,

$$\varphi(t) = \int_{-\infty}^{\infty} \delta(\tau - t)\varphi(\tau)\, d\tau. \tag{2.119}$$

In particular, if $\varphi(t) \equiv 1$ and $t = 0$, then this gives

$$\int_{-\infty}^{\infty} \delta(\tau)\, d\tau = 1. \tag{2.120}$$

Although the delta distribution is not a mathematical function in the ordinary sense, it is a meaningful statement to say that for $t \neq 0$, $\delta(t) = 0$. The derivative $\dot{\delta}(t) = d\delta(t)/dt$ of the impulse function is defined by

$$\dot{\varphi}(0) = -\int_{-\infty}^{\infty} \dot{\delta}(t)\,\varphi(t)\, dt, \tag{2.121}$$

and the $n$th derivative $\delta^{(n)}(t) = d^n \delta(t)/dt^n$ by

$$\varphi^{(n)}(0) = (-1)^n \int_{-\infty}^{\infty} \delta^{(n)}(t)\, \varphi(t)\, dt. \tag{2.122}$$

The impulse function $\delta(t)$ is even and the derivative $\delta^{(n)}(t)$ is even if $n$ is even, and odd if $n$ is odd [17].

The function

$$\text{sgn}(t) = 2U(t) - 1 = \begin{cases} 1 & \text{for } t > 0 \\ 0 & \text{for } t = 0 \\ -1 & \text{for } t < 0 \end{cases} \tag{2.123}$$

is an odd function, whose derivative is $2\delta(t)$. In Eq. (2.123), we have introduced the (Heaviside) unit step function $U(t)$, which equals 1 for $t > 0$ and 0 for $t < 0$. Note that $\dot{U}(t) = \delta(t)$.

### 2.5.2 Impulse Response: Time-Invariant System

Let us next assume that the input to a linear system is an impulse at time $t_1$. By this, we mean that the input function is $u(t) = \delta(t - t_1)$. The system's output $y(t) = h(t, t_1)$, which corresponds to the impulse input, is called the *impulse response*. By means of Eq. (2.119), we may write an arbitrary input as

$$u(t) = \int_{-\infty}^{\infty} u(t_1)\, \delta(t - t_1)\, dt_1, \tag{2.124}$$

which may be interpreted as a superposition of impulse inputs $u(t_1)\, \delta(t - t_1)\, dt_1$. Because the system is linear, we may superpose the corresponding outputs $u(t_1)\, h(t, t_1)\, dt_1$. Thus, the resulting output is

$$y(t) = \int_{-\infty}^{\infty} u(t_1)\, h(t, t_1)\, dt_1 \equiv L\{u(t)\}. \tag{2.125}$$

If we know the impulse response $h(t, t_1)$, we can use Eq. (2.125) to find the output corresponding to an arbitrary given input.

A linear system in which the impulse response $h(t, t_1)$ depends on $t$ and $t_1$ only through the time difference $t - t_1$ is called a time-invariant linear system. Then the integral in Eq. (2.125) becomes a convolution integral

$$y(t) = \int_{-\infty}^{\infty} u(t_1)\, h(t - t_1)\, dt_1 \equiv u(t) * h(t). \tag{2.126}$$

Note that convolution is commutative, i.e., $h(t) * u(t) = u(t) * h(t)$. [See Eq. (2.145), which proves this statement of commutativity.]

In many cases, a system which is linear and time invariant may be represented by a set of simultaneous first-order differential equations with constant coefficients, as discussed in Section 2.4. Let us, as an example of such a case, consider a SISO system, for which Eq. (2.115) simplifies to

$$y(t) = \int_0^t \mathbf{C}e^{\mathbf{A}(t-\tau)}\mathbf{B}u(\tau)\,d\tau, \tag{2.127}$$

where the input matrix $\mathbf{B}$ is of dimension $n \times 1$ and the output matrix $\mathbf{C}$ of dimension $1 \times n$, and where we have chosen $t_0 = 0$. Thus, for this special case, the impulse response is given by

$$h(t) = \begin{cases} 0 & \text{for } t < 0 \\ \mathbf{C}e^{\mathbf{A}t}\mathbf{B} & \text{for } t > 0, \end{cases} \tag{2.128}$$

as is easily seen if we compare Eq. (2.127) with Eq. (2.126). For this example, the system is causal.

We say that a linear time-invariant system is causal if the impulse response vanishes for negative times—that is, if

$$h(t) = 0 \quad \text{for } t < 0. \tag{2.129}$$

Let us now return to the general case of a linear, time-invariant, SISO system. If the input is $u_e(t) = e^{i\omega t}$, then, with $y_e(t) = L\{e^{i\omega t}\}$, we have

$$\begin{aligned} y_e(t_1 + t) = L\{e^{i\omega(t_1+t)}\} &= L\{e^{i\omega t_1}e^{i\omega t}\} \\ &= e^{i\omega t_1}L\{e^{i\omega t}\} = u_e(t_1)\,y_e(t), \end{aligned} \tag{2.130}$$

where $t_1$ is an arbitrary constant (of dimension time). Here we utilise the fact that the system is time invariant and linear. Inserting $t = 0$ and $t_1 = t$ into Eq. (2.130), we obtain

$$y_e(t) = y_e(0)\,e^{i\omega t} \equiv H(\omega)e^{i\omega t}, \tag{2.131}$$

which shows that if the input is an exponential function of time, then the output equals the input multiplied by a time-independent coefficient, which we have denoted $H(\omega)$. If the input is

$$u(t) = \tfrac{1}{2}(\hat{u}e^{i\omega t} + \hat{u}^*e^{-i\omega t}), \tag{2.132}$$

then the output is

$$y(t) = \tfrac{1}{2}(\hat{y}e^{i\omega t} + \hat{y}^*e^{-i\omega t}), \tag{2.133}$$

with

$$\hat{y} = H(\omega)\hat{u}, \quad \hat{y}^* = H^*(\omega)\hat{u}^*. \tag{2.134}$$

Note that $H(\omega)$, which characterises the system, may be complex and dependent on $\omega$ (see Figure 2.13). We shall see in the next section that $H(\omega)$, which we call the system's *transfer function*, is the Fourier transform of the impulse response $h(t)$.

Figure 2.13: System with input signal $u(t)$ and output signal $y(t)$. The corresponding Fourier transforms are $U(\omega)$ and $Y(\omega)$ respectively. The linear system's impulse response $h(t)$ is the inverse Fourier transform of the transfer function $H(\omega)$.

Returning again to the example of forced oscillations in the system shown in Figure 2.1, where we consider the external force $F(t)$ as input and the velocity $u(t)$ as output, we see from Eq. (2.53) that the transfer function is $H(\omega) = 1/Z$, where $Z = Z(\omega)$ is the mechanical impedance, defined by Eq. (2.51).

## 2.6 Fourier Transform and Other Integral Transforms

We have previously studied forced sinusoidal oscillations (Section 2.1) and we have introduced complex amplitudes and phasors as convenient means to analyse sinusoidal oscillations (Section 2.2). In cases in which linear theory is applicable, the obtained results may also be useful with quantities which do not necessarily vary sinusoidally with time. This follows from the following two facts. Firstly, functions of a rather general class may be decomposed into harmonic components according to Fourier analysis. Secondly, the superposition principle is applicable when linear theory is valid. A condition for the success of this method of approach is that the physical systems considered are time invariant. This implies that the inherent characteristics of the system remain the same at any time (see Section 2.5).

Instead of studying how the complex amplitude of a physical quantity varies with frequency, we shall now study the physical quantity's variation with time. The main content of this section includes a short review of Fourier analysis and of the connection between causality and the Kramers–Kronig relations.

### 2.6.1 Fourier Transformation in Brief

It is assumed that the reader is familiar with Fourier analysis. For convenience, however, let us collect some of the main formulas related to the Fourier transformation. For a more rigorous treatment, the reader may consult many textbooks, such as those by Papoulis [17] or Bracewell [18].

If the function $f(t)$ belongs to a certain class of reasonably well-behaved functions, its Fourier transform $F(\omega)$ is defined by

$$F(\omega) = \int_{-\infty}^{\infty} f(t)\, e^{-i\omega t}\, dt \equiv \mathcal{F}\{f(t)\}, \tag{2.135}$$

and the inverse transform is

$$f(t) = \frac{1}{2\pi} \int_{-\infty}^{\infty} F(\omega)\, e^{i\omega t}\, d\omega \equiv \mathcal{F}^{-1}\{F(\omega)\}. \tag{2.136}$$

Table 2.2: Some relations between functions of time and their corresponding Fourier transforms.

| Function of Time | Fourier Transform |
|---|---|
| $f(t)$ | $F(\omega)$ |
| $a_1 f_1(t) + a_2 f_2(t)$ | $a_1 F_1(\omega) + a_2 F_2(\omega)$ |
| $F(t)$ | $2\pi f(-\omega)$ |
| $f(at)$ ($a$ real) | $(1/|a|)F(\omega/a)$ |
| $f(t - t_0)$ | $F(\omega)\exp(-it_0\omega)$ |
| $f(t)\exp(i\omega_0 t)$ | $F(\omega - \omega_0)$ |
| $df(t)/dt = \dot{f}(t)$ | $i\omega F(\omega)$ |
| $\int_{-\infty}^{t} f(\tau)\, d\tau$ | $F(\omega)/i\omega + \pi F(0)\delta(\omega)$ |
| $\delta(t)$ | $1$ |
| $1$ | $2\pi\delta(\omega)$ |
| $\hat{u}\exp(i\omega_0 t) + \hat{u}^*\exp(-i\omega_0 t)$ | $\hat{u}2\pi\delta(\omega - \omega_0) + \hat{u}^*2\pi\delta(\omega + \omega_0)$ |
| $\mathrm{sgn}(t)$ | $2/i\omega$ |
| $U(t) = [\mathrm{sgn}(t) + 1]/2$ | $1/i\omega + \pi\delta(\omega)$ |
| $d\delta(t)/dt = \dot{\delta}(t)$ | $i\omega$ |

Note that these integrals, as well as other Fourier integrals in this book, must be interpreted as Cauchy principal value integrals when necessary (Papoulis [17], p. 10).

    Transformations (2.135) and (2.136) are linear. Other theorems concerning symmetry, time scaling, time shifting, frequency shifting, time differentiation and time integration are summarised in Table 2.2. The table also includes some transformation pairs relating to the delta function and its derivative (both of which belong to the class of generalised functions, also termed 'distributions'), the unit step function (Heaviside function) and the signum function.

    If $f(t)$ is a real function, then

$$F^*(-\omega) = F(\omega). \tag{2.137}$$

The real and imaginary parts of the Fourier transform are even and odd functions, respectively:

$$R(\omega) = \mathrm{Re}\{F(\omega)\} = \int_{-\infty}^{\infty} f(t)\cos(\omega t)\, dt = R(-\omega), \tag{2.138}$$

$$X(\omega) = \mathrm{Im}\{F(\omega)\} = -\int_{-\infty}^{\infty} f(t)\sin(\omega t)\, dt = -X(-\omega). \tag{2.139}$$

Moreover, if the real $f(t)$ is an even function, $f(t) = f_e(t)$, say—that is, $f_e(-t) = f_e(t)$—its Fourier transform is real and even,

$$F(\omega) = R(\omega) = 2\int_{0}^{\infty} f_e(t)\cos(\omega t)\, dt, \tag{2.140}$$

and the inverse transform may be written as

$$f_e(t) = \frac{1}{\pi} \int_0^\infty R(\omega) \cos(\omega t) \, d\omega. \tag{2.141}$$

In contrast, if the real $f(t)$ is an odd function, $f(t) = f_o(t)$, say—that is, $f_o(-t) = -f_o(t)$—its Fourier transform is purely imaginary

$$F(\omega) = iX(\omega) = -2i \int_0^\infty f_o(t) \sin(\omega t) \, dt, \tag{2.142}$$

and its inverse transform may be written as

$$f_o(t) = -\frac{1}{\pi} \int_0^\infty X(\omega) \sin(\omega t) \, d\omega. \tag{2.143}$$

Frequently used in connection with Fourier analysis is the convolution theorem. The function

$$f(t) = \int_{-\infty}^\infty f_1(t - \tau) f_2(\tau) \, d\tau \equiv f_1(t) * f_2(t) \tag{2.144}$$

is termed the convolution (convolution product) of the two functions $f_1(t)$ and $f_2(t)$, as we have seen in Eq. (2.126). A very useful relation is given by the convolution theorem, which states that the Fourier transform of the convolution is

$$\mathcal{F}\{f_1(t) * f_2(t)\} = F_1(\omega) F_2(\omega) = \mathcal{F}\{f_1(t)\} \mathcal{F}\{f_2(t)\}. \tag{2.145}$$

Thus convolution in the time domain corresponds to ordinary multiplication in the frequency domain. It is implied that the convolution is commutative—that is, $f_1(t) * f_2(t) = f_2(t) * f_1(t)$. If we multiply Eq. (2.145) by $i\omega$, it is the Fourier transform of

$$f_1(t) * \frac{df_2(t)}{dt} = f_2(t) * \frac{df_1(t)}{dt}. \tag{2.146}$$

By using the symmetry theorem

$$\mathcal{F}\{F(t)\} = 2\pi f(-\omega) \tag{2.147}$$

(see Table 2.2), we can show that

$$\mathcal{F}\{f_1(t) f_2(t)\} = \frac{1}{2\pi} \int_{-\infty}^\infty F_1(\omega - y) F_2(y) \, dy = \frac{1}{2\pi} F_1(\omega) * F_2(\omega). \tag{2.148}$$

This is the frequency convolution theorem.

Let us next consider the relationship between Fourier integrals and Fourier series, as follows. Let $f_1(t) = 0$ for $|t| > T/2$. Then from Eq. (2.135), its Fourier transform is

$$F_1(\omega) = \int_{-T/2}^{T/2} f_1(t) e^{-i\omega t} \, dt. \tag{2.149}$$

Now we define a periodic function

$$f_T(t) = f_T(t + T) \tag{2.150}$$

for all $t$ ($-\infty < t < \infty$), where

$$f_T(t) = f_1(t), \quad -T/2 < t < T/2. \tag{2.151}$$

We may write the periodic function as a Fourier series,

$$f_T(t) = \sum_{n=-\infty}^{\infty} c_n \exp(in\omega_0 t), \tag{2.152}$$

where

$$\omega_0 = 2\pi/T \tag{2.153}$$

and

$$c_n = \frac{1}{T} \int_{-T/2}^{T/2} f_T(t) \exp(-in\omega_0 t) \, dt = \frac{1}{T} F_1(n\omega_0). \tag{2.154}$$

If $f_T$ is real, then

$$c_{-n} = c_n^*, \tag{2.155}$$

which corresponds to Eq. (2.137). The Fourier transform of $f_T(t)$ is

$$\begin{aligned} F_T(\omega) &= \int_{-\infty}^{\infty} \sum_{n=-\infty}^{\infty} c_n \exp\left[i(n\omega_0 - \omega)t\right] dt \\ &= 2\pi \sum_{n=-\infty}^{\infty} c_n \delta(\omega - n\omega_0) \\ &= \frac{2\pi}{T} \sum_{n=-\infty}^{\infty} F_1(n\omega_0) \, \delta(\omega - n\omega_0). \end{aligned} \tag{2.156}$$

Thus, it follows that the Fourier transform of a periodic function is discontinuous and is made up of impulses weighted by the Fourier transform of the function $f_1(t)$ for $\omega = n\omega_0$, ($n = 1, 2, \dots$). Note that $|F_1(n\omega_0)|/T$ may be interpreted as the amplitude of Fourier component number $n$ of $f_T(t)$. See Eq. (2.154).

## 2.6.2 Time-Invariant Linear System

Fourier analysis may be applied to the study of linear systems. Let the input signal to a linear system be $u(t)$ and the corresponding output signal (response) be $y(t)$. We shall assume that the system is linear and time invariant; that is, it has the same characteristics now as in the past and in the future. We shall mostly consider systems which are causal, which means that there is no output response before a causing signal has been applied to the input. However, it is sometimes of practical interest to consider noncausal systems as well (see Sections 4.9 and 5.3.2).

The Fourier transform of the impulse response $h(t)$ [see Eq. (2.126)] is

$$H(\omega) = R(\omega) + iX(\omega) = \int_{-\infty}^{\infty} h(t)\, e^{-i\omega t} dt, \tag{2.157}$$

which we shall call the transfer function of the system. [If the system is not causal, that is, if the condition (2.129) is not satisfied, some authors use the term 'frequency-response function', instead of 'transfer function'.] The real and imaginary parts of the transfer function are denoted by $R(\omega)$ and $X(\omega)$, respectively.

The Fourier transform of the output response $y(t)$, as given by the convolution product (2.126), is

$$Y(\omega) = H(\omega)\, U(\omega), \tag{2.158}$$

according to the convolution theorem (2.145), where $U(\omega)$ is the Fourier transform of the input signal $u(t)$. A system (block diagram) interpretation of Eq. (2.158) is given by the right-hand part of Figure 2.13.

For the simple case in which the transfer function is independent of frequency,

$$H(\omega) = H_0, \tag{2.159}$$

where $H_0$ is constant, the impulse response is

$$h(t) = H_0 \delta(t) \tag{2.160}$$

(see Table 2.2). For an arbitrary input $u(t)$, the response then becomes

$$y(t) = \int_{-\infty}^{\infty} h(t-\tau)u(\tau)d\tau = \int_{-\infty}^{\infty} H_0 \delta(t-\tau)u(\tau)d\tau = H_0 u(t), \tag{2.161}$$

which—apart for the constant $H_0$—is a distortion-free reproduction of the input signal.

As another example, let us consider a harmonic input function

$$u(t) = \tfrac{1}{2}\hat{u}e^{i\omega_0 t} + \tfrac{1}{2}\hat{u}^* e^{-i\omega_0 t}, \tag{2.162}$$

where $\hat{u}$ is the complex amplitude and $\hat{u}^*$ its conjugate. The Fourier transform is

$$U(\omega) = \pi\hat{u}\delta(\omega - \omega_0) + \pi\hat{u}^*\delta(\omega + \omega_0) \tag{2.163}$$

(see Table 2.2). The response is given by

$$\begin{aligned} Y(\omega) &= H(\omega)\, U(\omega) \\ &= \pi\hat{u}\delta(\omega - \omega_0)\, H(\omega) + \pi\hat{u}^*\delta(\omega + \omega_0)\, H(\omega) \\ &= \pi\hat{u}\delta(\omega - \omega_0)\, H(\omega_0) + \pi\hat{u}^*\delta(\omega + \omega_0)\, H(-\omega_0). \end{aligned} \tag{2.164}$$

Now, since $h(t)$ is real, we have $H(-\omega_0) = H^*(\omega_0)$ in accordance with Eq. (2.137). Hence, we may write

$$y(t) = \tfrac{1}{2}\hat{y}e^{i\omega_0 t} + \tfrac{1}{2}\hat{y}^* e^{-i\omega_0 t}, \tag{2.165}$$

where

$$\hat{y} = H(\omega_0)\hat{u}, \quad \hat{y}^* = H^*(\omega_0)\hat{u}^*, \tag{2.166}$$

in agreement with Eq. (2.134). It follows that the complex amplitude of the response equals the product of the transfer function and the complex amplitude of the input signal. According to Eq. (2.166), we may consider $\hat{u}$ as input to a linear system in the frequency domain, where $H$ is the transfer function and $\hat{y}$ the output response.

Assuming that the impulse response $h(t)$ is a real function of time, we next wish to express the impulse response in terms of the real and imaginary parts of the transfer function:

$$\begin{aligned}
h(t) &= \frac{1}{2\pi} \int_{-\infty}^{\infty} H(\omega)e^{i\omega t}\, d\omega \\
&= \frac{1}{2\pi} \int_{-\infty}^{\infty} [R(\omega)\cos(\omega t) - X(\omega)\sin(\omega t)]\, d\omega \\
&\quad + \frac{i}{2\pi} \int_{-\infty}^{\infty} [R(\omega)\sin(\omega t) + X(\omega)\cos(\omega t)]\, d\omega.
\end{aligned} \tag{2.167}$$

If $h(t)$ is real, Eqs. (2.137)–(2.139) hold. Thus, because $R(\omega)$ and $\cos(\omega t)$ are even functions of $\omega$, whereas $X(\omega)$ and $\sin(\omega t)$ are odd functions of $\omega$, the imaginary part in Eq. (2.167) vanishes. Moreover, we can rewrite the (real) impulse response as

$$h(t) = h_e(t) + h_o(t), \tag{2.168}$$

where

$$h_e(t) = \frac{1}{\pi} \int_0^{\infty} R(\omega)\cos(\omega t)\, d\omega, \tag{2.169}$$

$$h_o(t) = -\frac{1}{\pi} \int_0^{\infty} X(\omega)\sin(\omega t)\, d\omega. \tag{2.170}$$

Notice that $h_e(t)$ and $h_o(t)$ are even and odd functions of $t$, respectively. Thus, according to Eq. (2.168), the real impulse response $h(t)$ is split into even and odd parts. The corresponding transforms are even and odd functions of $\omega$, respectively:

$$\mathcal{F}\{h_e(t)\} = R(\omega), \quad \mathcal{F}\{h_o(t)\} = iX(\omega). \tag{2.171}$$

### 2.6.3 Kramers–Kronig Relations and Hilbert Transform

The function $h(t)$ is the response upon application of an input impulse at $t = 0$. Hence, for a causal system, Eq. (2.129) holds because there can be no response before an input signal has been applied. Then Eq. (2.126) and the commutativity of the convolution product give

$$y(t) = h(t) * u(t) = \int_{-\infty}^{t} h(t - \tau)u(\tau)\, d\tau = \int_{0}^{\infty} h(\tau)u(t - \tau)\, d\tau. \qquad (2.172)$$

Further, if also the input is causal (i.e., if $u(t) = 0$ for $t < 0$), then

$$y(t) = h(t) * u(t) = \int_{0}^{t} h(t - \tau)u(\tau)\, d\tau = \int_{0}^{t} h(\tau)u(t - \tau)\, d\tau. \qquad (2.173)$$

This is the version of the convolution theorem used in the theory of the (one-sided) Laplace transform, which may be applied to causal systems assumed to be dead (quiescent) previous to an initial instant, $t = 0$. Using the Fourier transform when $h(t) = 0$ for $t < 0$ and inserting $s$ for $i\omega$, we obtain the following relation between the Laplace transform $H_L$ and the transfer function $H$:

$$H_L(s) \equiv \int_{0}^{\infty} h(t)e^{-st}\, dt = H(s/i) = H(-is). \qquad (2.174)$$

Using the convolution theorem (2.145), we obtain the Laplace transform of Eq. (2.173) as

$$Y_L(s) = H_L(s)\, U_L(s), \qquad (2.175)$$

where $U_L(s)$ is the Laplace transform of $u(t)$, which is assumed to be vanishing for $t < 0$.

For a constant transfer function $H(\omega) = H_0$, the impulse response is given by Eq. (2.160) as $H_0\delta(t)$. Note that this is an example of a causal impulse response. In this case, the response to a causal input $u(t)$ is $y(t) = H_0\, u(t)$, which means that $y(t) = 0$ for $t < 0$ if $u(t) = 0$ for $t < 0$. This example of a causal impulse response, $H_0\delta(t)$, is an even function. Another example of a causal impulse response is $h(t) = H_{00}\dot{\delta}(t)$, for which the transfer function is $H(\omega) = i\omega H_{00}$ (cf. Table 2.2). This impulse response is an odd function.

In general, however, any causal impulse response $h(t)$ which 'has a memory of the past' is neither an even nor an odd function, because $h(t)$ vanishes for all negative $t$ but not for all positive $t$. Next, let us make some observations for this more general situation. We shall first exclude cases for which $H(\infty) \neq 0$, such as in the preceding example, where $H(\infty) = H_0$.

For a causal function $h(t)$ decomposed into even and odd parts according to Eq. (2.168), we have

$$h(t) = \begin{cases} 2h_e(t) = 2h_o(t) & \text{for } t > 0 \\ h_e(0) & \text{for } t = 0 \\ 0 & \text{for } t < 0 \end{cases} \qquad (2.176)$$

and, in general (for $t \neq 0$),

$$h_o(t) = h_e(t)\, \text{sgn}(t), \quad h_e(t) = h_o(t)\, \text{sgn}(t), \qquad (2.177)$$

because the condition $h(t) = 0$ for $t < 0$ must be satisfied. Using Eqs. (2.169), (2.170), and (2.176) gives, for $t > 0$,

$$h(t) = 2h_e(t) = \frac{2}{\pi} \int_0^\infty R(\omega) \cos(\omega t) \, d\omega, \tag{2.178}$$

$$h(t) = 2h_o(t) = -\frac{2}{\pi} \int_0^\infty X(\omega) \sin(\omega t) \, d\omega. \tag{2.179}$$

It should be emphasised that these expressions apply for $t > 0$ only, whereas for $t < 0$, we have $h(t) = 0$. Note that the two alternative expressions (2.178) and (2.179) imply a relationship between $R(\omega)$ and $X(\omega)$, the real and imaginary parts of the Fourier transform of a causal, real function.

Using

$$\mathcal{F}\{\text{sgn}(t)\} = 2/i\omega \tag{2.180}$$

(see Table 2.2) and the frequency convolution theorem (2.148), we obtain the Fourier transforms of Eqs. (2.177):

$$\mathcal{F}\{h_e(t)\} = \mathcal{F}\{h_o(t)\text{sgn}(t)\} = R(\omega)$$
$$= \frac{1}{2\pi} i X(\omega) * \frac{2}{i\omega} = \frac{1}{\pi\omega} * X(\omega), \tag{2.181}$$

$$\mathcal{F}\{h_o(t)\} = \mathcal{F}\{h_e(t) \, \text{sgn}(t)\} = iX(\omega)$$
$$= \frac{1}{2\pi} R(\omega) * \frac{2}{i\omega} = -\frac{i}{\pi\omega} * R(\omega). \tag{2.182}$$

Writing convolutions explicitly in terms of integrals, we have

$$R(\omega) = \frac{1}{\pi} \int_{-\infty}^\infty \frac{X(y)}{\omega - y} \, dy, \tag{2.183}$$

$$X(\omega) = -\frac{1}{\pi} \int_{-\infty}^\infty \frac{R(y)}{\omega - y} \, dy. \tag{2.184}$$

Note that the integrand is singular for $y = \omega$ and that the integrals should be understood as principal value integrals (cf. footnote on p. 10 in Papoulis [17]). Relations (2.183) and (2.184) are called Kramers–Kronig relations [19, 20]. If for a causal function the real/imaginary part of the Fourier transform is known for all frequencies, then the remaining imaginary/real part of the transform is given by a principal-value integral. For the integrals to exist, it is necessary that

$$R(\omega) + iX(\omega) = H(\omega) \to 0 \quad \text{when } \omega \to \infty. \tag{2.185}$$

If this is not true, we may conveniently subtract the singular part of $H(\omega)$. Let us consider the case in which $h(t)$ has an impulse singularity. Note (from Table 2.2) that

$$\mathcal{F}\{\delta(t)\} = 1, \quad \mathcal{F}\{\delta(t - t_0)\} = \exp(-i\omega t_0). \tag{2.186}$$

These Fourier transforms do not satisfy condition (2.185). If $H(\infty) \neq 0$, we define a modified transfer function

$$H'(\omega) = H(\omega) - H(\infty) \tag{2.187}$$

and a new corresponding causal impulse response function

$$h'(t) = h(t) - H(\infty)\,\delta(t). \tag{2.188}$$

For instance, if

$$H(\infty) = R(\infty) \neq 0, \quad X(\infty) = 0, \tag{2.189}$$

we have

$$X(\omega) = -\frac{1}{\pi} \int_{-\infty}^{\infty} \frac{R(y) - R(\infty)}{\omega - y}\, dy, \tag{2.190}$$

$$R(\omega) - R(\infty) = \frac{1}{\pi} \int_{-\infty}^{\infty} \frac{X(y)}{\omega - y}\, dy. \tag{2.191}$$

Multiplying the former of the two equations by $i$ and summing yield the Hilbert transform

$$H'(\omega) = \frac{1}{i\pi} \int_{-\infty}^{\infty} \frac{H'(y)}{\omega - y}\, dy. \tag{2.192}$$

An alternative formulation of the Kramers–Kronig relations is obtained by noting that because $h(t)$ is real, we have

$$R(-\omega) = R(\omega), \quad X(-\omega) = -X(\omega) \tag{2.193}$$

[cf. Eqs. (2.138)–(2.139)]. Then

$$-\pi X(\omega) = \int_{-\infty}^{\infty} \frac{R(y)}{\omega - y}\, dy = \int_{-\infty}^{0} \frac{R(z)}{\omega - z}\, dz + \int_{0}^{\infty} \frac{R(y)}{\omega - y}\, dy. \tag{2.194}$$

Because

$$\int_{-\infty}^{0} \frac{R(z)}{\omega - z}\, dz = \int_{0}^{\infty} \frac{R(-y)}{\omega + y}\, dy = \int_{0}^{\infty} \frac{R(y)}{\omega + y}\, dy \tag{2.195}$$

and

$$\frac{1}{\omega + y} + \frac{1}{\omega - y} = \frac{2\omega}{\omega^2 - y^2}, \tag{2.196}$$

we obtain

$$X(\omega) = -\frac{2\omega}{\pi} \int_{0}^{\infty} \frac{R(y)}{\omega^2 - y^2}\, dy. \tag{2.197}$$

Similarly, we find

$$
\begin{aligned}
\pi R(\omega) &= \int_{-\infty}^{\infty} \frac{X(y)}{\omega - y}\, dy = \int_{0}^{\infty} \frac{X(-y)}{\omega + y}\, dy + \int_{0}^{\infty} \frac{X(y)}{\omega - y}\, dy \\
&= \int_{0}^{\infty} X(y)\left(\frac{1}{\omega - y} - \frac{1}{\omega + y}\right) dy,
\end{aligned}
\tag{2.198}
$$

which gives

$$
R(\omega) = \frac{2}{\pi} \int_{0}^{\infty} \frac{yX(y)}{\omega^2 - y^2}\, dy.
\tag{2.199}
$$

### 2.6.4 An Energy Relation for Non-sinusoidal Oscillation

In Section 2.3, we derived the formula (2.76) for the time-average power or rate of work associated with sinusoidal oscillations in a mechanical system, namely

$$
P = \tfrac{1}{4}\hat{f}\hat{u}^* + \tfrac{1}{4}\hat{f}^*\hat{u}
\tag{2.200}
$$

(note that we here denote force by the variable $f$ instead of $F$, as we did previously). For a more general oscillation which is not periodic in time, it is more convenient to consider the total work done than the time-average of the rate of work. The instantaneous rate of work is $P(t) = f(t)u(t)$—that is, the product of the instantaneous values of force $f$ and velocity $u$. Hence, the total work done is

$$
W = \int_{-\infty}^{\infty} f(t)u(t)\, dt.
\tag{2.201}
$$

By applying Parseval's theorem, or the frequency convolution theorem (2.148) with $\omega = 0$, we find that Eq. (2.201) gives

$$
W = \frac{1}{2\pi} \int_{-\infty}^{\infty} F(\omega)U(-\omega)\, d\omega,
\tag{2.202}
$$

where $F(\omega)$ and $U(\omega)$ are the Fourier transforms of the force $f(t)$ and the velocity $u(t)$, respectively. We shall assume that $F(\omega)$ and $U(\omega)$ are related through the transfer function

$$
Y(\omega) = 1/Z(\omega) = U(\omega)/F(\omega),
\tag{2.203}
$$

where

$$
Z(\omega) = R(\omega) + iX(\omega)
\tag{2.204}
$$

is a mechanical impedance. Its real and imaginary parts are the mechanical resistance $R(\omega)$ and the mechanical reactance $X(\omega)$, respectively. It may, for instance, be given by Eq. (2.51).

If $f(t)$ and $u(t)$ are real functions, then Eq. (2.137) is applicable to $F(\omega)$ and $U(\omega)$. Further, using also Eqs. (2.203) and (2.204), Eq. (2.202) becomes

$$
\begin{aligned}
W &= \frac{1}{2\pi} \int_0^\infty (F(\omega)U^*(\omega) + F^*(\omega)U(\omega))\, d\omega \\
&= \frac{1}{\pi} \int_0^\infty \mathrm{Re}\{F(\omega)U^*(\omega)\}\, d\omega \\
&= \frac{1}{\pi} \int_0^\infty \mathrm{Re}\{Z(\omega)\, U(\omega)U^*(\omega)\}\, d\omega \\
&= \frac{1}{\pi} \int_0^\infty R(\omega)|U(\omega)|^2\, d\omega.
\end{aligned}
\tag{2.205}
$$

It may be interesting (see Problem 2.14) to compare this result with Eqs. (2.76) and (2.77), where the time-average mechanical power is expressed in terms of complex amplitudes.

## Problems

### Problem 2.1: Free Oscillation

Show that the equation $m\ddot{x} + R\dot{x} + Sx = 0$ has the general solution

$$
x = (C_1 \cos \omega_d t + C_2 \sin \omega_d t)\, e^{-\delta t}.
$$

Further, show that the integration constants $C_1$ and $C_2$, as given by

$$
C_1 = x_0, \qquad C_2 = (u_0 + x_0\delta)/\omega_d,
$$

satisfy the initial conditions $x(0) = x_0$ and $\dot{x}(0) = u_0$.

### Problem 2.2: Quality Factor or $Q$ Value for Resonator

Using definition (2.8), derive the equations

$$
Q = \frac{\omega_0}{2\delta}\left(1 + \frac{\delta}{\omega_0} - \frac{1}{6}\frac{\delta^2}{\omega_0^2} + \mathcal{O}\left\{\frac{\delta^3}{\omega_0^3}\right\}\right)
$$

$$
\approx \frac{\omega_0}{2\delta} = \frac{\omega_0 m}{R} = \frac{S}{\omega_0 R} = \frac{(Sm)^{1/2}}{R}
$$

and

$$
\frac{\delta}{\omega_0} = \frac{1}{2Q}\left(1 + \frac{1}{2Q} + \frac{5}{24Q^2} + \mathcal{O}\left\{Q^{-3}\right\}\right) \approx \frac{1}{2Q}.
$$

[Hint: use Taylor series for an exponential function and for a binomial, and/or use the method of successive approximations.]

### Problem 2.3: Forced Oscillation

Show that the equation

$$m\ddot{x} + R\dot{x} + Sx = F(t) = F_0 \cos(\omega t + \varphi_F)$$

has a particular solution

$$x(t) = x_0 \cos(\omega t + \varphi_x),$$

where the phase difference $\varphi = \varphi_F - \varphi_u = \varphi_F - \varphi_x - \pi/2$ is an angle in quadrant 1 or 4, and satisfies

$$\tan \varphi = (\omega m - S/\omega)/R.$$

Also show that

$$x_0 = \frac{u_0}{\omega} = \frac{F_0}{|Z|\omega},$$

where $|Z| = \sqrt{R^2 + (\omega m - S/\omega)^2}$.

### Problem 2.4: Excursion Response Maximum

Using the results of Problem 2.3, discuss the nondimensionalised excursion ratio $|\xi| \equiv Sx_0/F_0$ versus the frequency ratio $\gamma \equiv \omega/\omega_0 = \omega\sqrt{m/S}$. Find the frequency at which $|\xi|$ is maximum. Determine also the inequality which $\delta = R/2m$ has to satisfy so that $1 < |\xi|_{max} < \sqrt{2}$. Observe that $|\xi|_{max} > |\xi(\gamma = 1)|$ and that $|\xi|_{max} \to |\xi(\gamma = 1)|$, as $\delta/\omega_0 \to 0$.

### Problem 2.5: Amplitude and Phase Constant

Determine the numerical values of the amplitude $A$ and the phase constant $\varphi$ for the oscillation

$$x = A \cos(62.8t + \varphi)$$

when the initial conditions at $t = 0$ are as follows: position $x(0) = 50$ mm, velocity $\dot{x}(0) = 0.8$ m/s. Determine the acceleration at $t = 0$. Draw a phasor diagram for the complex amplitudes of position (in scale 1:1), velocity (1 m/s $\hat{=}$ 20 mm) and acceleration (1 m/s$^2$ $\hat{=}$ 0.5 mm).

### Problem 2.6: Complex Representations of Harmonic Oscillation

The following harmonic oscillation is given:

$$u(t) = 100 \sin(\omega t) - 50 \cos(\omega t).$$

Rewrite $u(t)$ as

(a) a cosine function with phase constant,
(b) a sine function with phase constant,
(c) the real part of a complex quantity,
(d) the imaginary part of a complex quantity, and
(e) the sum of two complex conjugate quantities.

## Problem 2.7: Superposed Oscillations of the Same Frequency

An oscillation is given as a superposition of individual oscillations:

$$x(t) = 3\cos(\omega t + \pi/6) + 6\sin(\omega t + 3\pi/2)$$
$$- 3.5\sin(\omega t + \pi/3) + 2.2\sin(\omega t - \pi/9).$$

(a) Numerically determine the amplitude and the phase constant of the resultant oscillation.
(b) Rewrite $x(t)$ in complex form.
(c) Draw phasors for the four individual oscillations in the same diagram, and construct the resultant phasor. Compare it with the preceding numerical computation.

## Problem 2.8: Resonance Bandwidth

According to Eq. (2.53), the inverse of the mechanical impedance $Z$ may be interpreted as the transfer function of a system in which applied force $\hat{F}$ is the input and velocity $\hat{u}$ the output. The maximum modulus $|Y|_{max}$ of this transfer function $Y = 1/Z$ is $1/R$, corresponding to the resonance frequency $\omega_0 = \sqrt{S/m}$. Derive an expression for the upper and lower frequencies, $\omega_u$ and $\omega_l$, respectively, at which $|Y|/|Y|_{max} = R|Y| = 1/\sqrt{2}$, in terms of $\omega_0$ and $\delta = R/2m$. Also determine the resonance bandwidth $(\Delta\omega)_{res} = \omega_u - \omega_l$. Compare the relative bandwidth $(\omega_u - \omega_l)/\omega_0$ with the inverse of the quality factor $Q$ defined in Section 2.1.1.

## Problem 2.9: Solving System of Linear Differential Equations

Show that the solution

$$\mathbf{x}(t) = e^{\mathbf{A}(t-t_0)}\mathbf{x}(t_0) + \int_{t_0}^{t} e^{\mathbf{A}(t-\tau)}\mathbf{B}\mathbf{u}(\tau)\,d\tau$$

satisfies the vectorial differential equation $\dot{\mathbf{x}} = \mathbf{A}\mathbf{x} + \mathbf{B}\mathbf{u}$.

## Problem 2.10: State-Space Description of Oscillation

Let matrices $\mathbf{A}$, $\mathbf{B}$, $\mathbf{C}$ and $\mathbf{D}$ be given by Eqs. (2.110)–(2.111). Determine the eigenvalues $\Lambda_1$ and $\Lambda_2$ in terms of the coefficients $\delta$ and $\omega_d$ given by Eq. (2.5). Assume that $\Lambda_2 \neq \Lambda_1$ and that $\Lambda_1 - \Lambda_2$ either has a positive imaginary part, or otherwise that $\Lambda_1 - \Lambda_2$ is real and positive. Show that $\mathbf{A}$ has the linearly independent eigenvectors $\mathbf{m}_1$ and $\mathbf{m}_2$ with transposed vectors $\mathbf{m}_i^T = c_i \begin{bmatrix} 1 & \Lambda_i \end{bmatrix}$ $(i = 1, 2)$, where $c_1$ and $c_2$ are arbitrary constants. Determine the inverse of matrix $\mathbf{M} = \begin{bmatrix} \mathbf{m}_1 & \mathbf{m}_2 \end{bmatrix}$, and show that $\mathbf{A}'$ defined by the similarity transformation (2.102) is a diagonal matrix. Next perform the transformations (2.106) and (2.107); find $\mathbf{x}'$, $\mathbf{B}'$ and $\mathbf{C}'$; and use Eq. (2.99) to obtain the solution of Eqs. (2.108) and (2.109). In the final answer, set $t_0 = 0$, and replace $y_1$, $x_1$, $x_2$ and $u_1$ with $x$, $\dot{x}$, $u$ and $F$, respectively. Finally, write down the solution for free oscillations—i.e., for $F(t) = 0$. Compare the result with Eqs. (2.4) and (2.6).

## Problem 2.11: Critically Damped Oscillation

Discuss the solution of Problem 2.10 for the case when $\Lambda_1 = \Lambda_2 = \Lambda$. In this case, $\mathbf{A}$ does not have two linearly independent eigenvectors, and the transformed matrix $\mathbf{A}'$ is not diagonal. For the similarity transformation, choose the matrix

$$\mathbf{M} = \begin{bmatrix} 1 & 0 \\ \Lambda & 1 \end{bmatrix},$$

and perform the various tasks as in Problem 2.10. A special challenge is to determine matrix exponential $e^{\mathbf{A}'t}$ by using definition (2.98).

## Problem 2.12: Mechanical Impedance and Power

A mass of $m = 6$ kg is suspended by a spring of stiffness $S = 100$ N/m. The oscillating system has a mechanical resistance of $R = 3$ Ns/m. The system is excited by an alternating force $F(t) = |\hat{F}| \cos(\omega t)$, where the force amplitude is $|\hat{F}| = 1$ N and the frequency $f = \omega/2\pi = 1$ Hz.

(a) Determine the mechanical impedance $Z = R + iX = |Z|e^{i\varphi}$. State numerical values for $R$, $X$, $|Z|$ and $\varphi$.

(b) Determine the complex amplitudes $\hat{F}$ for the force, $\hat{s}$ for the position, $\hat{u}$ for the velocity and $\hat{a}$ for the acceleration. Draw the complex amplitudes as vectors in a phasor diagram.

(c) Find the frequency $f_0 = \omega_0/2\pi$ for which the mechanical reactance vanishes.

(d) Find the time-averaged mechanical power $P$ which the force $|\hat{F}| = 1$ N supplies to the system, at the two frequencies $f$ and $f_0$.

## Problem 2.13:  Convolution with Sinusoidal Oscillation

Show that if $f(t)$ varies sinusoidally with an angular frequency $\omega_0$, then the convolution product $g(t) = h(t) * f(t)$ may be written as a linear combination of $f(t)$ and $\dot{f}(t)$. Assuming that the transfer function $H(\omega)$ is known, determine the coefficients of the linear combination.

## Problem 2.14:  Work in Terms of Complex Amplitude or Fourier Integral

Consider the particular case of harmonic oscillations. Discuss the mathematical relationship between Eq. (2.205) for the total work and the equation

$$P = \tfrac{1}{4}(\hat{F}\hat{u}^* + \hat{F}^*\hat{u}) = \tfrac{1}{2}\mathrm{Re}\{\hat{F}\hat{u}^*\}$$

for the power of a mechanical system performing harmonical oscillation [cf. Eq. (2.76)]. Consider also the dimensions (or SI units) of the various physical quantities.

## Problem 2.15:  Mechanical Impedance at Zero Frequency

When we, by Eq. (2.51), defined the mechanical impedance

$$Z(\omega) = R + i(\omega m - S/\omega),$$

we tacitly assumed that $\omega \neq 0$. We define the transfer functions $Y(\omega) = 1/Z(\omega)$, $G(\omega) = i\omega Z(\omega)$, and $H(\omega) = 1/G(\omega) = Y(\omega)/i\omega$. Show that the corresponding impulse response functions $y(t)$, $g(t)$ and $h(t)$ are causal. [Hint: for transfer functions with poles, apply the method of contour integration when considering the inverse Fourier transform.] If the variables concerned are $s(t)$, $u(t)$ and $F(t)$, choose input and output variables for the four linear systems concerned. Further, show that in order to make $z(t) = \mathcal{F}^{-1}\{Z(\omega)\}$ a causal impulse response function, it is necessary to add a term to the impedance such that

$$Z(\omega) = R + i\omega m + S[1/i\omega + \pi\delta(\omega)].$$

Explain the physical significance of the last term.

# CHAPTER THREE

# Interaction between Oscillations and Waves

There are many different types of waves in nature. Apart from the visible waves on the surface of oceans and lakes, there are, for instance, sound waves, light waves and other electromagnetic waves. This chapter gives a brief description of waves in general and compares surface waves on water with other types of waves. It also presents a simple generic discussion on the interaction between waves and oscillations. One phenomenon is generated waves radiated from an oscillator, and another phenomenon is oscillations excited by a wave incident upon the oscillating system. We shall define the radiation resistance in terms of the power associated with the wave generated by the oscillator. The 'added mass' is related to added energy associated with the wave-generating process, not to kinetic energy alone but to the difference between kinetic and potential energies.

## 3.1 Comparison of Waves on Water with Other Waves

Waves on water propagate along a surface. Acoustic waves in a fluid and electromagnetic waves in free space may propagate in any direction in a three-dimensional space. Waves on a stretched string propagate along a line (in a one-dimensional 'space'). The same may be said about waves on water in a canal and about guided acoustic waves or guided electromagnetic waves along cylindrical structures, although in these cases the physical quantities (pressure, velocity, electric field, magnetic field, etc.) may vary in directions transverse to the direction of wave propagation.

 As was mentioned in Chapter 2, there is an exchange of kinetic energy and potential energy in a mechanical oscillator (or magnetic energy and electric energy in the electric analogue). In a propagating wave, too, there is interaction between different forms of energy—for instance, magnetic and electric energy with electromagnetic waves, and kinetic and potential energy with mechanical waves, such as acoustic waves and water waves. With an acoustic wave, the potential energy is associated with the elasticity of the medium in which the wave propagates. The potential energy with a water wave is due to gravity and surface tension. The contribution from the elasticity of water is negligible, because waves on water propagate rather slowly as compared with the velocity of sound in water. Gravity is responsible for the potential energy associated with the lifting of water from the wave troughs to the wave crests. As the waves increase the area of the interface between water and air, the work done against surface tension is converted to potential energy. For wavelengths in the range of $10^{-3}$ m to $10^{-1}$ m, both types of potential energy are important. For shorter waves, so-called capillary waves, the effect of gravity may be neglected (see Problems 3.2 and 4.1). For longer waves, so-called gravity waves, surface tension may be neglected, as we shall do in the following, since we restrict our study to water waves of wavelengths exceeding 0.25 m.

Oscillations are represented by physical quantities which vary with time. For waves, the quantities also vary with the spatial coordinates. In the present chapter, we shall consider waves which vary sinusoidally with time. Such waves are called 'harmonic' or 'monochromatic'. When daling with sea waves, they are also characterised as 'regular' if their time variation is sinusoidal.

Let

$$p = p(x, y, z, t) = \text{Re}\{\hat{p}(x, y, z)e^{i\omega t}\} \tag{3.1}$$

represent a general harmonic wave, and let $p$ denote the dynamic pressure in a fluid. (The total pressure is $p_{tot} = p_{stat} + p$, where the static pressure $p_{stat}$ is independent of time.) The complex pressure amplitude $\hat{p}$ is a function of the spatial coordinates $x$, $y$ and $z$.

For a plane acoustic wave propagating in a direction $x$, we have

$$p = p(x, t) = \text{Re}\{Ae^{i(\omega t - kx)} + Be^{i(\omega t + kx)}\}, \tag{3.2}$$

where $k = 2\pi/\lambda$ is the angular repetency (wave number) and $\lambda$ is the wavelength. The first and second terms represent waves propagating in the positive and negative $x$ direction, respectively. Assume that an observer moves with a velocity $v_p = \omega/k$ in the positive $x$ direction. Then he or she will experience a constant phase $(\omega t - kx)$ of the first right-hand term in Eq. (3.2). If the observer moves with same speed in the opposite direction, he or she will experience a constant phase $(\omega t + kx)$ of the last term in Eq. (3.2). For this reason, $v_p = \omega/k$ is called the phase velocity. At a certain instant, the phase is constant on all planes perpendicular to the direction of wave propagation. For this reason, the wave is called plane. In contrast, an acoustic wave

$$p(r, t) = \text{Re}\{(C/r)e^{i(\omega t - kr)}\} \tag{3.3}$$

radiated from a spherical loudspeaker in open air may be called a spherical wave, because the phase $(\omega t - kr)$ is the same everywhere on an envisaged sphere with a radius $r$ from the centre of the loudspeaker. Note that in this geometrical case, the pressure amplitude $|C|/r$ decreases with the distance from the loudspeaker.

The pressure amplitudes $|A|$ and $|B|$ are constant for the two oppositely propagating waves corresponding to the two right-hand terms in Eq. (3.2). An acoustic wave in a fluid is a longitudinal wave, because the fluid oscillates only in the direction of wave propagation. We shall not discuss acoustic waves in detail here but just mention that if the sound pressure is given by Eq. (3.2), then the fluid velocity is

$$v = v_x = \frac{1}{\rho c} \mathrm{Re}\left\{Ae^{i(\omega t - kx)} - Be^{i(\omega t + kx)}\right\}, \tag{3.4}$$

where $c$ is the sound velocity and $\rho$ is the (static) mass density of the fluid. The readers who are interested in the derivation of Eqs. (3.2) and (3.4) and the wave equation for acoustic waves are referred to textbooks in acoustics [21].

For a plane harmonic wave that propagates along a water surface, an expression similar to Eq. (3.2) applies. However, then $A$ and $B$ cannot be constants; they depend on the vertical coordinate $z$ (chosen to have positive direction upwards). On 'deep water', $A$ and $B$ are then proportional to $e^{kz}$, as will be shown later in Chapter 4. Then $p = p(x, z, t)$, and

$$\hat{p} = \hat{p}(x, z) = A(z)e^{-ikx} + B(z)e^{ikx}. \tag{3.5}$$

In the case of a plane and linearly polarised electromagnetic wave propagating in free space, the electric field is

$$E(x, t) = \mathrm{Re}\left\{Ae^{i(\omega t - kx)} + Be^{i(\omega t + kx)}\right\}, \tag{3.6}$$

where $A$ and $B$ are again complex constants.

## 3.2 Dispersion, Phase Velocity and Group Velocity

Both acoustic waves and electromagnetic waves are non-dispersive, which means that the phase velocity is independent of the frequency. The dispersion relation (the relationship between $\omega$ and $k$) is

$$\omega = ck, \tag{3.7}$$

where $c$ is the constant speed of sound or light, respectively.

Gravity waves on water are, in general, dispersive. As will be shown later, in Chapter 4 (Section 4.2), the relationship for waves on deep water is

$$\omega^2 = gk, \tag{3.8}$$

where $g$ is the acceleration of gravity. Using this dispersion relationship, we find that the phase velocity is

$$v_p \equiv \omega/k = g/\omega = \sqrt{g/k}. \tag{3.9}$$

Note that in the study of the propagation of dispersive waves (for which the phase velocity depends on frequency), we have to distinguish between phase velocity and group velocity. Let us assume that the dispersion relationship may be written as $F(\omega, k) = 0$, where $F$ is a differentiable function of two variables. Then the group velocity is defined as

$$v_g = \frac{d\omega}{dk} = -\frac{\partial F/\partial k}{\partial F/\partial \omega}. \tag{3.10}$$

As shown in Problem 3.1, if several propagating waves with slightly different frequencies are superimposed on each other, the result may be interpreted as a group of waves, each of them moving with the phase velocity, whereas the amplitude of the group of individual waves are modulated by an envelope moving with the group velocity. Here let us just consider the following simpler example of two superimposed harmonic waves of angular frequency $\omega \pm \Delta\omega$ and angular repetency $k \pm \Delta k$, namely

$$\begin{aligned} p(x,t) &= D\cos\{(\omega - \Delta\omega)t - (k - \Delta k)x\} + D\cos\{(\omega + \Delta\omega)t - (k + \Delta k)x\} \\ &= 2D\cos\{(\Delta\omega)t - (\Delta k)x\}\cos(\omega t - kx), \end{aligned} \tag{3.11}$$

where an elementary trigonometric identity has been used in the last step. If $\Delta\omega \ll \omega$, this last expression for $p(x,t)$ is the product of a fast varying function, representing a wave with propagation speed $\omega/k$ (the phase velocity), and a slowly varying function, representing an amplitude envelope, moving in the positive $x$ direction with a propagation speed $\Delta\omega/\Delta k$, which tends to the group velocity if $\Delta\omega$ (and, correspondingly, $\Delta k$) tend to zero.

Using Eq. (3.8), we find that for a wave on deep water, the group velocity is

$$v_g \equiv d\omega/dk = g/(2\omega) = v_p/2. \tag{3.12}$$

We shall see later, in Chapter 4, that this group velocity may be interpreted as the speed with which energy is transported by a deep-water wave. Although electromagnetic waves are non-dispersive in free space, such waves propagating along telephone lines or along optical fibres have, in general, some dispersion. In this case, it is of interest to know that not only the energy but also the information carried by the wave are usually propagated with a speed equal to the group velocity.

## 3.3 Wave Power and Energy Transport

Next let us consider the energy, power and intensity associated with waves. Intensity $I$ is the time-average energy transport per unit time and per unit area in the direction of wave propagation. Whereas the dimension in SI units is J (joule) for energy and W (watt) for power, it is $J/(s\,m^2) = W/m^2$ for intensity. For surface waves on water and for acoustic waves, the intensity is

$$I = \overline{p_{tot}v} = \overline{(p_{stat} + p)v}, \tag{3.13}$$

where the total pressure $p_{tot}$ is the sum of the static pressure $p_{stat}$ and the dynamic pressure $p$, and where $v = v_x$ is the fluid particle velocity component in the direction of wave propagation. (The overbar denotes time average.) Because the time-average particle velocity is zero, $\bar{v} = 0$, we have

$$I = p_{stat}\bar{v} + \overline{pv} = \overline{pv}. \tag{3.14}$$

For a harmonic wave, we have

$$p = \text{Re}\{\hat{p}e^{i\omega t}\}, \quad v_x = \text{Re}\{\hat{v}_x e^{i\omega t}\}, \tag{3.15}$$

where $\hat{p} = \hat{p}(x, y, z)$ and $\hat{v}_x = \hat{v}_x(x, y, z)$ are complex amplitudes at $(x, y, z)$ of the dynamic pressure $p$ and of the $x$ component of the fluid particle velocity, respectively. In analogy with the derivation of Eq. (2.76), we then have

$$I = I_x = I_x(x, y, z) = \overline{pv_x} = \tfrac{1}{2}\text{Re}\{\hat{p}\hat{v}_x^*\}. \tag{3.16}$$

Strictly speaking, $v$ and, hence, $I$ are vectors:

$$\vec{I} = \vec{I}(x, y, z) = \overline{p\vec{v}} = \tfrac{1}{2}\text{Re}\{\hat{p}\hat{\vec{v}}^*\}. \tag{3.17}$$

For an electromagnetic wave the intensity may be defined as the time-average of the so-called Poynting vector, which is well known in electromagnetics (see, e.g., Panofsky and Phillips [22]).

For a plane acoustic wave propagating in the positive $x$ direction, the sound pressure is as given by Eq. (3.2), and the oscillating fluid velocity is as given by Eq. (3.4), with $B = 0$ and the constant $|A|$ being equal to the pressure amplitude. Then the sound intensity, as given by Eq. (3.16), is constant (independent of $x$, $y$ and $z$).

For a plane gravity wave on water propagating in the positive $x$ direction, the dynamic pressure is as given by Eq. (3.5), with $B(z) \equiv 0$. Refer to Chapter 4 for a discussion of the oscillating fluid velocity $\vec{v}$ associated with this wave. It is just mentioned here that if the water is 'deep', then both $p$ and $\vec{v}$ are proportional to $e^{kz}$. Hence, the intensity is proportional to $e^{2kz}$, meaning that the intensity decreases exponentially with the distance downward from the water surface. Thus,

$$I_x = I_0 e^{2kz}, \tag{3.18}$$

where $I_0$ is the intensity at the (average) water surface, $z = 0$. By integrating $I_x = I_x(z)$ from $z = -\infty$ to $z = 0$, we arrive at the wave-energy transport

$$J = \int_{-\infty}^{0} I_x(z)\, dz = I_0 \int_{-\infty}^{0} e^{2kz}\, dz = I_0/2k, \tag{3.19}$$

which is the wave energy transported per unit time through an envisaged vertical strip of unit width parallel to the wave front—that is, parallel to the planes of constant phase of the propagating wave.

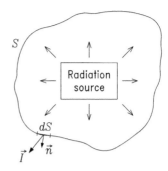

Figure 3.1: General source of radiation surrounded by an envisaged closed surface $S$.

Let us now consider an arbitrary, envisaged, closed surface with a radiation source (or wave generator) inside, as indicated in Figure 3.1. The radiation source could be an oscillating body immersed in water or another kind of wave generator. (In acoustics, the source could be a loudspeaker; in radio engineering, a transmitting antenna; and in optics, a light-emitting atom.)

The radiated power (energy per unit time) passing through the closed surface $S$ may be expressed as an integral of the intensity over the surface,

$$P_r = \oiint \vec{I} \cdot \vec{n}\, dS = \oiint \vec{I} \cdot \vec{dS}, \tag{3.20}$$

where $\vec{n}$ is the unit normal and $\vec{dS} \equiv \vec{n}\, dS$. When a spherical loudspeaker radiates isotropically in open air, the sound intensity $I(r)$ is independent of direction, and then the radiated power through an envisaged spherical surface of radius $r$ is

$$P_r = I(r)\, 4\pi r^2. \tag{3.21}$$

Here we have neglected reflection and absorption of the acoustic wave from ground and other obstacles. If we assume that the air does not absorb acoustic wave energy, $P_r$ must be independent of $r$. Hence, $I(r)$ is inversely proportional to the square of the distance $r$:

$$I(r) = I(a)\, (a/r)^2, \tag{3.22}$$

where $I(a)$ is the intensity for $r = a$. With this result in mind, it is easier to accept the statement, as implied in Eq. (3.3) for the spherical wave, that the amplitude of the sound pressure is inversely proportional to the distance $r$.

Assume now that an axisymmetric body immersed in water of depth $h$ is performing vertical oscillations. Then an axisymmetric wave generation will take place, and trains of circular waves will radiate along the water surface, outwards from the oscillating body. The power radiated through an envisaged vertical cylinder of large radius $r$ may, according to Eq. (3.20), be written as

$$P_r = \int_{-h}^{0} I(r, z)\, 2\pi r\, dz = J_r 2\pi r, \tag{3.23}$$

where

$$J_r = \int_{-h}^{0} I(r, z)\, dz \tag{3.24}$$

is the radiated wave-energy transport (per unit width of the wave front). If there is no loss of wave energy in the water, $P_r$ is independent of the distance $r$ from the axis of the oscillating body. Consequently, $J_r$ is inversely proportional to $r$:

$$J_r(r) = J_r(a)\,(a/r), \tag{3.25}$$

where $J_r(a)$ is the wave-energy transport at $r = a$. From this, we would expect that the dynamic pressure and other physical quantities associated with the radiated wave have amplitudes that are inverse to the square root of $r$. As we shall discuss in more detail later, in Chapters 4 and 5, this result is true, provided $r$ is large enough. (There may be significant deviation from this result if the distance from the oscillating body is shorter than one wavelength.)

## 3.4  Radiation Resistance and Radiation Impedance

If the mass $m$ indicated in Figure 2.1 is the membrane of a loudspeaker, an acoustic wave will be generated due to the oscillation of the system. Or, if the mass $m_m$ is immersed in water, as indicated in Figure 3.2, a water wave will be generated. Let us now consider a wave-tank laboratory where such an immersed body of mass $m_m$ is suspended through a spring $S_m$ and a mechanical resistance $R_m$, as indicated in Figure 2.1. Alternatively, the body could be a membrane suspended in a frame inside a loudspeaker cabinet. Assume that an external force

$$F(t) = \mathrm{Re}\{\hat{F} e^{i\omega t}\} \tag{3.26}$$

is applied to the body, resulting in a forced oscillatory motion with velocity

$$u(t) = \mathrm{Re}\{\hat{u} e^{i\omega t}\}. \tag{3.27}$$

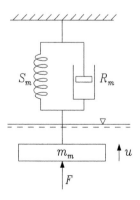

Figure 3.2: A body of mass $m_m$ suspended in water through a spring $S_m$ and a damper $R_m$.

The power consumed by the mechanical damper is (in time average)

$$P_m = \tfrac{1}{2} R_m |\hat{u}|^2, \tag{3.28}$$

in agreement with Eq. (2.79). The oscillating body generates a wave which carries away a radiated power $P_r$ [cf. Figure 3.1 and Eq. (3.20)]. In analogy with Eq. (3.28), we write

$$P_r = \tfrac{1}{2} R_r |\hat{u}|^2, \tag{3.29}$$

which defines the so-called *radiation resistance $R_r$*.

Due to the radiated wave, a reaction force $F_r$ acts on the body in addition to the externally applied force $F$. With the assumption of linear theory, $F_r$ is also varying as a harmonic oscillation—that is, $F_r = \mathrm{Re}\{\hat{F}_r e^{i\omega t}\}$. The dynamics of the system is then described by the following extension of Eq. (2.63) or Eq. (2.64):

$$i\omega m_m \hat{u} + R_m \hat{u} + (S_m/i\omega)\hat{u} = \hat{F} + \hat{F}_r, \tag{3.30}$$

or, in terms of the mechanical impedance $Z_m$:

$$Z_m \hat{u} = \hat{F} + \hat{F}_r. \tag{3.31}$$

Setting

$$\hat{F}_r = -Z_r \hat{u}, \tag{3.32}$$

we define an added impedance, or the so-called *radiation impedance $Z_r$*.

In general, $Z_r$ is a complex function of $\omega$:

$$Z_r = Z_r(\omega) = R_r(\omega) + iX_r(\omega), \tag{3.33}$$

which also depends on the geometry of the radiating system. Now we have from Eqs. (3.31) and (3.32) that

$$(Z_m + Z_r)\hat{u} = \hat{F}, \tag{3.34}$$

which gives the complex velocity amplitude

$$\hat{u} = \frac{\hat{F}}{Z_m + Z_r} = \frac{\hat{F}}{(R_m + R_r) + i(\omega m_m + X_r - S_m/\omega)}. \tag{3.35}$$

Comparing this result with Eq. (2.53), we observe that the oscillatory motion is modified because the oscillating mass has been immersed in water. The motion of the immersed body results in motion of the water surrounding the body. Some energy, represented by the radiated power (3.29), is carried away. Moreover, some energy is stored as kinetic energy, due to the velocity of the water, and as potential energy, due to gravity when the water surface is deformed and water is lifted from troughs to crests. The energy stored in the water is added to the energy stored in the mechanical system itself. Referring to Eq. (2.88), we may thus relate the *radiation reactance $X_r(\omega)$* to the difference between the average values of the added kinetic energy and the added potential energy. The radiation reactance $X_r(\omega)$ is frequently written as $\omega m_r$, where

$$m_r = m_r(\omega) = X_r(\omega)/\omega \tag{3.36}$$

is the so-called *added mass*, which is usually positive. There are, however, exceptional cases in which the added potential energy is larger than the added kinetic energy, and in such cases, the added mass becomes negative [23].

Combining Eqs. (3.33) and (3.36), we may write the radiation impedance as

$$Z_r(\omega) = R_r(\omega) + i\omega m_r(\omega). \tag{3.37}$$

Here, as well as in acoustics [21], radiation impedance has the dimension of force divided by velocity, and hence, the SI unit is $[Z_r] = [R_r] = \mathrm{Ns/m} = \mathrm{kg/s}$. Analogously, in theory for radio antennae, the (electric) radiation impedance has dimension voltage divided by current, and the corresponding SI unit is $\Omega = \mathrm{V/A}$. The term 'radiation impedance' is commonly used in connection with microphones and loudspeakers in acoustics [21] and also in connection with receiving and transmitting antennae in electromagnetics. A few authors [24, 25] have also adopted the term 'radiation impedance' in hydrodynamics, in connection with the generation and absorption of gravity waves on water. This term will also be used in the subsequent text.

## 3.5 Resonance Absorption

In the preceding section, we assumed that an external force $F$ was given [cf. Eq. (3.26) and Figure 3.2]. This external force could have been applied through a motor or some other mechanism, not shown in Figure 3.2. In addition to the external force, a reaction force $F_r$ was taken into consideration in Eq. (3.30), and we assumed in Eq. (3.32) that this reaction force is linear in the velocity $u$. If the motion had been prevented (by choosing at least one of the parameters $S_m$, $R_m$ and $m_m$ sufficiently large), then only the external force $F$ would remain.

Let us now assume that this force is applied through an incident wave. We shall adopt the term 'excitation force' for the wave force $F_e$ which acts on the immersed body when it is not moving—that is, when $u = 0$. For this case, we replace $\hat{F}$ in Eqs. (3.30), (3.31), (3.34) and (3.35) by $\hat{F}_e$. According to Eq. (3.35), the body's velocity is then given by

$$\hat{u} = \frac{\hat{F}_e}{R_m + R_r + i[\omega(m_m + m_r) - S_m/\omega]}. \tag{3.38}$$

The power absorbed in the mechanical damper resistance $R_m$ is [cf. Eq. (3.28)]

$$P_a = \frac{R_m}{2}|\hat{u}|^2 = \frac{(R_m/2)|\hat{F}_e|^2}{(R_m + R_r)^2 + (\omega m_m + \omega m_r - S_m/\omega)^2}. \tag{3.39}$$

Note that $R_m$ could, in an ideal case, represent a load resistance and that $P_a$ correspondingly represents useful power being consumed by the load resistance. We note that $P_a = 0$ for $R_m = 0$ and for $R_m = \infty$, and that $P_a > 0$ for $0 < R_m < \infty$. Thus, there is a maximum of absorbed power when $\partial P_a / \partial R_m = 0$, which occurs if

$$R_m = \left[ R_r^2 + (\omega m_m + \omega m_r - S_m/\omega)^2 \right]^{1/2} \equiv R_{m,\text{opt}}, \tag{3.40}$$

for which we have the maximum absorbed power

$$P_{a,\text{max}} = \frac{|\hat{F}_e|^2/4}{R_r + \left[ R_r^2 + (\omega m_m + \omega m_r - S_m/\omega)^2 \right]^{1/2}}. \tag{3.41}$$

See Problems 3.7 and 3.8.

Furthermore, we see by inspection of Eq. (3.39) that if we, for arbitrary $R_m$, can choose $m_m$ and $S_m$ such that

$$\omega m_m + \omega m_r - S_m/\omega = 0, \tag{3.42}$$

then the absorbed power has the maximum value

$$P_a = \frac{R_m |\hat{F}_e|^2/2}{(R_m + R_r)^2}. \tag{3.43}$$

If we now choose $R_m$ in accordance with condition (3.40), which now becomes

$$R_m = R_r \equiv R_{m,\text{OPT}}, \tag{3.44}$$

the maximum absorbed power is

$$P_{a,\text{MAX}} = |\hat{F}_e|^2/(8R_r), \tag{3.45}$$

and in this case, Eq. (3.38) simplifies to

$$\hat{u} = \hat{F}_e/(2R_r) \equiv \hat{u}_{\text{OPT}}. \tag{3.46}$$

When condition (3.42) is satisfied, we have resonance. We see from Eq. (3.38) that the oscillation velocity is in phase with the excitation force, since the ratio between the complex amplitudes $\hat{u}$ and $\hat{F}$ is then real. We may refer to Eq. (3.42) as the 'resonance condition' or the 'optimum phase condition'. Note that this condition is independent of the chosen value of the mechanical damper resistance $R_m$, and the maximum absorbed power is as given by Eq. (3.43).

If the optimum phase condition cannot be satisfied, then the maximum absorbed power is as given by Eq. (3.41), provided the 'optimum amplitude condition' (3.40) is satisfied.

If the optimum phase condition and the optimum amplitude condition can be satisfied simultaneously, then the maximum absorbed power is as given by

Eq. (3.45), and the optimum oscillation is as given by Eq. (3.46). In later chapters (Chapters 6–8), we encounter situations analogous to this optimum case of power absorption from a water wave. Note, however, that the discussion in the present section is also applicable to absorption of energy from an acoustic wave by a microphone or from an electromagnetic wave by a receiving antenna. (In this latter case, $u$ is the electric current flowing in the electric terminal of the antenna, and $F_e$ is the excitation voltage—that is, the voltage which is induced by the incident electromagnetic wave at the terminal when $u = 0$.)

Let us now, for simplicity, neglect the frequency dependence of the radiation resistance $R_r$ and of the added mass $m_r$. Introducing the natural angular frequency (eigenfrequency)

$$\omega_0 = \sqrt{S_m/(m_m + m_r)} \tag{3.47}$$

into Eq. (3.39), we rewrite the absorbed power as

$$P_a(\omega) = \frac{R_m|\hat{F}_e(\omega)|^2}{2(R_m + R_r)^2} \frac{1}{1 + (\omega_0/2\delta)^2(\omega/\omega_0 - \omega_0/\omega)^2}, \tag{3.48}$$

where

$$\delta = \frac{R_m + R_r}{2(m_m + m_r)} = \frac{(R_m + R_r)\omega_0^2}{2S_m} \tag{3.49}$$

is the so-called damping coefficient of the oscillator. Note that Eqs. (3.47) and (3.49) are extensions of definitions given in Eq. (2.5).

Referring to Eq. (2.20), we see that the relative absorbed-power response

$$\frac{P_a(\omega)/|\hat{F}_e(\omega)|^2}{P_a(\omega_0)/|\hat{F}_e(\omega_0)|^2} = \frac{1}{1 + (\omega_0/2\delta)^2(\omega/\omega_0 - \omega_0/\omega)^2}, \tag{3.50}$$

which has its maximum value of 1 at resonance ($\omega = \omega_0$), exceeds $\frac{1}{2}$ in a frequency interval $\omega_l < \omega < \omega_u$, where

$$\omega_u - \omega_l = (\Delta\omega)_{\text{res}} = 2\delta = \frac{R_m + R_r}{m_m + m_r} = \frac{(R_m + R_r)\omega_0^2}{S_m}. \tag{3.51}$$

The relative absorbed-power response versus frequency is plotted in Figure 3.3 for two different values of the damping factor

$$\frac{\delta}{\omega_0} = \frac{R_m + R_r}{2\omega_0(m_m + m_r)} = \frac{R_m + R_r}{2\sqrt{S_m(m_m + m_r)}} = \frac{(R_m + R_r)\omega_0}{2S_m}. \tag{3.52}$$

The last equalities in Eqs. (3.49), (3.51) and (3.52) have been obtained by using Eq. (3.47) to eliminate $(m_m + m_r)$. Observe that increasing (decreasing) the spring stiffness $S_m$ increases (decreases) the natural frequency $\omega_0$ and decreases (increases) the relative bandwidth $(\Delta\omega)_{\text{res}}/\omega_0$. On the other hand, increasing (decreasing) the mass $m_m$ decreases (increases) both $\omega_0$ and $(\Delta\omega)_{\text{res}}/\omega_0$.

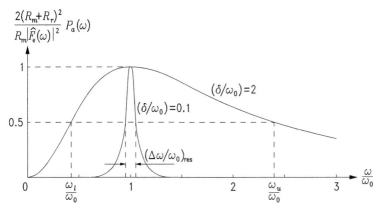

Figure 3.3: Frequency response of absorbed power for two different values of damping factor $\delta/\omega_0$.

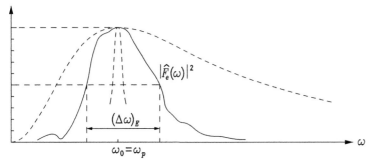

Figure 3.4: Representation of the wave spectrum (solid curve) compared with power absorption responses for the two cases given in Figure 3.3 (dashed curves).

Note that $|\hat{F}_e(\omega)|^2$ is a representation of the spectrum of the incident wave. An example is indicated in Figure 3.4, where $|\hat{F}_e(\omega)|^2$ is maximum at some angular frequency $\omega_p$ and exceeds half of its maximum in an interval of length $(\Delta\omega)_E$. Assume that an absorbing system has been chosen, for which $\omega_0 = \omega_p$. If a sufficiently large damping factor $\delta/\omega_0$ is chosen, we have that $(\Delta\omega)_{\text{res}} > (\Delta\omega)_E$. If we wish to absorb as much wave energy as possible, we should choose $R_m \geq R_r$ according to Eqs. (3.40) and (3.44). Then, from Eq. (3.51), we have

$$(\Delta\omega)_{\text{res}} \geq \frac{2R_r}{m_m + m_r}. \tag{3.53}$$

The two dashed curves in Figure 3.4 represent two different wave-absorbing oscillators, one with a narrow bandwidth, the other with a wide one. Evidently, the narrow-bandwidth oscillator can absorb efficiently from only a small part of the indicated wave spectrum.

## Problems

### Problem 3.1: Group Velocity

(a) Show that the sum $s = s_1 + s_2$ of two waves of equal amplitudes and different frequencies, $s_j(x, t) = A \cos(\omega_j t - k_j x)$ (for $j = 1, 2$), may be written as

$$s(x, t) = 2A \cos\left(\frac{\omega_1 - \omega_2}{2}t - \frac{k_1 - k_2}{2}x\right) \cos\left(\frac{\omega_1 + \omega_2}{2}t - \frac{k_1 + k_2}{2}x\right)$$

either by using a trigonometric formula or by applying the method of complex representation of harmonic waves. Set $\omega_1 = \omega + \Delta\omega/2$ and $\omega_2 = \omega - \Delta\omega/2$ and discuss the case $\Delta\omega \to 0$.

(b) Further, consider the more general case

$$s(x, t) = \tfrac{1}{2}\sum_j A_j \exp\left[i(\omega_j t - k_j x)\right] + \text{c. c.}$$

Assume that $|A_j|$ has a maximum value $|A_m|$ for $j = m$ ($m \gg 1$) and that $|A_j|$ is negligible when $\omega_j$ deviates from $\omega_m$ by more than a relatively small amount $\Delta\omega$ ($\Delta\omega \ll \omega_m$, 'narrow spectrum'). Show that the 'signal' $s(x, t)$ may be interpreted as a 'carrier wave', of angular frequency $\omega_m$ and angular repetency $k_m$, modulated by an 'envelope wave' propagating with the group velocity $(d\omega/dk)_m$, provided the dispersion curve $\omega = \omega(k)$ or $k = k(\omega)$ may be approximated by its tangent at $(k_m, \omega_m)$ in the interval where $|A_j|$ is not negligible.

### Problem 3.2: Capillary-Gravity Surface Wave

If the contribution to the wave's potential energy from capillary forces, in addition to the contribution from gravitational forces, is taken into consideration, then the phase velocity for waves on deep water is given by

$$v_p = \sqrt{g/k + \gamma k/\rho}$$

instead of by Eq. (3.9), where $\gamma = 0.07\,\text{N/m}$ is the surface tension on the air–water interface. Find the corresponding dispersion relationship which replaces Eq. (3.8). Derive also an expression for the group velocity. Find the numerical value of the phase velocity for a ripple of wavelength $2\pi/k = 10\,\text{mm}$ (assume $g = 9.8\,\text{m/s}^2$ and $\rho = 1.0 \times 10^3\,\text{kg/m}^3$). Finally, derive expressions and numerical values for the wavelength and the frequency of the ripple which has the minimum phase velocity.

### Problem 3.3: Optical Dispersion in Gas

For a gas, the refraction index $n = c/v_p$ (the ratio between the speed of light in vacuum $c$ and the phase velocity $v_p$) is to a good approximation

$$n = 1 + p/(\omega_0^2 - \omega^2)$$

for frequencies where $n$ is not significantly different from 1. Here, $p$ is a positive constant and $\omega_0$ is a resonance angular frequency for the gas. Verify that the group velocity $v_g = d\omega/dk$ for an electromagnetic wave (light wave) in the gas (contrary to the phase velocity $v_p = \omega/k$) is smaller than $c$, for $\omega \ll \omega_0$ as well as for $\omega \gg \omega_0$.

## Problem 3.4: Radiation Impedance for Spherical Loudspeaker

An acoustic wave is radiated from (generated by) a pulsating sphere of radius $a$. The surface of the sphere oscillates with a radial velocity $u = \hat{u}e^{i\omega t}$. For $r > a$, the wave may be represented by the sound pressure

$$p = (A/r)\, e^{i(\omega t - kr)}$$

and the (radially directed) particle speed

$$v = v_r = [1 + 1/(ikr)]\,(p/\rho c),$$

where the constants $\rho$ and $c = \omega/k$ are the fluid density and the speed of sound, respectively [21, pp. 114 and 163].

Use the boundary condition $u = [v]_{r=a}$ to determine the unknown coefficient $A$. Then derive expressions for the radiation impedance $Z_r$, radiation resistance $R_r$ and added mass $m_r$ as functions of the angular repetency $k$. (Base the derivation of $Z_r$ on the reaction force from the wave on the surface of the sphere. Check the result for $R_r$ by deriving an expression for the radiated power.)

## Problem 3.5: Acoustic Point Absorber

Let us consider a pulsating sphere as a microphone. Let the radius $a$ of the sphere be so small ($ka \ll 1$) that the microphone may be considered as an acoustic point absorber. When $ka \ll 1$, the radiation resistance is approximately

$$R_r \approx 4\pi a^4 k^2 \rho c.$$

The spherical shell of the microphone has a mass $m$ and a stiffness $S$ against radial displacements. Moreover, it has a mechanical resistance $R$ representing conversion of absorbed acoustical power to electric power.

A plane harmonic wave

$$p_i = A_0\, e^{i(\omega t - kx)}$$

is incident upon the sphere. For a point absorber, we may disregard reflected or diffracted waves. At a certain frequency ($\omega = \omega_0$) we have resonance. Find the value of $R$ for which the absorbed power $P$ has its maximum $P_{max}$ and express $P_{max}$ in terms of $A_0$, $a$, $k$ and $\rho c$. Also show that the maximum absorption cross

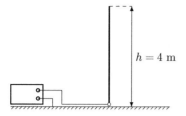

Figure 3.5: Vertical antenna connected to an electric circuit for transmission or reception of electromagnetic waves.

$h = 4$ m

section of the spherical microphone is $\lambda^2/4\pi$. (This might be compared with the absorption cross section $A_a = \lambda^2/2\pi$ for a microphone of the type of a plane vibrating piston in a plane stiff wall, as shown, for instance, in [26]. The absorption cross section is defined as the absorbed power divided by the sound intensity $I_i = \frac{1}{2}|A_0|^2/\rho c$ of the incident wave.)

### Problem 3.6: Short Dipole Antenna

A vertical grounded antenna of height $h = 4$ m has, at frequency $v = \omega/2\pi = 3$ MHz (or wavelength $\lambda = 100$ m), an effective height $h_{\mathrm{eff}} = 2$ m. The input port for the antenna is at the ground plane and the antenna is coupled to an electric circuit as indicated in Figure 3.5. Show that the radiation resistance is $R_r = 0.63\ \Omega$. The antenna is being used for reception of a plane electromagnetic wave with vertically polarised electric field of amplitude $|E_i| = 10^{-3}$ V/m.

The antenna circuit is resistance matched as well as tuned to resonance. Calculate the power which is absorbed by the antenna circuit. Also calculate the absorption cross section for the antenna.

[Hint: for the grounded antenna, the effective dipole length is $l_{\mathrm{eff}} = 2h_{\mathrm{eff}}$. For a Hertz dipole, of infinitesimal length $l_{\mathrm{eff}}$, the radiation resistance is $R_r = (kl_{\mathrm{eff}})^2 Z_0/6\pi$, where $Z_0 = (\mu_0/\epsilon_0)^{1/2} = 377\ \Omega$. The intensity (average power per unit area) of the incident electromagnetic wave is $\frac{1}{2}|E_i|^2/Z_0$.]

### Problem 3.7: Optimum Load Resistance

Derive Eqs. (3.40) and (3.41) for the optimum load resistance $R_m$ and the corresponding absorbed power $P_{a,\max}$.

### Problem 3.8: Maximum Absorbed Power

Show that Eq. (3.39) for the absorbed power may be reformulated as

$$\frac{8R_r}{|\hat{F}_e|^2}P_a = \frac{4R_m R_r}{(R_m + R_r)^2 + X^2} = 1 - \frac{(R_m - R_r)^2 + X^2}{(R_m + R_r)^2 + X^2} = 1 - \left|1 - \frac{2R_r\hat{u}}{\hat{F}_e}\right|^2,$$

where $X = \omega m_m + \omega m_r - S_m/\omega$. From this, Eqs. (3.42)–(3.46) may be obtained simply by inspection.

## Problem 3.9:  Power Radiated from Oscillating Submerged Body

Assume that the mass $m$ of Problem 2.12 is submerged in water, where it generates a wave. Further, assume that the radiation resistance $R_r = 1$ Ns/m is included as one-third of the total resistance $R = 3$ Ns/m. Similarly, assume that the added mass $m_r$ is included in the total mass $m = 6$ kg. Here we neglect the fact that $R_r$ and $m_r$ vary with frequency. How large is the radiated power at frequencies $f$ and $f_0$ defined in Problem 2.12?

# CHAPTER FOUR

# Gravity Waves on Water

The subject of this chapter is the study of waves on an ideal fluid, namely a fluid which is incompressible and in which wave motion takes place without loss of mechanical energy. It is also assumed that the fluid motion is irrotational and that the wave amplitude is so small that linear theory is applicable. Starting from basic hydrodynamics, we shall derive the dispersion relationship for waves on water which is deep or otherwise has a constant depth. Plane and circular waves are discussed, and the transport of energy and momentum associated with wave propagation is considered. The final parts of the chapter introduce some concepts and derive some mathematical relations, which turn out to be very useful when, in subsequent chapters, interactions between waves and oscillating systems are discussed.

## 4.1 Basic Equations: Linearisation

Let us start with two basic hydrodynamic equations which express conservation of mass and momentum, namely the continuity equation

$$\frac{\partial \rho}{\partial t} + \nabla \cdot (\rho \vec{v}) = 0 \tag{4.1}$$

and the Navier–Stokes equation

$$\frac{D\vec{v}}{Dt} \equiv \frac{\partial \vec{v}}{\partial t} + (\vec{v} \cdot \nabla)\vec{v} = -\frac{1}{\rho}\nabla p_{\text{tot}} + \nu\nabla^2\vec{v} + \frac{1}{\rho}\vec{f}. \tag{4.2}$$

Here $\rho$ is the mass density of the fluid, $\vec{v}$ is the velocity of the flowing fluid element, $p_{\text{tot}}$ is the pressure of the fluid and $\nu = \eta/\rho$ is the kinematic viscosity coefficient, which we shall neglect by assuming the fluid to be ideal. Hence, we set $\nu = 0$. Finally, $\vec{f}$ is external force per unit volume. Here we consider only gravitational force—that is,

$$\vec{f} = \rho\vec{g}, \tag{4.3}$$

where $\vec{g}$ is the acceleration due to gravity. With the introduced assumptions, Eq. (4.2) becomes

$$\frac{\partial \vec{v}}{\partial t} + (\vec{v} \cdot \nabla)\vec{v} = -\frac{1}{\rho}\nabla p_{tot} + \vec{g}. \tag{4.4}$$

For an incompressible fluid, $\rho$ is constant, and Eq. (4.1) gives

$$\nabla \cdot \vec{v} = 0. \tag{4.5}$$

Further, we assume that the fluid is irrotational, which mathematically means that

$$\nabla \times \vec{v} = 0. \tag{4.6}$$

Because of the vector identity $\nabla \times \nabla \phi \equiv 0$ for any scalar function $\phi$, we can then write

$$\vec{v} = \nabla \phi, \tag{4.7}$$

where $\phi$ is the so-called *velocity potential*. Inserting into Eq. (4.4) and using the vector identity

$$\vec{v} \times (\nabla \times \vec{v}) \equiv \frac{1}{2}\nabla v^2 - (\vec{v} \cdot \nabla)\vec{v} \tag{4.8}$$

and the relation $\vec{g} = -\nabla(gz)$ (the $z$-axis is pointing upwards) give

$$\nabla \left( \frac{\partial \phi}{\partial t} + \frac{v^2}{2} + \frac{p_{tot}}{\rho} + gz \right) = 0. \tag{4.9}$$

Integration gives

$$\frac{\partial \phi}{\partial t} + \frac{v^2}{2} + \frac{p_{tot}}{\rho} + gz = C, \tag{4.10}$$

where $C$ is an integration constant or some function of time. This is (a non-stationary version of) the so-called Bernoulli equation.

Because $v^2 = \nabla \phi \cdot \nabla \phi$ according to Eq. (4.7), the scalar equation (4.10) for the scalar quantity $\phi$ replaces the vectorial equation (4.4) for the vectorial quantity $\vec{v}$. This is a mathematical convenience which is a benefit resulting from the assumption of irrotational flow.

For the static case, when the fluid is not in motion, $\vec{v} = 0$ and $\phi = $ constant. Then Eq. (4.10) gives

$$p_{tot} = p_{stat} = -\rho gz + \rho C. \tag{4.11}$$

At the free surface, $z = 0$, we have $p_{tot} = p_{atm}$, where $p_{atm}$ is the atmospheric air pressure. Note that we here neglect surface tension on the air–fluid interface. This gives $C = p_{atm}/\rho$ and

$$p_{stat} = -\rho gz + p_{atm}. \tag{4.12}$$

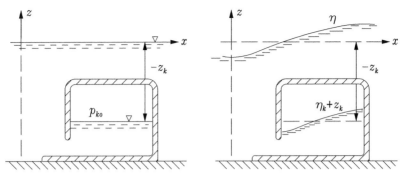

Figure 4.1: Submerged chamber for an OWC with an equilibrium water level below the mean sea level and containing entrapped air with elevated static pressure $p_{k0}$.

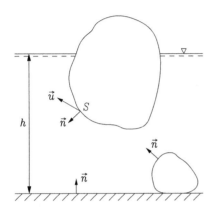

Figure 4.2: Unit normals $\vec{n}$ on wet body surfaces $S$ (interfaces between water and rigid bodies) directed into the fluid domain. Body velocity $\vec{u}$ may be a function of the considered point of $S$.

The hydrostatic pressure increases linearly with the vertical displacement below the surface of the fluid.

The static air pressure in a submerged air chamber (Figure 4.1) is

$$p_{k0} = \rho g(-z_k) + p_{\text{atm}},\qquad(4.13)$$

where the water surface below the entrapped air is at depth $-z_k$.

The conditions of incompressibility (4.5) and of irrotational motion (4.7) require that Laplace's equation

$$\nabla^2\phi = 0\qquad(4.14)$$

must be satisfied throughout the fluid. Solutions of this partial differential equation must satisfy certain boundary conditions, which are considered in the following paragraphs.

On a solid-body boundary moving with velocity $\vec{u}$ (Figure 4.2), we have

$$u_n \equiv \vec{u}\cdot\vec{n} = v_n = \nabla\phi\cdot\vec{n} \equiv \partial\phi/\partial n,\qquad(4.15)$$

as there is no fluid flow through the boundary. In an ideal fluid which is inviscid ($v = 0$), no condition is required on the tangential component of $\vec{v} = \nabla\phi$. On a solid surface not in motion, we have

$$\partial\phi/\partial n = 0. \tag{4.16}$$

On a horizontal bottom of depth $h$, we have, in particular,

$$\partial\phi/\partial z = 0 \quad \text{at } z = -h. \tag{4.17}$$

Notice that the acceleration of gravity $g$, which is an important quantity for ocean waves, does not enter into Laplace's equation (4.14) or the body boundary condition (4.15). Moreover, these equations do not contain any derivative with respect to time. However, the fact that gravity waves can exist on water is associated with the presence of the quantity $g$, as well as of a time derivative, in the following free-surface boundary condition.

On the free surface $z = \eta(x,y,t)$, which is the interface between water and open air, the pressure in the fluid equals the air pressure (if we neglect capillary forces—which are, however, considered in Problem 4.1). Using $[p_{\text{tot}}]_{z=\eta} = p_{\text{atm}}$ in the Bernoulli equation (4.10) gives

$$\left[\frac{\partial\phi}{\partial t} + \frac{v^2}{2}\right]_{z=\eta} + g\eta = C - \frac{p_{\text{atm}}}{\rho}. \tag{4.18}$$

In the static case, the left-hand side vanishes. Hence, $C = p_{\text{atm}}/\rho$, and with $\vec{v} = \nabla\phi$ [cf. Eq. (4.7)] we have the free-surface boundary condition

$$g\eta + \left[\frac{\partial\phi}{\partial t} + \frac{1}{2}\nabla\phi \cdot \nabla\phi\right]_{z=\eta} = 0. \tag{4.19}$$

In a wave-power converter of the oscillating-water-column (OWC) type with pneumatic power takeoff, the air pressure is not constant above the OWC. Let the dynamic part of the air pressure be $p_k$. The total air pressure is

$$p_{\text{air}} = p_{k0} + p_k = \rho g(-z_k) + p_{\text{atm}} + p_k, \tag{4.20}$$

where we have used Eq. (4.13). (Except for submerged OWCs, the equilibrium water level is at $z_k = 0$, which is also the case if air pressure fluctuation due to wind is considered.) Let $\eta_k = \eta_k(x,y,t)$ denote the vertical deviation of the water surface from its equilibrium position below the entrapped air (see Figure 4.1). Thus, $\eta = \eta_k$ when $z_k = 0$. From the Bernoulli equation (4.10), we now have

$$\left[\frac{\partial\phi}{\partial t} + \frac{v^2}{2}\right]_{z=\eta_k+z_k} + g\eta_k + gz_k = C - \left[\frac{p_{\text{tot}}}{\rho}\right]_{z=\eta_k+z_k}$$

$$= C - \frac{p_{\text{air}}}{\rho} = C - (-gz_k) - \frac{p_{\text{atm}} + p_k}{\rho}. \tag{4.21}$$

Using $C - p_{atm}/\rho = 0$ and Eq. (4.7) gives

$$g\eta_k + \frac{p_k}{\rho} + \left[ \frac{\partial \phi}{\partial t} + \frac{1}{2} \nabla \phi \cdot \nabla \phi \right]_{z=\eta_k+z_k} = 0. \tag{4.22}$$

This is the so-called *dynamic boundary condition*, which we linearise to

$$g\eta_k + \frac{p_k}{\rho} + \left[ \frac{\partial \phi}{\partial t} \right]_{z=z_k} = 0 \tag{4.23}$$

when we assume that the dynamic variables like $\phi$, $\eta$, $\eta_k$ and all their derivatives are small, and we neglect small terms of second or higher order, such as $v^2 = \nabla \phi \cdot \nabla \phi$. Also the difference

$$\left[ \frac{\partial \phi}{\partial t} \right]_{z=\eta_k+z_k} - \left[ \frac{\partial \phi}{\partial t} \right]_{z=z_k} = \eta_k \left[ \frac{\partial^2 \phi}{\partial z \partial t} \right]_{z=z_k} + \cdots \tag{4.24}$$

may then be neglected. [Note that the right-hand side of Eq. (4.24) is a Taylor expansion where only the first non-vanishing term is explicitly written. It is a product of two small quantities, $\eta_k$ and a derivative of $\phi$.]

In addition to the dynamic boundary condition, there is also a *kinematic boundary condition* on the interface between water and air. Physically, it is the condition that a fluid particle on the interface stays on the interface as this undulates as a result of wave motion. For the linearised case, the kinematic boundary condition is simply

$$\left[ \frac{\partial \phi}{\partial z} \right]_{z=z_k} = [v_z]_{z=z_k} = \frac{\partial \eta_k}{\partial t}, \tag{4.25}$$

where $\eta_k$ may be replaced by $\eta$ for the open-air case. (Readers interested in learning about the kinematic boundary condition for the more general nonlinear case may consult, e.g., [1, chapter 1].) Taking the time derivative of Eq. (4.23) and inserting it into (4.25) give

$$\left[ \frac{\partial^2 \phi}{\partial t^2} + g \frac{\partial \phi}{\partial z} \right]_{z=z_k} = -\frac{1}{\rho} \frac{\partial p_k}{\partial t}. \tag{4.26}$$

With zero dynamic air pressure (for $z = 0$), we get, in particular,

$$\left[ \frac{\partial^2 \phi}{\partial t^2} + g \frac{\partial \phi}{\partial z} \right]_{z=0} = 0. \tag{4.27}$$

Note that time $t$ enters explicitly only into the free-surface boundary conditions. (It does not enter into the partial differential equation $\nabla^2 \phi = 0$ and the remaining boundary conditions.) Thus, without a free surface, the solution $\phi = \phi(x, y, z, t)$ could not represent a wave.

We may also note that the boundary condition (4.15) has to be satisfied on the wet surface of the moving body. However, if the body is oscillating with a small amplitude, we may make the linear approximation that the boundary condition (4.15) is to be applied at the time-average (or equilibrium) position of the wet surface of the oscillating body.

Any solution $\phi = \phi(x, y, z, t)$ for the velocity potential has to satisfy Laplace's equation (4.14) in the fluid domain and the inhomogeneous boundary conditions (4.15) and (4.26), which may sometimes or somewhere simplify to the homogeneous boundary conditions (4.16) and (4.27), respectively. For cases in which the fluid domain is of infinite extent, later sections (Sections 4.3 and 4.6) supplement the preceding boundary conditions with a 'radiation condition' at infinite distance.

From the velocity potential $\phi(x, y, z, t)$, which is an auxiliary mathematical function, we can derive the following physical quantities. Everywhere in the fluid domain, we can derive the fluid velocity from Eq. (4.7),

$$\vec{v} = \vec{v}(x, y, z, t) = \nabla \phi, \tag{4.28}$$

and the *hydrodynamic pressure* from the dynamic part of Eq. (4.10),

$$p = p(x, y, z, t) = -\rho \left( \frac{\partial \phi}{\partial t} + \frac{v^2}{2} \right) \approx -\rho \frac{\partial \phi}{\partial t}, \tag{4.29}$$

where in the last step we have neglected the small term of second order. Moreover, the elevation of the interface between water and entrapped air is given by the linearised dynamic boundary condition (4.23) as

$$\eta_k = \eta_k(x, y, t) = -\frac{1}{g} \left[ \frac{\partial \phi}{\partial t} \right]_{z=z_k} - \frac{1}{\rho g} p_k. \tag{4.30}$$

The *wave elevation* (the elevation of the interface between the water and the open, constant-pressure air) is

$$\eta = \eta(x, y, t) = -\frac{1}{g} \left[ \frac{\partial \phi}{\partial t} \right]_{z=0}. \tag{4.31}$$

## 4.2 Harmonic Waves on Water of Constant Depth

Except when otherwise stated, let us in the following consider the case of a plane horizontal sea bottom. If the water is sufficiently deep, the sea bottom does not influence the waves on the water surface. Then the shape of the bottom is of no concern for the waves. If the water is not sufficiently deep, and the sea bottom is not horizontal, an analysis differing from the following analysis is required.

With sinusoidal time variation, we write

$$\phi = \phi(x, y, z, t) = \text{Re}\{\hat{\phi}(x, y, z)e^{i\omega t}\}, \tag{4.32}$$

where $\hat{\phi}$ is the complex amplitude of the velocity potential at the space point $(x, y, z)$. Similarly, we define the complex amplitudes $\hat{\vec{v}} = \hat{\vec{v}}(x, y, z), \hat{p} = \hat{p}(x, y, z),$

$\hat{\eta}_k = \hat{\eta}_k(x, y)$ and $\hat{\eta} = \hat{\eta}(x, y)$. The linearised basic equations, the Laplace equation (4.14) and the boundary conditions (4.15) and (4.26), now become

$$\nabla^2 \hat{\phi} = 0 \tag{4.33}$$

everywhere in the water,

$$\left[ \partial \hat{\phi} / \partial n \right]_S = \hat{u}_n \tag{4.34}$$

on the wet surface of solid bodies (Figure 4.2) and

$$\left[ -\omega^2 \hat{\phi} + g \frac{\partial \hat{\phi}}{\partial z} \right]_{z=z_k} = -\frac{i\omega}{\rho} \hat{p}_k \tag{4.35}$$

on the water–air surfaces (Figure 4.1). Here, $\hat{u}_n$ and $\hat{p}_k$ are also complex amplitudes. Note that the normal component $u_n$ of the motion of the wet body surface (Figure 4.2) is a function of the point considered on that surface, whereas the dynamic pressure $p_k$ is assumed to have the same value everywhere inside the volume of entrapped air (Figure 4.1). Equations (4.28)–(4.31) for determining the physical variables become, in terms of complex amplitudes,

$$\hat{\vec{v}} = \nabla \hat{\phi}, \tag{4.36}$$

$$\hat{p} = -i\omega \rho \hat{\phi}, \tag{4.37}$$

$$\hat{\eta}_k = -\frac{i\omega}{g} \left[ \hat{\phi} \right]_{z=z_k} - \frac{1}{\rho g} \hat{p}_k, \tag{4.38}$$

$$\hat{\eta} = -\frac{i\omega}{g} \left[ \hat{\phi} \right]_{z=0}. \tag{4.39}$$

The remaining part of this section (and also Sections 4.3–4.6) discusses some particular solutions which satisfy Laplace's equation and the homogeneous boundary conditions

$$\left[ \frac{\partial \hat{\phi}}{\partial z} \right]_{z=-h} = 0, \tag{4.40}$$

$$\left[ -\omega^2 \hat{\phi} + g \frac{\partial \hat{\phi}}{\partial z} \right]_{z=0} = 0. \tag{4.41}$$

These solutions will thus satisfy the boundary conditions on a (non-moving) horizontal bottom of a sea of depth $h$ and the boundary condition at the free water surface (the interface between water and air), where the air pressure is constant. Additional (inhomogeneous) boundary conditions will be imposed later (e.g., in Section 4.7).

Using the method of separation of variables, we seek a particular solution of the form

$$\hat{\phi}(x, y, z) = H(x, y)Z(z). \tag{4.42}$$

Inserting this into the Laplace equation (4.33) and dividing by $\hat{\phi}$, we get

$$0 = \frac{\nabla^2 \hat{\phi}}{\hat{\phi}} = \frac{1}{H}\left[\frac{\partial^2 H}{\partial x^2} + \frac{\partial^2 H}{\partial y^2}\right] + \frac{1}{Z}\frac{d^2 Z}{dz^2} \tag{4.43}$$

or

$$-\frac{1}{Z}\frac{d^2 Z}{dz^2} = \frac{1}{H}\left[\frac{\partial^2 H}{\partial x^2} + \frac{\partial^2 H}{\partial y^2}\right] \equiv \frac{1}{H}\nabla_H^2 H. \tag{4.44}$$

The left-hand side of Eq. (4.44) is a function of $z$ only. The right-hand side is a function of $x$ and $y$. This is impossible unless both sides equal a constant, $-k^2$, say. Thus, from Eq. (4.44), we get two equations:

$$\frac{d^2 Z(z)}{dz^2} = k^2 Z(z), \tag{4.45}$$

$$\nabla_H^2 H(x, y) = -k^2 H(x, y). \tag{4.46}$$

We note here that the separation constant $k^2$ has the dimension of inverse length squared. Later, when discussing a solution of the two-dimensional Helmholtz equation (4.46), we shall see that $k$ may be interpreted as the angular repetency of a propagating wave.

Let us, however, first discuss the following solution of Eq. (4.45):

$$Z(z) = c_+ e^{kz} + c_- e^{-kz}. \tag{4.47}$$

Here $c_+$ and $c_-$ are two integration constants. Because we have two new integration constants when solving Eq. (4.46) for $H$, we may choose $Z(0) = 1$, which means that $c_+ + c_- = 1$. Another relation to determine the integration constants is provided by the bottom boundary condition

$$\frac{\partial \hat{\phi}}{\partial z} = \frac{dZ(z)}{dz}H(x, y) = 0 \quad \text{for } z = -h. \tag{4.48}$$

It is then easy to show that (see Problem 4.2)

$$c_\pm = \frac{e^{\pm kh}}{e^{kh} + e^{-kh}}. \tag{4.49}$$

From Eq. (4.47) it now follows that

$$Z(z) = \frac{e^{k(z+h)} + e^{-k(z+h)}}{e^{kh} + e^{-kh}}. \tag{4.50}$$

Hence, we have the following particular solution of the Laplace equation:

$$\hat{\phi} = H(x, y)e(kz), \tag{4.51}$$

where $H(x, y)$ has to satisfy the Helmholtz equation (4.46), and where

$$e(kz) = \frac{\cosh(kz + kh)}{\cosh(kh)} = \frac{e^{k(z+h)} + e^{-k(z+h)}}{e^{kh} + e^{-kh}} = \frac{1 + e^{-2k(z+h)}}{1 + e^{-2kh}}e^{kz}. \tag{4.52}$$

To be strict, since this is a function of two variables, we should perhaps have denoted the function by $e(kz, kh)$. We prefer, however, to use the simpler notation $e(kz)$, in particular because, for the deep-water case, $kh \gg 1$, it tends to the exponential function $e(kz) \approx e^{kz}$ (although it then approaches $2e^{kz}$ when $z$ approaches $-h$, a $z$-coordinate which is usually of little practical interest in the deep-water case). We may note that the solution (4.51) is applicable for the deep-water case even if the water depth is not constant, provided $kh_{\min} \gg 1$.

In order to satisfy the free-surface boundary condition (4.41), we require

$$\omega^2 = \omega^2 e(0) = g \left[ \frac{de(kz)}{dz} \right]_{z=0} = gk \frac{\sinh(kh)}{\cosh(kh)} \tag{4.53}$$

or

$$\omega^2 = gk \tanh(kh), \tag{4.54}$$

which for deep water ($kh \gg 1$) simplifies to

$$\omega^2 = gk, \tag{4.55}$$

a result which has already been presented in Eq. (3.8).

We have now derived the dispersion equation (4.54), which is a relation between the angular frequency $\omega$ and the angular repetency $k$. Rewriting Eq. (4.54) as

$$\omega^2/(gk) = \tanh(kh), \tag{4.56}$$

we observe that, in the interval $0 < k < +\infty$, the right-hand side is monotonically increasing from 0 to 1, while the left-hand side, for a given $\omega$, is monotonically decreasing from $+\infty$ to 0. Hence, there is one, and only one, positive $k$ which satisfies (4.56), and there is correspondingly one, and only one, negative solution, since both sides of the equation are odd functions of $k$. Thus, there is only one possible positive value of the separation constant $k^2$ in Eqs. (4.45)–(4.46). We may raise the question whether negative or even complex values are possible.

Replacing $k^2$ by $\lambda_n$ and $\hat{\phi}(x, y, z) = Z(z)H(x, y)$ by

$$\hat{\phi}_n(x, y, z) = Z_n(z)H_n(x, y), \tag{4.57}$$

we rewrite Eqs. (4.45)–(4.46) as

$$\nabla_H^2 H_n(x, y) = -\lambda_n H_n(x, y), \tag{4.58}$$
$$Z_n''(z) = \lambda_n Z_n(z), \tag{4.59}$$

where the integer subscript $n$ is used to label the various possible solutions. Since $\hat{\phi}_n(x, y, z)$ has to satisfy the boundary conditions (4.40)–(4.41), the functions $Z_n(z)$ are subject to the boundary conditions

$$Z_n'(-h) = 0, \tag{4.60}$$

$$Z'_n(0) = \frac{\omega^2}{g} Z_n(0). \tag{4.61}$$

In Eq. (4.59), the separation constant $\lambda_n$ is an eigenvalue and $Z_n(z)$ is the corresponding eigenfunction. We shall now show that all eigenvalues have to be real. Let us consider two possible eigenvalues $\lambda_n$ and $\lambda_m$ with corresponding eigenfunctions $Z_n(z)$ and $Z_m(z)$. Replacing $n$ by $m$ in Eq. (4.59)—assuming $\omega$ is real—and taking the complex conjugate give

$$Z_m^{*\prime\prime}(z) = \lambda_m^* Z_m^*(z). \tag{4.62}$$

Now, let us multiply Eq. (4.59) by $Z_m^*(z)$ and Eq. (4.62) by $Z_n(z)$. Subtracting and integrating then give

$$I \equiv \int_{-h}^{0} [Z_m^*(z) Z_n''(z) - Z_n(z) Z_m^{*\prime\prime}(z)]\, dz = (\lambda_n - \lambda_m^*) \int_{-h}^{0} Z_m^*(z) Z_n(z)\, dz. \tag{4.63}$$

Noting that the integrand in the first integral of Eq. (4.63) may be written as

$$Z_m^*(z) Z_n''(z) - Z_n(z) Z_m^{*\prime\prime}(z) = \frac{d}{dz} [Z_m^*(z) Z_n'(z) - Z_n(z) Z_m^{*\prime}(z)] \tag{4.64}$$

and that the boundary conditions (4.60)–(4.61) apply to $Z_m^*$ as well, we find that the integral $I$ vanishes. This is true because

$$
\begin{aligned}
I &= [Z_m^*(z) Z_n'(z) - Z_n(z) Z_m^{*\prime}(z)]_{-h}^{0} \\
&= Z_m^*(0) \frac{\omega^2}{g} Z_n(0) - Z_n(0) \frac{\omega^2}{g} Z_m^*(0) - 0 - 0 \\
&= 0.
\end{aligned}
\tag{4.65}
$$

We have here tacitly assumed (as we shall do throughout) that $\omega$ is real—that is, $\omega^* = \omega$. Hence, we have shown that for all $m$ and $n$,

$$(\lambda_n - \lambda_m^*) \int_{-h}^{0} Z_m^*(z) Z_n(z)\, dz = 0. \tag{4.66}$$

For $m = n$, we have for non-trivial solutions $|Z_n(z)| \neq 0$ that $(\lambda_n - \lambda_n^*) = 0$. This means that $\lambda_n$ is real for all $n$.

For $\lambda_n \neq \lambda_m$, Eq. (4.66) gives the following orthogonality condition for the eigenfunctions $\{Z_n(z)\}$:

$$\int_{-h}^{0} Z_m^*(z) Z_n(z)\, dz = 0. \tag{4.67}$$

We have shown that the eigenvalues are real and that there is only one positive eigenvalue $k^2$ satisfying the dispersion equation (4.54). The other eigenvalues have to be negative. We shall label the eigenvalues in decreasing order as

$$\lambda_0 = k^2 > \lambda_1 > \lambda_2 > \lambda_3 > \ldots > \lambda_n > \ldots \tag{4.68}$$

Thus, $\lambda_n$ is negative if $n \geq 1$. A negative eigenvalue may be conveniently written as

$$\lambda_n = -m_n^2, \tag{4.69}$$

where $m_n$ is real. Let us now in Eq. (4.47) replace $k$ with $-im_n$ and $Z(z)$ with $Z_n(z)$. One of the two integration constants may then be eliminated by using boundary condition (4.60). The resulting eigenfunction is (see Problem 4.3)

$$Z_n(z) = N_n^{-1/2} \cos(m_n z + m_n h), \tag{4.70}$$

where $N_n^{-1/2}$ is an arbitrary integration constant. This result also follows from Eqs. (4.50) and (4.52) if we observe that $\cosh\left[-im_n(z+h)\right] = \cos\left[-m_n(z+h)\right] = \cos\left[m_n(z+h)\right]$. The free-surface boundary condition (4.61) is satisfied if (see Problem 4.3)

$$\omega^2/(gm_n) = -\tan(m_n h). \tag{4.71}$$

This equation also follows from Eq. (4.56) if $k$ is replaced by $-im_n$. The left-hand side of Eq. (4.71) decreases monotonically from $+\infty$ to 0 in the interval $0 < m_n < +\infty$. The right-hand side decreases monotonically from $+\infty$ to 0 in the interval $(n - 1/2)\pi/h < m_n < n\pi/h$. Hence, Eq. (4.71) has a solution in this latter interval. Noting that $n = 1, 2, 3, \ldots$, we find that there is an infinite, but numerable, number of solutions for $m_n$. For the solutions, we have (see Problem 4.3)

$$m_n \to \frac{n\pi}{h} - \frac{\omega^2}{n\pi g} \to \frac{n\pi}{h} \quad \text{as } n \to \infty. \tag{4.72}$$

Further (as is also shown in Problem 4.3), if

$$\frac{1}{h} \int_{-h}^{0} |Z_n(z)|^2 dz = 1 \tag{4.73}$$

is chosen as a normalisation condition, then the integration constant in Eq. (4.70) is given by

$$N_n = \frac{1}{2} \left[ 1 + \frac{\sin(2m_n h)}{2m_n h} \right]. \tag{4.74}$$

For $n = 0$, we set $m_0 = ik$ and then we have, in particular,

$$N_0 = \frac{1}{2} \left[ 1 + \frac{\sinh(2kh)}{2kh} \right]. \tag{4.75}$$

We may note (see Problem 4.2) that

$$Z_0(z) = \sqrt{\frac{2kh}{D(kh)}} \, e(kz), \tag{4.76}$$

where $e(kz)$ is given by Eq. (4.52) and

$$D(kh) = \left[1 + \frac{2kh}{\sinh(2kh)}\right]\tanh(kh).\qquad(4.77)$$

The orthogonal set of eigenfunctions $\{Z_n(z)\}$ is complete if the function $Z_0(z)$ is included in the set—that is, if $n = 0, 1, 2, 3, \ldots$. The completeness follows from the fact that the eigenvalue problem (4.59)–(4.61) is a Sturm–Liouville problem [27].

We have now discussed possible solutions of Eq. (4.45) or (4.59) with the homogeneous boundary conditions (4.40) and (4.41), or (4.60) and (4.61). It remains for us to discuss solutions of the Helmholtz equation (4.46) or (4.58). In the discussion that follows, we start by considering the case with $n = 0$—that is, $m_n = m_0 = ik$. Then we take Eq. (4.46) as the starting point.

## 4.3 Plane Waves: Propagation Velocities

We consider a two-dimensional case with no variation in the $y$-direction. Setting $\partial/\partial y = 0$, we have from Eqs. (4.44) and (4.46)

$$d^2 H(x)/dx^2 = -k^2 H(x),\qquad(4.78)$$

which has the general solution

$$H(x) = ae^{-ikx} + be^{ikx},\qquad(4.79)$$

where $a$ and $b$ are arbitrary integration constants. Hence, from (4.51),

$$\hat{\phi} = e(kz)(ae^{-ikx} + be^{ikx})\qquad(4.80)$$

and from (4.32),

$$\phi = \mathrm{Re}\{\hat{\phi}e^{i\omega t}\} = \mathrm{Re}\{ae^{i(\omega t - kx)} + be^{i(\omega t + kx)}\}e(kz),\qquad(4.81)$$

which demonstrates that $k$ is the angular repetency (wave number).

The first and second terms represent plane waves propagating in the positive and negative $x$-directions, respectively, with a phase velocity

$$v_p \equiv \frac{\omega}{k} = \frac{g}{\omega}\tanh(kh) = \sqrt{\frac{g}{k}\tanh(kh)},\qquad(4.82)$$

which is obtained from the dispersion relation (4.54). Note that $v_p$ is the velocity by which the wave crest (or a line of constant phase) propagates (in a direction perpendicular to the wave crest). Later (in Section 4.4), we prove that the wave energy associated with a harmonic plane wave is transported with the group velocity $v_g = d\omega/dk$ [cf. Eq. (3.10)], which, in general, differs from the phase velocity. The quantity $v_g$ bears its name because it is the velocity by which a wave group, composed of harmonic waves with slightly different frequencies, propagates (see Problem 3.1). For instance, a swell wave, which originates from a storm centre a distance $L$ from a coast line, reaches this coast line a time $L/v_g$ later (see Problem 4.4).

In Eq. (4.80), we might interpret the first term as an incident wave and the second term as a reflected wave. Introducing $\Gamma = b/a$, which is a complex reflection coefficient, and $A = -i\omega a/g$, which is the complex elevation amplitude at $x = 0$ in the case of no reflection, we may rewrite Eq. (4.80) as

$$\hat{\phi} = -\frac{g}{i\omega} A\, e(kz) \left(e^{-ikx} + \Gamma e^{ikx}\right). \tag{4.83}$$

Correspondingly, the wave elevation $\eta = \eta(x,t)$ has a complex amplitude [cf. Eqs. (4.39) and (4.52)]

$$\hat{\eta} = \hat{\eta}(x) = A\left(e^{-ikx} + \Gamma e^{ikx}\right). \tag{4.84}$$

Setting

$$\hat{\eta}_f = Ae^{-ikx}, \tag{4.85}$$

$$\hat{\eta}_b = Be^{ikx} = A\Gamma e^{ikx}, \tag{4.86}$$

we have

$$\hat{\eta} = \hat{\eta}_f + \hat{\eta}_b, \tag{4.87}$$

$$\hat{\phi} = -\frac{g}{i\omega}\hat{\eta}e(kz), \tag{4.88}$$

$$\hat{p} = \rho g \hat{\eta} e(kz), \tag{4.89}$$

where we have also made use of Eq. (4.37). Note that the hydrodynamic pressure $p$ decreases monotonically (exponentially for deep water) with the distance $(-z)$ below the mean free surface, $z = 0$. For the fluid velocity, we have, from Eqs. (4.36) and (4.52),

$$\hat{v}_x = \frac{\partial\hat{\phi}}{\partial x} = g\frac{k}{\omega}e(kz)(\hat{\eta}_f - \hat{\eta}_b) = \omega\frac{\cosh(kz+kh)}{\sinh(kh)}(\hat{\eta}_f - \hat{\eta}_b), \tag{4.90}$$

$$\hat{v}_z = \frac{\partial\hat{\phi}}{\partial z} = g\frac{ik}{\omega}e'(kz)(\hat{\eta}_f + \hat{\eta}_b) = i\omega\frac{\sinh(kz+kh)}{\sinh(kh)}(\hat{\eta}_f + \hat{\eta}_b), \tag{4.91}$$

where

$$e'(kz) = \frac{de(kz)}{d(kz)} = \frac{\sinh(kz+kh)}{\cosh(kh)} = e(kz)\tanh(kz+kh). \tag{4.92}$$

We have used the dispersion relation (4.54) to obtain the last expressions for $\hat{v}_x$ and $\hat{v}_z$.

If we have a progressive wave in the forward (positive $x$) direction ($\hat{\eta}_b = 0$) on deep water ($kh \gg 1$), the motion of fluid particles is circularly polarised

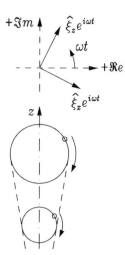

Figure 4.3: Phasor diagram of the $x$- and $z$-components of the fluid particle displacement (top) and fluid particle trajectories (bottom).

with a negative sense of rotation (in the clockwise direction in the $xz$-plane), notably

$$\hat{v}_x = \omega \hat{\eta}_f e^{kz}, \tag{4.93}$$

$$\hat{v}_z = i\omega \hat{\eta}_f e^{kz}. \tag{4.94}$$

The displacement of the fluid particles (as shown in Figure 4.3) are given by

$$\hat{\vec{\xi}} = \hat{\vec{v}}/i\omega, \tag{4.95}$$

$$\hat{\xi}_x = -i\hat{\eta}_f e^{kz}, \tag{4.96}$$

$$\hat{\xi}_z = \hat{\eta}_f e^{kz}. \tag{4.97}$$

For a progressive wave in the backwards direction ($\hat{\eta}_f = 0$) on deep water, the motion is circularly polarised with a positive sense of rotation.

For a plane wave propagating in a direction making an angle $\beta$ with the $x$-axis, the wave elevation is given by

$$\hat{\eta} = A \exp\left[-ik(x\cos\beta + y\sin\beta)\right], \tag{4.98}$$

which can be easily shown by considering transformation between two coordinate systems $(x', y')$ and $(x, y)$ which differ by a rotation angle $\beta$. The corresponding velocity potential $\hat{\phi}$, still given by Eq. (4.88) but now with the new $\hat{\eta}$, is a solution of the Laplace equation (4.33), satisfying the boundary conditions on the sea bed, $z = -h$, and on the free water surface, $z = 0$ (see Problem 4.5).

For deep water ($kh \gg 1$), the dispersion relation (4.54) is approximately $\omega^2 = gk$, in agreement with Eq. (4.55) and the previous statement, Eq. (3.8). Note that

$$\tanh(kh) > \begin{cases} 0.95 & \text{for } kh > 1.83 \text{ or } h > 0.3\lambda, \\ 0.99 & \text{for } kh > 2.6 \text{ or } h > 0.42\lambda. \end{cases} \tag{4.99}$$

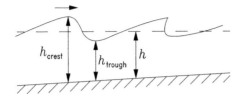

Figure 4.4: Increasing wave steepness as a wave propagates on shallow water.

Hence, depending on the desired accuracy, we may assume deep water when the depth is at least one third or one half of the wavelength, respectively. For deep water, the wavelength is

$$\lambda = 2\pi/k = 2\pi g/\omega^2 = (g/2\pi)T^2 = (1.56 \text{ m/s}^2)T^2. \tag{4.100}$$

As previously found [see Eqs. (3.9) and (3.12)], for deep water we have

$$2v_g = v_p = g/\omega = \sqrt{g/k}. \tag{4.101}$$

In contrast, for shallow water with horizontal bottom, $kh \ll 1$, a power series expansion of Eq. (4.54) gives

$$\omega^2 = gk(kh + \ldots) \approx ghk^2. \tag{4.102}$$

(Here the error is less than 1% or 5% if the water depth is less than 1/36 or 1/16 of the wavelength, respectively.) In this approximation a wave is not dispersive, because

$$v_g = v_p = \sqrt{gh} \tag{4.103}$$

is independent of $k$ and $\omega$. The group velocity $v_g$ differs, however, from the phase velocity $v_p$ in the general dispersive case, as is considered next. The present theory is linear, but let us nevertheless try to give a qualitative explanation of wave breaking on shallow water. The formula (4.103) indicates that the wave passes faster on the wave crest than on the wave through because $h_{\text{crest}} > h_{\text{trough}}$ (see Figure 4.4). Finally there will be a vertical edge and then, of course, linear theory does not apply since $\partial\eta/\partial x \to \infty$.

For constant water depth $h$, the phase velocity $v_p$ is, in the general case, given by Eq. (4.82). In order to obtain the group velocity $v_g$, we differentiate the dispersion equation (4.54):

$$\begin{aligned}
2\omega d\omega &= gdk \tanh(kh) + \frac{gk}{\cosh^2(kh)}hdk \\
&= \frac{dk}{k}gk \tanh(kh) + \frac{gk \tanh(kh)}{\cosh(kh)\sinh(kh)}hdk \\
&= \frac{dk}{k}\omega^2 + \frac{2\omega^2 hdk}{\sinh(2kh)}.
\end{aligned} \tag{4.104}$$

Hence,

$$v_g = \frac{d\omega}{dk} = \frac{\omega}{2k}\left[1 + \frac{2kh}{\sinh(2kh)}\right].$$ (4.105)

Since $v_p = \omega/k$, we then obtain

$$v_g = \frac{D(kh)}{2\tanh(kh)}v_p = \frac{g}{2\omega}D(kh),$$ (4.106)

where we have, for convenience, used the depth function $D(kh)$ defined by Eq. (4.77). Because we shall refer to this function frequently later, here we present some alternative expressions:

$$
\begin{aligned}
D(kh) &= \left[1 + \frac{2kh}{\sinh(2kh)}\right]\tanh(kh) \\
&= \tanh(kh) + \frac{kh}{\cosh^2(kh)} \\
&= \tanh(kh) + kh - kh\tanh^2(kh) \\
&= \left[1 - \left(\omega^2/gk\right)^2\right]kh + \omega^2/gk \\
&= (2\omega/g)v_g = (2k/g)v_pv_g.
\end{aligned}
$$ (4.107)

In the last equality, the somewhat complicated mathematical function $D(kh)$ has been more simply expressed by the physical quantities $v_g$ and $v_p$, the group and phase velocities. We shall frequently also need the relationship (see Problem 4.12)

$$2k\int_{-h}^{0} e^2(kz)\,dz = D(kh),$$ (4.108)

where the vertical eigenfunction $e(kz)$ is given by Eq. (4.52) and the depth function $D(kh)$ by Eq. (4.107).

Note that on deep water, $kh \gg 1$, we have $D(kh) \approx 1$ and $\tanh(kh) \approx 1$. For small values of $kh$, we have $\tanh(kh) = kh + \ldots$ and $D(kh) = 2kh + \ldots$. Whereas $\tanh(kh)$ is a monotonically increasing function, it can be shown that $D(kh)$ has a maximum $D_{max} = x_0$ for $kh = x_0$, where $x_0 = 1.1996786$ is a solution of the transcendental equation $x_0\tanh(x_0) = 1$ (see Problem 4.12). This maximum occurs for $\omega^2 h/g = 1$.

For arbitrary $\omega$, we can obtain $k$ from the transcendental dispersion equation (4.54) for instance by applying some numerical iteration procedure, and then we obtain $v_p$ and $v_g$ from Eqs. (4.82) and (4.106). The relationships are shown graphically in the curves of Figure 4.5.

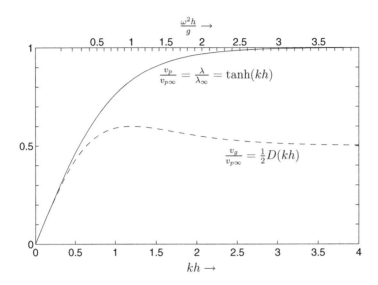

Figure 4.5: Phase velocity $v_p$, group velocity $v_g$ and wavelength $\lambda = 2\pi/k$ as functions of depth $h$ for a given frequency, or as functions of frequency for a given water depth. The subscript $\infty$ corresponds to (infinitely) deep water.

We have now discussed a propagating wave, which corresponds to the case with $n = 0$ in Eqs. (4.70), (4.71) and (4.74). Let now $n$ be an arbitrary non-negative integer. Setting again $\partial/\partial y = 0$, we find that Eq. (4.58), with Eq. (4.69) inserted, has a solution

$$H_n(x) = a_n e^{-m_n x} + b_n e^{m_n x}, \tag{4.109}$$

where $a_n$ and $b_n$ are integration constants. The corresponding complex amplitude of the velocity potential is, from Eq. (4.57),

$$\hat{\phi}_n(x, z) = (a_n e^{-m_n x} + b_n e^{m_n x}) Z_n(z). \tag{4.110}$$

Note that this is a particular solution, which satisfies the homogeneous Laplace equation (4.33) and the homogeneous boundary conditions (4.40) and (4.41). Hence, a superposition of such solutions as Eq. (4.110) is a solution:

$$\hat{\phi}(x, z) = \sum_{n=0}^{\infty} (a_n e^{-m_n x} + b_n e^{m_n x}) N_n^{-1/2} \cos(m_n z + m_n h), \tag{4.111}$$

where we have used Eq. (4.70).

Remember that for the term with $n = 0$ we set $m_n = ik$. Here we choose $k$ and $m_n$ (for $n \geq 1$) to be positive. Note that the term with $a_n$ decays with increasing $x$, or, for $n = 0$, is a wave progressing in the positive $x$-direction. The term with $b_n$ decays with decreasing $x$ or, for $n = 0$, is a wave progressing in the negative $x$-direction. If our region of interest is $0 < x < +\infty$, we thus have to set $b_n = 0$ for all $n$, except for the possibility that $b_0 \neq 0$ for a case where a wave is incident from infinite distance, $x = +\infty$. For $n \geq 1$, this is a consequence of avoiding infinite values. The *radiation condition* was briefly mentioned previously (in Section 4.1). We have the opportunity to apply such a condition here. When $b_0 = 0$, the radiation condition of an outgoing wave at infinite distance is satisfied. This means that there is no contribution of the form $e^{ikx}$ to the potential $\hat{\phi}$. Similarly, if our region of interest is $-\infty < x < 0$, then $a_n = 0$ for all $n$ if the radiation condition is satisfied as $x \to -\infty$. However, if a wave is incident from $x = -\infty$, then $a_0 \neq 0$. If our region of interest is finite, $x_1 < x < x_2$, for example, then we may have $a_n \neq 0$ and $b_n \neq 0$ for all $n$.

For the two-dimensional case, Eq. (4.111) represents a general plane-wave solution for the velocity potential in a uniform fluid of constant depth. The terms with $n = 0$ are propagating waves, whereas the terms with $n \geq 1$ are 'evanescent waves'. Propagation is along the $x$-axis. If $x$ in Eq. (4.111) is replaced by $(x \cos \beta + y \sin \beta)$, then the plane-wave propagation is along a direction which makes an angle $\beta$ with the $x$-axis [cf. Eq. (4.98)].

## 4.4 Wave Transport of Energy and Momentum

### 4.4.1 Potential Energy

Let us consider the potential energy associated with the elevation of water from the wave troughs to the wave crests. Per unit (horizontal) area, the potential energy relative to the sea bed equals the product of $\rho g(h + \eta)$, the water weight per unit area, and $(h + \eta)/2$, the height of the water mass centre above the sea bed:

$$(\rho g/2)(h + \eta)^2 = (\rho g/2)h^2 + \rho gh\eta + (\rho g/2)\eta^2. \tag{4.112}$$

The increase in relation to calm water is

$$\rho gh\eta + (\rho g/2)\eta^2, \tag{4.113}$$

where the first term has a vanishing average value.

Hence, the time-average potential energy per unit (horizontal) area is

$$E_p(x, y) = (\rho g/2)\overline{\eta^2(x, y, t)}. \tag{4.114}$$

In the case of a harmonic wave,

$$E_p(x, y) = (\rho g/4)|\hat{\eta}(x, y)|^2. \tag{4.115}$$

In particular, for the harmonic plane wave given by Eq. (4.87), we have

$$\begin{aligned} E_p(x) &= \frac{\rho g}{4}|\hat{\eta}_f + \hat{\eta}_b|^2 = \frac{\rho g}{4}\left(|\hat{\eta}_f|^2 + |\hat{\eta}_b|^2 + \hat{\eta}_f\hat{\eta}_b^* + \hat{\eta}_f^*\hat{\eta}_b\right) \\ &= \frac{\rho g}{4}\left(|A|^2 + |B|^2 + AB^*e^{-i2kx} + A^*Be^{i2kx}\right). \end{aligned} \tag{4.116}$$

If $AB \neq 0$, this expression for $E_p(x)$ contains a term which varies sinusoidally with $x$ with a 'wavelength' $\lambda/2 = 2\pi/2k = \pi/k$. This sinusoidal variation does not contribute if we average over an $x$ interval which is either very long or, alternatively, an integer multiple of $\lambda/2$. Denoting this average by $\langle E_p \rangle$, we have

$$\langle E_p \rangle = (\rho g/4) \left( |A|^2 + |B|^2 \right), \tag{4.117}$$

and for a progressive plane wave ($B = 0$),

$$\langle E_p \rangle = (\rho g/4)|A|^2. \tag{4.118}$$

In Section 7.1.8, we shall show that for an OWC the time-average potential energy per unit horizontal area includes an additional contribution from the dynamic air pressure $p_k(t)$ in the chamber above the water column—that is,

$$E_p(x, y) = (\rho g/2)\overline{\eta_k^2(x, y, t)} + \overline{p_k(t)\eta_k(x, y, t)} \tag{4.119}$$

on the air–water interface of the OWC.

### 4.4.2 Kinetic Energy

For simplicity, we consider a progressive, plane, harmonic wave on deep water. The fluid velocity is given by Eqs. (4.93)–(4.94). The average kinetic energy per unit volume is

$$\frac{1}{2}\rho\frac{1}{2}\text{Re}\left\{ |\hat{v}_x|^2 + |\hat{v}_z|^2 \right\} = \frac{\rho}{4}\left( |\hat{v}_x|^2 + |\hat{v}_z|^2 \right) = \frac{\rho}{2}\omega^2|\hat{\eta}_f|^2 e^{2kz} = \frac{\rho}{2}\omega^2|A|^2 e^{2kz}. \tag{4.120}$$

By integrating from $z = -\infty$ to $z = 0$, we obtain the average kinetic energy per unit (horizontal) area:

$$E_k = \frac{\rho}{2}\omega^2|A|^2 \int_{-\infty}^{0} e^{2kz}\,dz = \frac{\rho}{2}\frac{\omega^2}{2k}|A|^2. \tag{4.121}$$

Using $\omega^2 = gk$, we obtain

$$E_k = (\rho g/4)|A|^2. \tag{4.122}$$

It can be shown (cf. Problem 4.7) that this expression for $E_k$ is also valid for an arbitrary constant water depth $h$. Moreover, it can be shown (cf. Problem 4.8) that for a plane wave, as given by Eq. (4.87), the kinetic energy per unit horizontal surface, averaged both over time and over the horizontal plane, is

$$\langle E_k \rangle = (\rho g/4) \left( |A|^2 + |B|^2 \right). \tag{4.123}$$

### 4.4.3 Total Stored Energy

The total energy is the sum of potential energy and kinetic energy. For a progressive plane harmonic wave, the time-average stored energy per unit (horizontal) area is

$$E = E_k + E_p = 2E_k = 2E_p = (\rho g/2)|A|^2. \tag{4.124}$$

Averaging also over the horizontal plane, we have for a plane wave as given by Eq. (4.87)

$$\langle E \rangle = 2\langle E_k \rangle = 2\langle E_p \rangle = (\rho g/2)\left(|A|^2 + |B|^2\right). \tag{4.125}$$

### 4.4.4 Wave-Energy Transport

Consider the energy transport of a plane harmonic wave propagating in the $x$-direction. Per unit (vertical) area, the time-average power propagating in the positive $x$-direction equals the intensity [cf. Eq. (3.16)]

$$I = \overline{pv_x} = \tfrac{1}{2}\mathrm{Re}\{\hat{p}\hat{v}_x^*\}. \tag{4.126}$$

Note that since $\hat{p}\hat{v}_z^*$ is purely imaginary [cf. Eqs. (4.89) and (4.91)], the intensity has no $z$-component.

Inserting for $\hat{p}$ and $\hat{v}_x$ from Eqs. (4.89)–(4.90), we need the product

$$\left(\hat{\eta}_f + \hat{\eta}_b\right)\left(\hat{\eta}_f - \hat{\eta}_b\right)^* = |\hat{\eta}_f|^2 - |\hat{\eta}_b|^2 + \left(\hat{\eta}_f^*\hat{\eta}_b - \hat{\eta}_f\hat{\eta}_b^*\right). \tag{4.127}$$

Since the last term here is purely imaginary, we get

$$I = \frac{k\rho g^2}{2\omega}\left(|\hat{\eta}_f|^2 - |\hat{\eta}_b|^2\right) e^2(kz). \tag{4.128}$$

Integrating from $z = -h$ to $z = 0$ gives the transported wave power per unit width of the wave front:

$$\begin{aligned}
J = \int_{-h}^{0} I\, dz &= \frac{\rho g^2}{4\omega}\left(|\hat{\eta}_f|^2 - |\hat{\eta}_b|^2\right) 2k \int_{-h}^{0} e^2(kz)\, dz \\
&= \frac{\rho g^2 D(kh)}{4\omega}\left(|\hat{\eta}_f|^2 - |\hat{\eta}_b|^2\right) \\
&= \frac{\rho g^2 D(kh)}{4\omega}\left(|A|^2 - |B|^2\right) = \frac{\rho g}{2}v_g(|A|^2 - |B|^2),
\end{aligned} \tag{4.129}$$

where we have also used relations (4.107)–(4.108).

For a purely progressive wave ($\hat{\eta}_b = 0$, $\hat{\eta}_f = Ae^{-ikx}$), this gives

$$J = \frac{\rho g^2 D(kh)}{4\omega}|\hat{\eta}_f|^2 = \frac{\rho g^2 D(kh)}{4\omega}|A|^2 = \frac{\rho g}{2}v_g|A|^2. \tag{4.130}$$

Introducing the period $T = 2\pi/\omega$ and the wave height $H = 2|A|$, we have for deep water ($kh \gg 1$, $D(kh) \approx 1$)

$$J = \frac{\rho g^2}{32\pi} TH^2 = (976\ \mathrm{Ws^{-1}m^{-3}}) TH^2, \tag{4.131}$$

with $\rho = 1020\ \mathrm{kg/m^3}$ for sea water. For $T = 10\ \mathrm{s}$ and $H = 2\ \mathrm{m}$, this gives

$$J = 3.9 \times 10^4\ \mathrm{W/m} \approx 40\ \mathrm{kW/m}. \tag{4.132}$$

In the case of a reflecting fixed wall, for instance at a plane $x = 0$, we have $B = A$—which, according to Eq. (4.129), means that $J = 0$, since the incident power is cancelled by the reflected power. Imagine that a device at $x = 0$ extracts all the incident wave energy. Then it is necessary that the device radiates a wave which cancels the otherwise reflected wave. In this situation, a net energy transport as given by Eq. (4.130) would result, for $x < 0$.

We shall call the quantity $J$ the *wave-energy transport*. (Note that some authors call this quantity 'wave-energy flux' or 'wave-power flux'. This terminology will be avoided here, because of the confusion resulting from improper discrimination between 'flux' and 'flux density' in different branches of physics.) An alternative term for $J$ could be *wave-power level* [7].

### 4.4.5 Relation between Energy Transport and Stored Energy

For a progressive plane harmonic wave, the energy transport $J$ (energy per unit time and unit width of wave frontage) is given by Eq. (4.130), and the stored energy per unit horizontal area is $E$ as given by Eq. (4.124). We may use this to define an energy transport velocity $v_E$ by

$$J = v_E E. \tag{4.133}$$

Thus,

$$v_E = J/E = gD(kh)/(2\omega). \tag{4.134}$$

Comparing with the expression (4.106) for the group velocity $v_g$, we see that $v_E = v_g$. Hence,

$$J = v_g E. \tag{4.135}$$

Note that this simple relationship is valid for a purely progressive plane harmonic wave.

For a wave given by Eq. (4.87), we have, from Eqs. (4.129) and (4.125),

$$J = v_g \langle E \rangle \frac{|A|^2 - |B|^2}{|A|^2 + |B|^2}. \tag{4.136}$$

Thus, in this case, measurement of

$$\langle E \rangle = \rho g \langle \eta^2(x, y, t) \rangle \tag{4.137}$$

does not alone determine wave-energy transport $J$. Indiscriminate use of Eq. (4.135) would then result in an overestimation of $J$.

### 4.4.6 Momentum Transport and Momentum Density of a Wave

As we have seen, waves are associated with energy transport $J$ and with stored energy density $E$, which is equally divided between kinetic energy density $E_k$ and potential energy density $E_p$. With a plane progressive wave $\hat{\eta} = Ae^{-ikx}$, these relations are stated mathematically by Eqs. (4.124), (4.130) and (4.133)–(4.135). Moreover, the wave is associated with a momentum and, hence, a mean mass transport. A propagating wave induces a mean drift of water, the so-called Stokes drift (see [28], p. 251).

The $x$-component of the momentum (the rate of the mass transport) per unit volume is

$$\rho v_x = \rho \frac{\partial \phi}{\partial x}. \tag{4.138}$$

Per unit area of free water surface, the momentum is given by

$$M_x = \int_{-h}^{\eta} \rho v_x \, dz. \tag{4.139}$$

Integration over $-h < z < 0$ (instead of $-h < z < \eta$) would yield a quantity with zero time-average because $v_x$ varies sinusoidally with time. Hence, in lowest-order approximation, the time-average momentum density (per unit area of the free surface) is

$$\overline{M_x} = \overline{\int_0^{\eta} \rho v_x \, dz} = \overline{\eta \, [\rho v_x]_{z=0}} + \cdots \approx \overline{\eta \, [\rho v_x]_{z=0}}. \tag{4.140}$$

Thus, in linearised theory, $J$, $E$, $E_k$, $E_p$ and $\overline{M_x}$ are quadratic in the wave amplitude. In analogy with Eq. (2.77), we have, with sinusoidal variation with time,

$$\overline{M_x} = \overline{\eta \, [\rho v_x]_{z=0}} = \frac{\rho}{2} \mathrm{Re}\{\hat{\eta}^* [\hat{v}_x]_{z=0}\}. \tag{4.141}$$

Let us now consider a plane wave with complex amplitudes of elevation, velocity potential and horizontal fluid velocity given by Eqs. (4.87), (4.88) and (4.90), respectively. Using these equations together with Eq. (4.141) gives

$$\begin{aligned}
\overline{M_x} &= \frac{\rho g k}{2\omega} \mathrm{Re}\left\{\left(Ae^{-ikx} - Be^{ikx}\right)\left(A^*e^{-ikx} - B^*e^{ikx}\right)\right\} \\
&= \frac{\rho g k}{2\omega} \mathrm{Re}\left\{|A|^2 - |B|^2 + \left(AB^*e^{-i2kx} - A^*Be^{i2kx}\right)\right\}.
\end{aligned} \tag{4.142}$$

Noting that the last term is purely imaginary, we obtain

$$\overline{M_x} = \frac{\rho g k}{2\omega} \left(|A|^2 - |B|^2\right). \tag{4.143}$$

If we compare this result with Eq. (4.129) and use Eqs. (4.82) and (4.106), we find the following relationship between the average momentum density and the wave-energy transport:

$$J = v_p v_g \overline{M}_x. \tag{4.144}$$

For a progressive wave, Eq. (4.135) is applicable and then

$$E = J/v_g = v_p \overline{M}_x. \tag{4.145}$$

Let us next consider the momentum transport through a vertical plane normal to the direction of wave propagation. Per unit width of the wave front the momentum transport in the $x$-direction due to the wave is [29]

$$\mathcal{I}_x = \overline{\int_{-h}^{\eta} p_{\text{tot}} + \rho v_x^2 \, dz} - \left( -\int_{-h}^{0} \rho g z \, dz \right), \tag{4.146}$$

where the second term is the momentum transport in the absence of waves. The integrals in Eq. (4.146) may be split into the following parts:

$$\mathcal{I}_x = \int_{-h}^{0} \overline{p_{\text{tot}}} \, dz + \int_{-h}^{0} \overline{\rho v_x^2} \, dz + \overline{\int_{0}^{\eta} p_{\text{tot}} \, dz} + \int_{-h}^{0} \rho g z \, dz. \tag{4.147}$$

Because we keep only terms up to the second order, the second integral is taken to the mean free surface $z = 0$ instead of the instantaneous free surface $z = \eta$. Note that for the first and second integrals, since the limits of integration are constants, we can transfer the time average on the integrands.

Now, inserting the expression for the total pressure $p_{\text{tot}}$ as given by the Bernoulli equation (4.10):

$$p_{\text{tot}} = -\rho \frac{\partial \phi}{\partial t} - \frac{\rho}{2}(v_x^2 + v_z^2) - \rho g z, \tag{4.148}$$

we arrive at

$$\mathcal{I}_x = -\overline{\int_{0}^{\eta} \rho \left( g z + \frac{\partial \phi}{\partial t} \right) dz} + \int_{-h}^{0} \overline{\frac{\rho}{2}(v_x^2 - v_z^2)} \, dz \equiv \mathcal{I}_{x1} + \mathcal{I}_{x2}. \tag{4.149}$$

Because the upper limit of the first integral is a first-order quantity and we are evaluating the integral correct to second order, only terms up to the first order are kept in the integrand [cf. (4.140)]. Thus,

$$\mathcal{I}_{x1} = -\overline{\int_{0}^{\eta} \rho \left( g z + \left[ \frac{\partial \phi}{\partial t} \right]_{z=0} \right) dz} = -\overline{\int_{0}^{\eta} \rho \left( g z - g \eta \right) dz}$$

$$= -\frac{\rho g}{2} \overline{\left[ z^2 \right]_0^{\eta}} + \rho g \overline{\eta^2} = \frac{\rho g}{2} \overline{\eta^2}. \tag{4.150}$$

With the plane wave as given by Eqs. (4.87), (4.88), (4.90) and (4.91), we therefore have

$$\mathcal{I}_{x1} = \frac{\rho g}{4} \left( |A|^2 + |B|^2 + AB^* e^{-i2kx} + A^* B e^{i2kx} \right) \tag{4.151}$$

and

$$\begin{aligned}
\mathcal{I}_{x2} &= \frac{\rho}{4} \left( \frac{gk}{\omega} \right)^2 \left( |A|^2 + |B|^2 \right) \int_{-h}^{0} e^2(kz) \left[ 1 - \tanh^2(kz + kh) \right] dz \\
&\quad - \frac{\rho}{4} \left( \frac{gk}{\omega} \right)^2 \left( AB^* e^{-i2kx} + A^* B e^{i2kx} \right) \int_{-h}^{0} e^2(kz) \left[ 1 + \tanh^2(kz + kh) \right] dz \\
&= \frac{\rho g}{4} \left( |A|^2 + |B|^2 \right) \frac{2kh}{\sinh(2kh)} - \frac{\rho g}{4} \left( AB^* e^{-i2kx} + A^* B e^{i2kx} \right).
\end{aligned} \tag{4.152}$$

Taking the sum, we finally have

$$\mathcal{I}_x = \mathcal{I}_{x1} + \mathcal{I}_{x2} = \frac{\rho g}{4} \left( |A|^2 + |B|^2 \right) \left( 1 + \frac{2kh}{\sinh(2kh)} \right), \tag{4.153}$$

which may alternatively be expressed as

$$\mathcal{I}_x = \frac{\rho g^2 k D(kh)}{4\omega^2} \left( |A|^2 + |B|^2 \right) \tag{4.154}$$

using Eq. (4.107). For a progressive wave ($B = 0$), this simplifies to

$$\mathcal{I}_x = \frac{\rho g^2 k D(kh)}{4\omega^2} |A|^2. \tag{4.155}$$

Comparing this with Eqs. (4.143)–(4.145) gives

$$\mathcal{I}_x = \frac{g D(kh)}{2\omega} \overline{M}_x = v_g \overline{M}_x = \frac{J}{v_p} = \frac{E v_g}{v_p}. \tag{4.156}$$

Thus, for a purely progressive wave, the momentum transport equals the momentum density multiplied by the group velocity. This may be interpreted as follows: the momentum associated with the wave is propagated with a speed equal to the group velocity.

Defining

$$\overline{M}_{x,+} = \frac{\rho g k}{2\omega} |A|^2, \quad \overline{M}_{x,-} = \frac{\rho g k}{2\omega} |B|^2, \tag{4.157}$$

we may rewrite Eqs. (4.143) and (4.154) as

$$\overline{M}_x = \overline{M}_{x,+} - \overline{M}_{x,-}, \tag{4.158}$$

$$\mathcal{I}_x = v_g \overline{M}_{x,+} + v_g \overline{M}_{x,-}. \tag{4.159}$$

Note the minus sign associated with the wave propagating in the negative $x$-direction in the expression for $\overline{M}_x$. In the expression for $\mathcal{I}_x$ this minus sign is cancelled by the negative group velocity for this wave.

### 4.4.7 Drift Forces due to Absorption and Reflection of Wave Energy

If an incident wave $\hat{\eta} = Ae^{-ikx}$ is completely absorbed at the plane $x = 0$, then $\mathcal{I}_x = 0$ for $x > 0$, whereas $\mathcal{I}_x > 0$ for $x < 0$. Thus, the momentum transport is stopped at $x = 0$. In a time $\Delta t$, a momentum $\overline{M}_x \Delta x = F'_d \Delta t$ disappears. Here $\Delta x = v_g \Delta t$ is the distance of momentum transport during the time $\Delta t$. A force $F'_d$ per unit width of wave front is required to stop the momentum transport. This (time-average) drift force is given by

$$F'_d = \frac{\Delta x}{\Delta t}\overline{M}_x = v_g \overline{M}_x = \mathcal{I}_x = \frac{\rho g^2 k D(kh)}{4\omega^2}|A|^2. \tag{4.160}$$

Using the dispersion relationship (4.54), we find that this gives

$$F'_d = \frac{\rho g}{4}\frac{D(kh)}{\tanh(kh)}|A|^2. \tag{4.161}$$

Note that on deep water [see Eqs. (4.101) and (4.106)],

$$F'_d = (\rho g/4)|A|^2; \tag{4.162}$$

that is, the ratio between $F'_d$ and $|A|^2$ is independent of frequency.

The drift force must be taken up by the anchor or mooring system of the wave absorber. Note that the drift force is a time-average force usually much smaller than the amplitude of the oscillatory force. With complete absorption of a wave with amplitude $|A| = 1$ m on deep water, the drift force is

$$F'_d = (1020 \times 9.81/4)1^2 = 2500 \text{ N/m}. \tag{4.163}$$

The drift force is twice as large if the wave is completely reflected instead of completely absorbed, because of the equally large but oppositely directed momentum transport due to the reflected wave [cf. Eq. (4.154)].

A two-dimensional wave-energy absorber (or an infinitely long array of equally interspaced three-dimensional wave absorbers) may partly reflect and partly transmit the incident wave as indicated in Figure 4.6. In this case, the drift force is given by

$$F'_d = \mathcal{I}_{x\,\text{lhs}} - \mathcal{I}_{x\,\text{rhs}}, \tag{4.164}$$

where the two terms represent the momentum transport on the left-hand side (lhs) and right-hand side (rhs) of the absorbing system indicated in Figure 4.6. Thus,

$$F'_d = \frac{\rho g}{4}\frac{D(kh)}{\tanh(kh)}\left(|A|^2 + |B|^2 - |A_t|^2\right), \tag{4.165}$$

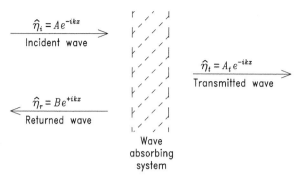

$\hat{\eta}_i = Ae^{-ikx}$
Incident wave

$\hat{\eta}_t = A_t e^{-ikx}$
Transmitted wave

$\hat{\eta}_r = Be^{+ikx}$
Returned wave

Wave
absorbing
system

Figure 4.6: Wave which is incident upon a wave-absorbing system is partly absorbed, partly transmitted downstream and partly reflected and/or radiated upstream.

where $A_t$ denotes the complex amplitude of the transmitted wave. Applying Eqs. (4.129) and (4.130) for the wave-energy transport and using the principle of conservation of energy, we obtain

$$
\begin{aligned}
F'_d &= \frac{\rho g^2 D(kh)}{4\omega}\frac{k}{\omega}\left(|A|^2 + |B|^2 - |A_t|^2\right) = \frac{k}{\omega}(J_i + J_r - J_t) \\
&= \frac{k}{\omega}(J_i - J_r - J_t + 2J_r) = \frac{k}{\omega}(P' + 2J_r),
\end{aligned}
\tag{4.166}
$$

where $P'$ is the absorbed power per unit width. Further, $J_i$, $J_r$ and $J_t$ are the wave-energy transport for the incident, reflected and transmitted waves, respectively. Assuming that $J_r$ is positive, we have $F'_d > (k/\omega)P' = F'_{d\,\mathrm{min}}$. As an example, let us consider absorption of 0.5 MW from a wave of period $T = 10$ s in deep water ($\omega/k = g/\omega$). This is associated with a drift force

$$
F_d > F_{d\,\mathrm{min}} = \frac{2\pi}{10}\frac{1}{9.81}5 \times 10^5 = 32 \times 10^3 \text{ N } (\hat{=} 3.2 \text{ tons}).
\tag{4.167}
$$

## 4.5 Real Ocean Waves

Harmonic waves were the main subject of the previous section. Such waves are also called 'regular waves', as opposed to the real 'irregular' waves of the ocean. The swells—that is, travelling waves which have left their regions of generation by winds—are closer to harmonic waves than the more irregular, locally wind-generated sea waves.

Irregular waves are of a stochastic nature, and they may be considered, at least approximately, as a superposition of many different frequencies. Usually only statistical information is, at best, available for the amplitude, the phase and the direction of propagation for each individual harmonic wave. Most of the energy content of ocean waves is associated with waves of periods in the interval from 5 s to 15 s. According to Eq. (4.100), the corresponding interval of deep-water wavelengths is from 40 m to 350 m.

The present section relates wave spectra to superposition of plane waves on deep water, or otherwise water of finite but constant depth. For more thorough studies, readers may consult other literature [30, 31].

For a progressive harmonic plane wave, the stored energy per unit surface is

$$E = E_k + E_p = 2E_k = 2E_p = \frac{\rho g}{2}|\hat{\eta}_f|^2 = \rho g \overline{\eta^2(x,y,t)} \tag{4.168}$$

according to linear wave theory. See Eqs. (4.85), (4.115) and (4.125). Note that for a harmonic plane wave, $\overline{\eta^2}$ is independent of $x$ and $y$. Still assuming linear theory, the superposition principle is applicable, which means that a real sea state may be described in terms of components of harmonic waves. Correspondingly, we may generalise and rewrite Eq. (4.168) as

$$E = \rho g \overline{\eta^2(x,y,t)} = \rho g \int_0^\infty S(f)\, df, \tag{4.169}$$

where

$$\overline{\eta^2(x,y,t)} = \int_0^\infty S(f)\, df \equiv \frac{H_s^2}{16}. \tag{4.170}$$

Here $S(f)$ is called the energy spectrum, or simply the spectrum. The so-called *significant wave height* $H_s$ is defined as four times the square root of the integral of the spectrum. Sometimes the wave spectrum is defined in terms of the angular frequency $\omega = 2\pi f$. The corresponding spectrum may be written $S_\omega(\omega)$, where

$$\int_0^\infty S_\omega(\omega)\, d\omega = \int_0^\infty S(f)\, df. \tag{4.171}$$

Thus,

$$S(f) = S_f(f) = 2\pi S_\omega(2\pi f) = 2\pi S_\omega(\omega), \tag{4.172}$$

where the units of $S$ and $S_\omega$ are m²/Hz and m²s/rad, respectively.

In a more precise spectral description, the direction of propagation and the phase of each harmonic component should also be taken into account. A harmonic plane wave with direction of incidence given by $\beta$ (the angle between the propagation direction and the $x$-axis) has an elevation given by

$$\eta(x,y,t) = \mathrm{Re}\{\eta_i \exp\left[i(\omega t - kx\cos\beta - ky\sin\beta)\right]\} \tag{4.173}$$

[cf. Eq. (4.98)]. Note that the complex amplitude $\eta_i = \eta_i(\omega) = |\eta_i(\omega)|e^{i\psi}$ contains information on the phase $\psi$ as well as the amplitude $|\eta_i|$. Consider a general sea state $\eta(x,y,t)$ decomposed into harmonic components

$$\eta(x,y,t) = \sum_m \sum_n \mathrm{Re}\{\eta_i(\omega_m,\beta_n) \exp\left[i(\omega_m t - k_m x\cos\beta_n - k_m y\sin\beta_n)\right]\}, \tag{4.174}$$

where $\omega_m$ and $k_m$ are related through the dispersion relationship (4.54). Although any frequency $\omega_m > 0$ and any angle of incidence $-\pi < \beta_n \le \pi$ are, in principle, possible in this summation, only certain finite intervals for $\omega_m$ and $\beta_n$ contribute significantly to the complete irregular wave in most cases. For convenience, we shall assume that in the sums over $\omega_m$ and $\beta_n$ we have $\omega_{m+1} - \omega_m = \Delta\omega$ independent of $m$, and $\beta_{n+1} - \beta_n = \Delta\beta$ independent of $n$.

Taking the square of the elevation (4.174) and then averaging with respect to time as well as the horizontal coordinates $x$ and $y$ results in

$$\langle \eta^2 \rangle = \sum_m \sum_n \frac{1}{2} |\eta_i(\omega_m, \beta_n)|^2. \tag{4.175}$$

Introducing the direction-resolved energy spectrum

$$s(f, \beta) = 2\pi s_\omega(\omega, \beta), \tag{4.176}$$

where

$$\Delta\omega \, \Delta\beta \, s_\omega(\omega_m, \beta_n) = \frac{1}{2} |\eta(\omega_m, \beta_n)|^2, \tag{4.177}$$

and replacing the sums by integrals, we rewrite the wave-elevation variance as

$$\langle \eta^2 \rangle = \int_0^\infty \int_{-\pi}^\pi s_\omega(\omega, \beta) \, d\beta \, d\omega = \int_0^\infty \int_{-\pi}^\pi s(f, \beta) \, d\beta \, df. \tag{4.178}$$

The (direction-integrated) energy spectrum is

$$S(f) = \int_{-\pi}^\pi s(f, \beta) \, d\beta. \tag{4.179}$$

In accordance with Eq. (4.168), the average stored energy per unit horizontal surface is

$$\langle E \rangle = \rho g \langle \eta^2 \rangle = \rho g \int_0^\infty \int_{-\pi}^\pi s(f, \beta) \, d\beta \, df = \rho g \int_0^\infty S(f) \, df. \tag{4.180}$$

From wave measurements, much statistical information has been obtained for the spectral functions $S(f)$ and $s(f, \beta)$ from various ocean regions. It has been found that wave spectra have some general characteristics which may be approximately described by semi-empirical mathematical relations. The most well-known functional relation is the Pierson–Moskowitz (PM) spectrum,

$$S(f) = (A/f^5) \exp(-B/f^4) \tag{4.181}$$

(for $f > 0$). Various parameterisations have been proposed for $A$ and $B$. One possible variant is

$$A = BH_s^2/4, \quad B = (5/4)f_p^4. \tag{4.182}$$

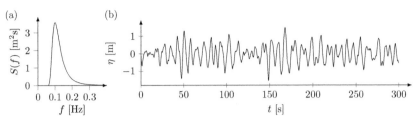

Figure 4.7: (a) Example of a Pierson–Moskowitz spectrum with $H_s = 2$ m and $T_p = 10$ s. (b) Single realisation of wave elevation at $(x, y) = (0, 0)$ simulated from the given spectrum.

Here $H_s$ is the significant wave height and $f_p$ is the 'peak frequency' (the frequency for which $S$ has its maximum). The corresponding wave period $T_p = 1/f_p$ is called the 'peak period'. Note that the energy period $T_J$ as defined in Eq. (4.189) is different from the peak period $T_p$. In general, $T_J < T_p$.

For the direction-resolved spectrum, we may write

$$s(f, \beta) = D(\beta, f) S(f), \tag{4.183}$$

where it is required that

$$\int_{-\pi}^{\pi} D(\beta, f) \, d\beta = 1 \tag{4.184}$$

[cf. Eq. (4.179)]. One proposal for the directional distribution (neglecting its possible frequency dependence) is

$$D(\beta) = \begin{cases} (2/\pi) \cos^2(\beta - \beta_0) & \text{for } |\beta - \beta_0| < \pi/2 \\ 0 & \text{otherwise}, \end{cases} \tag{4.185}$$

where $\beta_0$ is the predominant angle of incidence.

When a real irregular wave is simulated from a specified spectrum, a common procedure is to sum a finite number of wave components according to Eq. (4.174). Care must be exercised when selecting the complex amplitudes $\eta(\omega_m, \beta_n)$ and the number of wave components to ensure that the resulting wave has the correct statistical properties. Figure 4.7 shows an example of a simulated wave elevation at $(x, y) = (0, 0)$, which we have generated from a Pierson–Moskowitz spectrum defined in Eq. (4.181), with $H_s = 2$ m and $T_p = 10$ s.

### 4.5.1 Wave-Energy Transport of Irregular Waves

Because of the superposition principle, Eq. (4.135)—which is valid for a progressive plane harmonic wave—can be applied to a superposition of progressive plane harmonic waves. Thus,

$$J = \rho g \sum_m \frac{1}{2} |\eta_i(\omega_m)|^2 v_g(\omega_m) = \rho g \sum_m \Delta\omega \, S_\omega(\omega_m) v_g(\omega_m). \tag{4.186}$$

Replacing the sums by integrals, we have an expression of the energy transport for irregular plane waves:

$$J = \rho g \int_0^\infty S_\omega(\omega) v_g(\omega) \, d\omega = \rho g \int_0^\infty S(f) v_g(f) \, df. \tag{4.187}$$

On deep water, an expression corresponding to Eq. (4.131), which is applicable for regular waves, can be obtained for irregular waves by noting that $v_g = g/2\omega$ for deep water, and thus,

$$J = \frac{\rho g^2}{2} \int_0^\infty S_\omega(\omega) \, \omega^{-1} \, d\omega = \frac{\rho g^2}{4\pi} \int_0^\infty S(f) f^{-1} \, df. \tag{4.188}$$

Defining the *energy period* $T_J$ as

$$T_J = \frac{2\pi \int_0^\infty S_\omega(\omega) \, \omega^{-1} \, d\omega}{\int_0^\infty S_\omega(\omega) \, d\omega} = \frac{\int_0^\infty S(f) f^{-1} \, df}{\int_0^\infty S(f) \, df} \tag{4.189}$$

and using relationship (4.170), we have, for plane irregular waves on deep water,

$$J = \frac{\rho g^2}{64\pi} T_J H_s^2 = (488 \ \text{Ws}^{-1}\text{m}^{-3}) T_J H_s^2. \tag{4.190}$$

As an example, for an irregular wave with energy period $T_J = 10$ s and significant wave height $H_s = 2$ m, the energy transport is $J \approx 20$ kW/m.

## 4.6 Circular Waves

Using the method of separation of variables in a Cartesian $(x, y, z)$ coordinate system we have studied plane-wave solutions of the Laplace equation (4.33) with the free-surface and sea-bed homogeneous boundary conditions (4.41) and (4.40), respectively. Explicit expressions for plane-wave solutions are given, for instance, by Eq. (4.80) or by Eq. (4.88) with Eq. (4.98). These are examples of plane waves propagating along the $x$-axis, or in a direction which differs by an angle $\beta$ from the $x$-axis, respectively. Another plane-wave solution which includes evanescent plane waves is given by Eq. (4.111).

In the following paragraphs, let us discuss solutions of the problem by using the method of separation of variables in a cylindrical coordinate system $(r, \theta, z)$. The form of the Helmholtz equation (4.46) or (4.58) will be affected. The differential equation (4.45) or (4.59) with boundary conditions (4.60) and (4.61) remains, however, unchanged. Hence, we may still utilise the function $e(kz)$ defined by Eq. (4.52) and the orthogonal set of function $\{Z_n(z)\}$, which is defined by Eq. (4.70) and which is a complete set if $n = 0, 1, 2, \ldots$. In the following, we shall concentrate on the propagating wave (that is, $n = 0$) and leave the evanescent waves aside.

The coordinates in the horizontal plane are related by $(x, y) = (r \cos \theta, r \sin \theta)$, and we replace $H(x, y)$ by $H(r, \theta)$. The Helmholtz equation (4.46) now becomes

$$-k^2 H = \nabla_H^2 H = \frac{\partial^2 H}{\partial x^2} + \frac{\partial^2 H}{\partial y^2} = \frac{\partial^2 H}{\partial r^2} + \frac{1}{r}\frac{\partial H}{\partial r} + \frac{1}{r^2}\frac{\partial^2 H}{\partial \theta^2}. \tag{4.191}$$

Once more, we use the method of separation of variables to obtain a particular solution

$$H(r,\theta) = R(r)\,\Theta(\theta). \tag{4.192}$$

Inserting into Eq. (4.191) and dividing by $H/r^2$ give

$$\frac{r^2}{R}\left(\frac{d^2 R}{dr^2} + \frac{1}{r}\frac{dR}{dr} + k^2 R\right) = -\frac{1}{\Theta}\frac{d^2\Theta}{d\theta^2} = m^2, \tag{4.193}$$

where $m$ is a constant because the left-hand side is a function of $r$ only, while the right-hand side is independent of $r$. Hence,

$$\frac{d^2\Theta}{d\theta^2} = \Theta''(\theta) = -m^2\Theta, \tag{4.194}$$

$$\frac{d^2 R}{dr^2} + \frac{1}{r}\frac{dR}{dr} + \left(k^2 - \frac{m^2}{r^2}\right)R = 0. \tag{4.195}$$

The general solution of Eq. (4.194) may be written as

$$\Theta = \Theta_m = c_c \cos(m\theta) + c_s \sin(m\theta), \tag{4.196}$$

where $c_c$ and $c_s$ are integration constants. We require an unambiguous solution; that is

$$\Theta_m(\theta + 2\pi) = \Theta_m(\theta). \tag{4.197}$$

This requires $m$ to be an integer: $m = 0, 1, 2, 3, \dots$. Since there is also an integration constant in the solution for $R(r)$, we can set

$$\Theta_{max} = 1, \tag{4.198}$$

such that

$$c_c^2 + c_s^2 = 1. \tag{4.199}$$

Then we may write

$$\begin{aligned}\Theta_m(\theta) &= \cos(m\theta + \psi_m) \\ &= \cos(\psi_m)\cos(m\theta) - \sin(\psi_m)\sin(m\theta),\end{aligned} \tag{4.200}$$

where $\psi_m$ is an integration constant which is not currently specified.

Before considering the general solution of the Bessel differential equation (4.195), we note that in an ideal fluid which is non-viscous and, hence, loss-free, the power $P_r$ which passes an envisaged cylinder with large radius $r$ has to be independent of $r$; that is

$$P_r = J 2\pi r \tag{4.201}$$

is independent of $r$. Hence, the energy transport $J$ is inversely proportional to the distance $r$. For large $r$, the circular wave is approximately plane (the curvature of the wave front is of little importance). Then Eq. (4.130) for $J$ applies. This shows that the wave-elevation amplitude $|\hat{\eta}|$ is proportional to $r^{-1/2}$. Hence, $\hat{\phi}$ is proportional to $r^{-1/2}$ [because $\hat{\phi} = (-g/i\omega)e(kz)\hat{\eta}$ according to Eq. (4.88)], and thus, $|R(r)|$ is proportional to $r^{-1/2}$. However, $R(r)$ also has, for an outgoing wave, a phase which varies with $r$ as $-kr$. For large $r$, we then expect that

$$\hat{\phi} \propto r^{-1/2}e^{-ikr}e(kz),\tag{4.202}$$

where the coefficient of proportionality may be complex.

Bessel's differential equation (4.195) has two linearly independent solutions $J_m(kr)$ and $Y_m(kr)$, which are Bessel functions of order $m$. The function $J_m(kr)$ is of the first kind and $Y_m(kr)$ is of the second kind. The function $J_m(kr)$ is finite for $r = 0$, whereas $Y_m(kr) \rightarrow \infty$ when $r \rightarrow 0$. Other singularities do not exist for $r \neq \infty$. For large $r$, $J_m(kr)$ and $Y_m(kr)$ approximate cosine and sine functions multiplied by $r^{-1/2}$. In analogy with Euler's formulas

$$e^{ikr} = \cos(kr) + i\sin(kr),\tag{4.203}$$

$$e^{-ikr} = \cos(kr) - i\sin(kr),\tag{4.204}$$

the Hankel functions of order $m$ and of the first kind and second kind are defined by

$$H_m^{(1)}(kr) = J_m(kr) + iY_m(kr),\tag{4.205}$$

$$H_m^{(2)}(kr) = J_m(kr) - iY_m(kr),\tag{4.206}$$

respectively. Note that

$$H_m^{(1)}(kr) = \left[H_m^{(2)}(kr)\right]^*.\tag{4.207}$$

Asymptotically, we have [32]

$$H_m^{(2)}(kr) \approx \sqrt{\frac{2}{\pi kr}} \exp\left[-i\left(kr - \frac{m\pi}{2} - \frac{\pi}{4}\right)\right]\left(1 + \mathcal{O}\left\{\frac{1}{kr}\right\}\right)\tag{4.208}$$

as $kr \rightarrow \infty$. Hence, the solution of the differential equation (4.195) can be written as

$$R(r) = a_m H_m^{(2)}(kr) + b_m H_m^{(1)}(kr),\tag{4.209}$$

where $a_m$ and $b_m$ are integration constants. From the asymptotic expression (4.208), we see that $H_m^{(2)}(kr)$ represents an outgoing (divergent) wave and from Eq. (4.207) that $H_m^{(1)}(kr)$ represents an incoming (convergent) wave. [It might here be compared with the two terms in each of Eqs. (4.80), (4.81) and (4.83) for the case of propagating plane waves.]

In Section 4.1, it was mentioned that the boundary conditions might have to be supplemented by a *radiation condition*. Let us now introduce the radiation condition of outgoing wave at infinity, which means we have to set $b_m = 0$. Then we have from Eqs. (4.51), (4.192) and (4.209) the solution

$$\hat{\phi} = a_m H_m^{(2)}(kr)\Theta_m(\theta)e(kz) \tag{4.210}$$

for the complex amplitude of the velocity potential. A sum of similar solutions

$$\hat{\phi} = \sum_{m=0}^{\infty} a_m H_m^{(2)}(kr)\Theta_m(\theta)e(kz) \tag{4.211}$$

is also a solution which satisfies the radiation condition, the dispersion equation (4.54) and Laplace's equation (4.33). This solution satisfies the homogeneous boundary conditions (4.40) and (4.41) on the sea bed ($z = -h$) and on the free water surface ($z = 0$).

According to the asymptotic expression (4.208), the particular solution (4.211)—of the partial differential equation (4.191)—is, in the asymptotic limit $kr \gg 1$,

$$\hat{\phi} \approx A(\theta)e(kz)(kr)^{-1/2}e^{-ikr}, \tag{4.212}$$

where

$$A(\theta)\sqrt{\frac{2}{\pi}} \sum_{m=0}^{\infty} a_m \Theta_m(\theta) \exp\left[i\left(\frac{m\pi}{2} + \frac{\pi}{4}\right)\right] \tag{4.213}$$

is the so-called far-field coefficient of the outgoing wave. If we insert Eq. (4.200) for $\Theta_m(\theta)$, we may note that Eq. (4.213) is a Fourier-series representation of the far-field coefficient.

The outgoing wave could originate from a radiation source at or close to the origin $r = 0$. The $z$-variation according to factor $e(kz)$, as in the particular solution (4.211), can satisfy the inhomogeneous boundary conditions, such as Eq. (4.15) or (4.26), at the radiation source only in simple particular cases. A more general solution which satisfies Laplace's equation (4.33), the boundary conditions and the radiation condition as $r \to \infty$ is

$$\hat{\phi} = \hat{\phi}_l(r, \theta, z) + A(\theta)e(kz)(kr)^{-1/2}e^{-ikr}. \tag{4.214}$$

Here the last term represents the so-called *far field* while the  first term $\hat{\phi}_l$ represents the *near field*, a local velocity potential which connects the far field to the radiation source in such a way that the inhomogeneous boundary conditions at the radiation source are satisfied. The term $\hat{\phi}_l(r, \theta, z)$ may contain terms corresponding to evanescent waves. For large $r$, such terms decay exponentially as $\exp(-m_1 r)$ (or faster) with increasing $r$, where $m_1$ is the smallest positive solution of Eq. (4.71)—that is, $\pi/(2h) < m_1 < \pi/h$. (See Problems 4.3 and 5.6.) It can be shown that, in general, the local potential $\hat{\phi}_l$ decreases with distance $r$ at least as fast as $1/r$ when $r \to \infty$ (cf. Wehausen and Laitone [33], pp. 475–478).

The wave elevation corresponding to Eq. (4.214) is given by

$$\hat{\eta} = \hat{\eta}(r,\theta) = -\frac{i\omega}{g}\hat{\phi}_l(r,\theta,0) - \frac{i\omega}{g}A(\theta)(kr)^{-1/2}e^{-ikr}, \tag{4.215}$$

which is obtained by using Eq. (4.39). For large values of $kr$, where $\hat{\phi}_l$ is negligible, $(kr)^{-1/2}$ varies relatively little over one wavelength, which means that the curvature of the wave front is of minor importance. Then we may use the energy transport formula (4.130) for plane waves and there replace the wave-elevation amplitude $|A|$ by $|(\omega/g)A(\theta)(kr)^{-1/2}|$. The radiated energy transport per unit wave frontage then becomes

$$J(r,\theta) = \frac{1}{kr}\frac{D(kh)\rho g^2}{4\omega}\frac{\omega^2}{g^2}|A(\theta)|^2 = \frac{\omega\rho D(kh)}{4kr}|A(\theta)|^2. \tag{4.216}$$

The radiated power is

$$P_r = \int_0^{2\pi} J(r,\theta)r\,d\theta = \frac{\omega\rho D(kh)}{4k}\int_0^{2\pi}|A(\theta)|^2\,d\theta. \tag{4.217}$$

For a circularly symmetric radiation source, the far-field coefficient $A(\theta)$ must be independent of $\theta$. Hence, according to Eq. (4.213), we have

$$A(\theta) = \sqrt{2/\pi}a_0 e^{i\pi/4}. \tag{4.218}$$

Hence, the power associated with the outgoing wave is

$$P_r = \frac{\omega\rho D(kh)}{k}|a_0|^2 = \frac{\pi\omega\rho D(kh)}{2k}|A|^2, \tag{4.219}$$

where $A$ is now the $\theta$-independent far-field coefficient. This is a relation we may use in combination with Eq. (3.29) to obtain an expression for the radiation resistance for an oscillating body which generates an axisymmetric (circularly symmetric) wave. See also Section 5.5.3 and, in particular, Eq. (5.183).

## 4.7 A Useful Integral Based on Green's Theorem

This section is a general mathematical preparation for further studies in the following sections and chapters. We start this section by stating that Green's theorem

$$\oiint\left(\varphi_i\frac{\partial\varphi_j}{\partial n} - \varphi_j\frac{\partial\varphi_i}{\partial n}\right)dS = 0 \tag{4.220}$$

applies to two arbitrary differentiable functions $\varphi_i$ and $\varphi_j$, both of which satisfy the (three-dimensional) Helmholtz equation

$$\nabla^2\varphi = \lambda\varphi \tag{4.221}$$

within the volume $V$ contained inside the closed surface of integration. In the integrand, $\partial/\partial n = \vec{n} \cdot \nabla$ is the normal component of the gradient on the surface of integration. The 'eigenvalue' $\lambda$ is a constant, whereas $\varphi_i$ and $\varphi_j$ depend on the spatial coordinates. In particular, Green's theorem is applicable to two functions, $\phi_i$ and $\phi_j$, which satisfy Laplace's equation (corresponding to $\lambda = 0$).

Green's theorem follows from Gauss's divergence theorem

$$\iiint_V \nabla \cdot \vec{A} \, dV = \oiint A_n \, dS = \oiint \vec{n} \cdot \vec{A} \, dS \tag{4.222}$$

if we consider

$$\vec{A} = \varphi_i \nabla \varphi_j - \varphi_j \nabla \varphi_i. \tag{4.223}$$

Then

$$A_n = \varphi_i \frac{\partial \varphi_j}{\partial n} - \varphi_j \frac{\partial \varphi_i}{\partial n} \tag{4.224}$$

and

$$\begin{aligned} \nabla \cdot \vec{A} &= \varphi_i \nabla^2 \varphi_j + \nabla \varphi_i \cdot \nabla \varphi_j - \varphi_j \nabla^2 \varphi_i - \nabla \varphi_j \cdot \nabla \varphi_i \\ &= \varphi_i \nabla^2 \varphi_j - \varphi_j \nabla^2 \varphi_i. \end{aligned} \tag{4.225}$$

In view of the Helmholtz equation (4.221), we have

$$\nabla \cdot \vec{A} = \varphi_i \lambda \varphi_j - \varphi_j \lambda \varphi_i = 0, \tag{4.226}$$

from which Green's theorem follows as a corollary to Gauss's theorem.

Let us consider a finite region of the sea containing one or more structures, fixed or oscillating, as indicated in Figure 4.8. It is assumed that outside this region, the sea is unbounded in all horizontal directions. Moreover, the water is deep there, or otherwise the water has a constant depth $h$, outside the

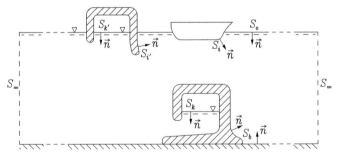

Figure 4.8: System of bodies and chambers for air-pressure distributions (OWCs) contained within an imaginary cylindrical control surface $S_\infty$. Wetted surfaces of oscillating bodies are indicated by $S_{i'}$ and $S_i$, whereas $S_k$ and $S_{k'}$ denote internal water surfaces. Fixed surfaces, including the sea bed, are given by $S_b$, and $S_0$ denotes the external free water surface. The arrows indicate unit normals pointing into the fluid region.

mentioned finite region. The structures, some of which may contain air chambers above OWCs, diffract incoming waves. The phenomenon of diffraction occurs when the incident plane wave alone, such as given by Eq. (4.88), violates the boundary conditions (4.34) and (4.35) on the structures. Furthermore, waves may be generated by oscillating bodies or by oscillating air pressures above the water, such as in the entrapped air within the chambers indicated in the figure.

Let us next apply Green's theorem to the fluid region shown in Figure 4.8. The fluid region is contained inside a closed surface composed of the following surfaces:

$S$, which is the sum (union) of all $N_k$ internal water surfaces $S_k$ above the OWCs and of the wet surface $S_i$ of all $N_i$ bodies, i.e., $S = \sum_{k=1}^{N_k} S_k + \sum_{i=1}^{N_i} S_i$

$S_0$, which is the free water surface (at $z = 0$) external to bodies and chamber structures for OWCs

$S_b$, which is the sum (union) of all wet surfaces of fixed rigid structures, including fixed OWC chamber structures and the sea bed (which is the plane $z = -h$ in case the sea bed is horizontal)

$S_\infty$, which is a 'control' surface, an envisaged vertical cylinder. If this cylinder is circular, we may denote its radius by $r$. In many cases, we consider the limit $r \to \infty$

If the extension of $S_0$ and of the fluid region is not infinite—that is, if the fluid is contained in a finite basin—the surfaces $S_0$ and $S_b$ intersect along a closed curve, and $S_\infty$ does not come into play. We may also consider a case in which an infinite coast line intersects the chosen control surface $S_\infty$.

Let $\phi_i$ and $\phi_j$ be two arbitrary functions which satisfy Laplace's equation

$$\nabla^2 \phi_{i,j} = 0 \tag{4.227}$$

and the homogeneous boundary conditions:

$$\frac{\partial}{\partial n} \phi_{i,j} = 0 \quad \text{on } S_b, \tag{4.228}$$

$$\left( \omega^2 - g \frac{\partial}{\partial z} \right) \phi_{i,j} = \left( \omega^2 + g \frac{\partial}{\partial n} \right) \phi_{i,j} = 0 \quad \text{on } S_0. \tag{4.229}$$

We define the very useful integral

$$I(\phi_i, \phi_j) \equiv \iint_S \left( \phi_i \frac{\partial \phi_j}{\partial n} - \frac{\partial \phi_i}{\partial n} \phi_j \right) dS. \tag{4.230}$$

Next, we shall show that, instead of integrating over $S$, the totality of wave-generating surfaces, we may integrate over $S_\infty$, a 'control' surface in the far-field region of waves radiated from $S$. Note that the integrand in Eq. (4.230) vanishes on $S_b$ and on $S_0$ because $(\partial \phi_{i,j}/\partial n) = 0$ on $S_b$ and

$$\phi_i \frac{\partial \phi_j}{\partial n} - \frac{\partial \phi_i}{\partial n} \phi_j = -\phi_i \frac{\omega^2}{g} \phi_j + \frac{\omega^2}{g} \phi_i \phi_j = 0 \quad \text{on } S_0.$$
(4.231)

Hence, it follows from Green's theorem (4.220) that

$$I(\phi_i, \phi_j) = -\iint_{S_\infty} \left( \phi_i \frac{\partial \phi_j}{\partial n} - \frac{\partial \phi_i}{\partial n} \phi_j \right) dS$$
(4.232)

or

$$I(\phi_i, \phi_j) = \iint_{S_\infty} \left( \phi_i \frac{\partial \phi_j}{\partial r} - \frac{\partial \phi_i}{\partial r} \phi_j \right) dS.$$
(4.233)

Note that if the control surface $S_\infty$ is a cylinder which is not circular, this formula still applies, provided $\partial/\partial r$ is interpreted as the normal component of the gradient on $S_\infty$ pointing in the outwards direction. Note that Eq. (4.230) is a definition, while Eq. (4.233) is a theorem.

Let us make the following comments:

(i) $\phi_i$ and $\phi_j$ may represent velocity potentials of waves generated by the oscillating rigid-body surfaces $S_i$ or oscillating internal water surfaces $S_k$. Alternatively, they may represent other kinds of waves. The constraints on $\phi_i$ and $\phi_j$ are that they have to satisfy the homogeneous boundary conditions (4.228) and (4.229) on $S_b$ and $S_0$.

(ii) $\phi_i$ and/or $\phi_j$ may be matrices provided they (that is, all matrix elements) satisfy the mentioned homogeneous conditions. If both $\phi_i$ and $\phi_j$ are matrices which do not commute, their order of appearance in products is important.

(iii) We may (for real $\omega$) replace $\phi_i$ and/or $\phi_j$ by their complex conjugate, because $\phi_i^*$ and $\phi_j^*$ also satisfy Laplace's equation and the homogeneous boundary conditions on $S_b$ and $S_0$.

(iv) If at least one of $\phi_i$ and $\phi_j$ is a scalar function, or if they otherwise commute, we have

$$I(\phi_j, \phi_i) = -I(\phi_i, \phi_j).$$
(4.234)

(v) Observe that we could include some finite part of surface $S_b$ to belong to surface $S$. For instance, the surface of a fixed piece of rock on the sea bed (or of the fixed sea bed-mounted OWC structure shown in Figure 4.8) could be included in the surface $S$ as a wet surface oscillating with zero amplitude.

(vi) Definition (4.230) and theorem (4.233) are also valid if the boundary condition (4.228) on $S_b$ is replaced by

$$\frac{\partial}{\partial n} \phi_{i,j} + C_n \phi_{i,j} = 0,$$
(4.235)

where $C_n$ is a complex constant or a complex function given along the surface $S_b$. For this to be true when $C_n$ is not real, however, the two functions $\phi_i$ and $\phi_j$ have to satisfy the same radiation condition; that means it would not be allowable to replace just one of the two functions by its complex conjugate.

We next consider the integral (4.233) when $\phi_i$ and $\phi_j$, in addition to satisfying the homogeneous boundary conditions (4.228) and (4.229), also satisfy a radiation condition as $r \to \infty$.

Waves satisfying the radiation condition of outgoing waves at infinite distance from the wave source, such as diffracted waves or radiated waves, have—in the three-dimensional case—an asymptotic expression of the type

$$\phi = \psi \sim A(\theta)e(kz)(kr)^{-1/2}e^{-ikr} \quad \text{as } kr \to \infty \tag{4.236}$$

[see Eqs. (4.212) and (4.214)]. In the remaining part of Section 4.7, and in Section 4.8, let us use the symbol $\psi$ to denote the velocity potential of a general wave which satisfies the radiation condition. We have

$$\frac{\partial \psi}{\partial r} \sim -ik\psi \tag{4.237}$$

when we include only the dominating term of the asymptotic expansion in the far-field region. Thus, for the two waves $\psi_i$ and $\psi_j$, we have

$$\lim_{r \to \infty} \iint_{S_\infty} \left( \psi_i \frac{\partial \psi_j}{\partial r} - \frac{\partial \psi_i}{\partial r} \psi_j \right) dS = 0. \tag{4.238}$$

Hence, according to Eq. (4.233),

$$I(\psi_i, \psi_j) = 0. \tag{4.239}$$

Note that also

$$I(\psi_i^*, \psi_j^*) = 0 \tag{4.240}$$

because

$$\frac{\partial \psi^*}{\partial r} \sim ik\psi^*. \tag{4.241}$$

However, since $\psi_i$ and $\psi_j^*$ satisfy opposite radiation conditions,

$$I(\psi_i, \psi_j^*) \neq 0. \tag{4.242}$$

We have

$$I(\psi_i, \psi_j^*) = \lim_{r \to \infty} \iint_{S_\infty} A_i(\theta)A_j^*(\theta)e^2(kz)\frac{2ik}{kr} dS$$

$$= 2i \int_{-h}^{0} e^2(kz)\, dz \int_{0}^{2\pi} A_i(\theta)A_j^*(\theta)\, d\theta. \tag{4.243}$$

Using Eq. (4.108) gives

$$I(\psi_i, \psi_j^*) = i\frac{D(kh)}{k} \int_{0}^{2\pi} A_i(\theta)A_j^*(\theta)\, d\theta. \tag{4.244}$$

Note that since the two terms of the integrand $\psi_i(\partial\psi_j^*/\partial r) - (\partial\psi_i/\partial r)\psi_j^*$ have equal contributions in the far-field region, we also have

$$I(\psi_i, \psi_j^*) = \lim_{r\to\infty} 2 \iint_{S_\infty} \psi_i \frac{\partial\psi_j^*}{\partial r} \, dS. \tag{4.245}$$

For the two-dimensional case, possible evanescent waves—set up as a result of inhomogeneous boundary conditions in the finite part of the fluid region (see Figure 4.8)—are negligible in the far-field region. Thus, all terms corresponding to $n \geq 1$ in Eq. (4.111) may be omitted there. The remaining terms for $n = 0$ correspond to Eq. (4.83). Consequently, for waves satisfying the radiation condition, we have the far-field asymptotic expression

$$\phi = \psi \sim -\frac{g}{i\omega} A^\pm e(kz) e^{\mp ikx} = -\frac{g}{i\omega} A^\pm e(kz) e^{-ik|x|}. \tag{4.246}$$

The upper signs apply for $x \to +\infty$ ($\theta = 0$), and the lower signs for $x \to -\infty$ ($\theta = \pi$). The constants $A^+$ and $A^-$ represent wave elevations.

Consider now a vertical cylinder $S_\infty$ with rectangular cross section of width $d$ in the $y$-direction and of an arbitrarily large length in the $x$-direction. Then

$$I(\psi_i, \psi_j^*) = \iint_{S_\infty} \left(\psi_i \frac{\partial\psi_j^*}{\partial r} - \frac{\partial\psi_i}{\partial r}\psi_j^*\right) dS$$

$$= \left(\frac{-g}{i\omega}\right)\left(\frac{-g}{-i\omega}\right)\left(A_i^+ A_j^{+*} + A_i^- A_j^{-*}\right) 2ik \int_{-h}^0 e^2(kz) \, dz \int_{-d/2}^{d/2} dy. \tag{4.247}$$

Per unit width in the $y$-direction, we have

$$I'(\psi_i, \psi_j^*) \equiv \frac{I(\psi_i, \psi_j^*)}{d} = i\left(\frac{g}{\omega}\right)^2 D(kh)\left(A_i^+ A_j^{+*} + A_i^- A_j^{-*}\right). \tag{4.248}$$

Also in the two-dimensional case it is easily seen that, for two waves satisfying the same radiation condition, we have

$$I'(\psi_i, \psi_j) = 0, \quad I'(\psi_i^*, \psi_j^*) = 0. \tag{4.249}$$

## 4.8 Far-Field Coefficients and Kochin Functions

In Section 4.6, we discussed circular waves and derived the asymptotic expression (4.212) for an outgoing wave. At very large distance ($r \to \infty$), all the waves diffracted and radiated from the structures shown in Figure 4.8 seem to originate from a region near the origin ($r = 0$). The resulting outgoing wave is, in linear theory, given by an asymptotic expression such as (4.212), where the far-field coefficient is a superposition of outgoing waves from the individual diffracting and radiating structures; that is,

$$A(\theta) = \sum_i A_i(\theta), \tag{4.250}$$

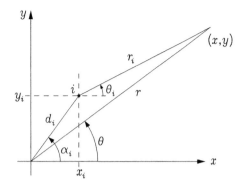

Figure 4.9: Horizontal coordinates, Cartesian $(x, y)$ and polar $(r, \theta)$. Global coordinates with respect to the common origin $O$. Local coordinates with respect to the local origin $O'$ of the (radiating and/or diffracting) structure number $i$.

where it is summed over all the sources for outgoing circular waves. Let

$$(x_i, y_i) = (d_i \cos \alpha_i, d_i \sin \alpha_i) \tag{4.251}$$

represent the local origin (or the vertical line through a reference point, e.g., centre of gravity) of source number $i$ for the outgoing circular wave. See Figure 4.9 for definition of the lengths $d_i$ and $r_i$ and of the angles $\alpha_i$, $\theta_i$ and $\theta$. The horizontal coordinates of an arbitrary point may be written as

$$(x, y) = (r \cos \theta, r \sin \theta) = (x_i + r_i \cos \theta_i, y_i + r_i \sin \theta_i). \tag{4.252}$$

We shall assume that all sources for outgoing waves are located in a bounded region in the neighbourhood of the origin, that is,

$$r \gg \max(d_i) \quad \text{for all } i, \tag{4.253}$$

and that

$$kr \gg 1. \tag{4.254}$$

Then for very large distances, each term of the velocity potential $\sum \phi_i = \sum \psi_i$ of the resulting outgoing wave is given by an asymptotic expression such as [cf. Eq. (4.212)]

$$\phi_i = \psi_i \sim B_i(\theta_i) e(kz)(kr_i)^{-1/2} \exp(-ikr_i) \tag{4.255}$$

or

$$\phi_i = \psi_i \sim A_i(\theta) e(kz)(kr)^{-1/2} \exp(-ikr). \tag{4.256}$$

Here, the complex function $B_i(\theta)$ is the far-field coefficient referring to the source's local origin, and $A_i(\theta)$ is the far-field coefficient, for the same wave source number $i$, referring to the common origin. For very distant field points $(x, y)$—that is, $r \gg d_i$—we have the following asymptotic approximations for the relation between local and global coordinates (see Figure 4.9):

$$\theta_i \approx \theta, \tag{4.257}$$

$$r_i \approx r - d_i \cos(\alpha_i - \theta), \tag{4.258}$$

$$(kr_i)^{-1/2} \approx (kr)^{-1/2}. \tag{4.259}$$

For the outgoing wave from source number $i$, we have

$$\begin{aligned}
\phi_i &\sim B_i(\theta_i)e(kz)(kr_i)^{-1/2}\exp(-ikr_i) \\
&\sim B_i(\theta)e(kz)(kr)^{-1/2}\exp[ikd_i\cos(\alpha_i - \theta)]\exp(-ikr).
\end{aligned} \tag{4.260}$$

Note that we use a higher-order approximation for $r_i$ in the phase than in the amplitude (modulus), as also usual in interference and diffraction theories, for instance, in optics.

Hence, we have the following relations between the far-field coefficients

$$\begin{aligned}
A_i(\theta) &= B_i(\theta)\exp[ikd_i\cos(\alpha_i - \theta)] \\
&= B_i(\theta)\exp[ik(x_i\cos\theta + y_i\sin\theta)].
\end{aligned} \tag{4.261}$$

For a single axisymmetric body oscillating in heave only, the far-field coefficient $B_0$, say, is independent of $\theta$. If the vertical symmetry axis coincides with the $z$-axis, also the corresponding $A_0$ is independent of $\theta$. However, if $d_i \neq 0$, $A_0$ varies with $\theta$, due to the exponential factor in Eq. (4.261).

Radiated waves and diffracted waves satisfy the radiation condition of outgoing waves at infinite distance. They are represented asymptotically by expressions such as (4.255) or (4.256), where the amplitude, the phase and the direction dependence are determined by the far-field coefficient.

An incident wave $\Phi$, for instance [see Eqs. (4.88) and (4.98)],

$$\Phi = \hat{\phi}_0 = -\frac{g}{i\omega}Ae(kz)\exp[-ik(x\cos\beta + y\sin\beta)], \tag{4.262}$$

does not satisfy the radiation condition. Let $\phi_i$ and $\phi_j$ be two arbitrary waves, where

$$\phi_{i,j} = \Phi_{i,j} + \psi_{i,j}. \tag{4.263}$$

Here $\psi_i$ and $\psi_j$ are two waves which satisfy the radiation condition (of outgoing waves at infinity), and $\Phi_i$ and $\Phi_j$ are two arbitrary incident waves. From the definition of the integral (4.230), we see that

$$I(\phi_i, \phi_j) = I(\Phi_i, \Phi_j) + I(\Phi_i, \psi_j) + I(\psi_i, \Phi_j) + I(\psi_i, \psi_j). \tag{4.264}$$

Since $\psi_i$ and $\psi_j$ satisfy the same radiation condition, we have from Eq. (4.239) that $I(\psi_i, \psi_j) = 0$. Further, since $\Phi_i$ and $\Phi_j$ satisfy the homogeneous boundary conditions on the planes $z = 0$ and $z = -h$, and since they satisfy Laplace's equation everywhere in the volume region between these planes, *including the* volume region occupied by the oscillator structures, it follows from Green's theorem (4.220) and Eq. (4.233) that also $I(\Phi_i, \Phi_j) = 0$. Hence,

$$I(\phi_i, \phi_j) = I(\Phi_i, \psi_j) + I(\psi_i, \Phi_j)$$
$$= I(\Phi_i, \psi_j) - I(\Phi_j, \psi_i), \qquad (4.265)$$

where we have in the last step also used Eq. (4.234), for which it is necessary to assume that $\Phi_j$ commutes with $\psi_i$.

Let us now consider the two incident waves

$$\Phi_{i,j} = -\frac{g}{i\omega} A_{i,j} e(kz) \exp\left[-ikr(\beta_{i,j})\right], \qquad (4.266)$$

where

$$r(\beta) \equiv x \cos\beta + y \sin\beta. \qquad (4.267)$$

Note that the two waves have different complex elevation amplitudes $A_i$ and $A_j$, and different angles of incidence $\beta_i$ and $\beta_j$. They have, however, equal period $T = 2\pi/\omega$ and, hence, equal wavelengths $\lambda = 2\pi/k$.

Following Newman [34], we define the so-called Kochin function

$$H_j(\beta) \equiv -\frac{k}{D(kh)} I\left\{e(kz)e^{ikr(\beta)}, \psi_j\right\}. \qquad (4.268)$$

Note that $H_j(\beta)$ depends on a chosen angle $\beta$. Otherwise, $H_j(\beta)$ is a property of the wave $\psi_j$ which satisfies the radiation condition. We shall show that $H_j(\theta)$ is proportional to the far-field coefficient $A_j(\theta)$ of the wave $\psi_j$. According to Eq. (4.267),

$$r(\beta \pm \pi) = -r(\beta). \qquad (4.269)$$

Hence,

$$H_j(\beta \pm \pi) = -\frac{k}{D(kh)} I\left\{e(kz)e^{-ikr(\beta)}, \psi_j\right\}. \qquad (4.270)$$

Using this in combination with the two chosen incident waves $\Phi_i$ and $\Phi_j$ given by Eq. (4.266), we have from Eq. (4.265),

$$I(\phi_i, \phi_j) = \frac{gD(kh)}{i\omega k} \left\{A_i H_j(\beta_i \pm \pi) - A_j H_i(\beta_j \pm \pi)\right\}. \qquad (4.271)$$

Note that this result is based on Eq. (4.265) and hence on Eq. (4.234). Thus, observe that $A_j$ has to commute with $H_i$ when Eq. (4.271) is being applied. Also note that the given complex wave-elevation amplitude $A_j$ here, and in Eq. (4.266), should not be confused with the function $A_j(\theta)$, which is the far-field coefficient.

Next, we shall show that

$$H_j(\theta) = \sqrt{2\pi} A_j(\theta) e^{i\pi/4}. \qquad (4.272)$$

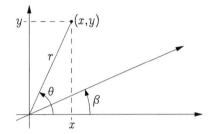

Figure 4.10: Cartesian and polar coordinates in the horizontal plane, with $\beta$ as the angle of wave incidence.

Using Eqs. (4.233) and (4.256) in Eq. (4.268) gives

$$H_j(\beta) = -\frac{k}{D(kh)} I \left\{ e(kz)e^{ikr(\beta)}, A_j(\theta)e(kz)(kr)^{-1/2}e^{-ikr} \right\}, \tag{4.273}$$

where the integration surface is now $S_\infty$ (cf. Figure 4.8). It should be observed here that $r$ is the (horizontal) radial coordinate (see Figure 4.9), whereas $r(\beta)$ is as given by Eq. (4.267).

We shall assume that the control surface $S_\infty$ is a circular cylinder of radius $r$. The integration variables are then $\theta$ and $z$ (cf. Figure 4.10), whereas $r$ is a constant which tends to infinity. Noting that $r(\beta) \equiv x\cos\beta + y\sin\beta = r\cos\theta \cos\beta + r\sin\theta\sin\beta = r\cos(\theta - \beta)$ and using Eq. (4.108), we have

$$H_j(\beta) = \lim_{kr \to \infty} \frac{i}{2} \int_0^{2\pi} \sqrt{kr} A_j(\theta)[1 + \cos(\theta - \beta)]\exp\{-ikr[1 - \cos(\theta - \beta)]\}d\theta. \tag{4.274}$$

We now substitute $\varphi = \theta - \beta$ as the new integration variable. Further, using the asymptotic expression (for $u \to \infty$)

$$\lim_{u \to \infty} \int_{-\beta}^{2\pi-\beta} f(\varphi)\exp\left[-iu(1-\cos\varphi)\right]d\varphi = \lim_{u \to \infty}\sqrt{\frac{\pi}{u}}\left[(1-i)f(0) + (1+i)e^{-i2u}f(\pi)\right], \tag{4.275}$$

which is derived in the next paragraph, we obtain

$$H_j(\beta) = \frac{i}{2}(1-i)\sqrt{\pi}A_j(\beta)(1+\cos 0) = (1+i)\sqrt{\pi}A_j(\beta) = \sqrt{2\pi}A_j(\beta)e^{i\pi/4}, \tag{4.276}$$

in accordance with the earlier statement (4.272) which was to be proven.

The previously mentioned mathematical relation (4.275) is derived as follows, by the 'method of stationary phase'. To be slightly more general, we shall first consider the asymptotic limit (as $u \to \infty$) of the integral

$$I_{ab} \equiv \int_a^b f(\varphi)e^{-iu\psi(\varphi)}d\varphi, \tag{4.277}$$

where $f(\varphi)$ and $\psi(\varphi)$ are analytic functions in the interval $a < \varphi < b$. Moreover, $\psi(\varphi)$ is real and has one, and only one, extremum $\psi(c)$ in the interval, such that $\psi'(c) = 0$ and $\psi''(c) \neq 0$, where $a < c < b$. If we take a constant $e^{-iu\psi(c)}$ outside the integral, and if we then let $u \to \infty$, the integrand oscillates infinitely fast with $\varphi$, except near the 'stationary' point $\varphi = c$, where the imaginary exponent varies slowly with $\varphi$. Hence, the contribution to the integral is negligible outside the interval $c - \epsilon < \varphi < c + \epsilon$, where $\epsilon$ is a small positive number. Using a Taylor expansion for $\psi(\varphi)$ around $\varphi = c$, we have

$$\lim_{u \to \infty} I_{ab} = \lim_{u \to \infty} f(c) e^{-iu\psi(c)} \int_{c-\epsilon}^{c+\epsilon} \exp\left[-iu\psi''(c)(\varphi - c)^2/2\right] d\varphi. \qquad (4.278)$$

We now take

$$\alpha = (\varphi - c)\sqrt{u\psi''(c) \, \mathrm{sgn}[\psi''(c)]/2} = (\varphi - c)\sqrt{u|\psi''(c)|/2} \qquad (4.279)$$

as the new integration variable. Note that the new integration limits are given by

$$\alpha = \pm\epsilon\sqrt{u|\psi''(c)|/2} \to \pm\infty \quad \text{as } u \to \infty. \qquad (4.280)$$

This gives

$$\lim_{u \to \infty} I_{ab} = \lim_{u \to \infty} f(c)\sqrt{\frac{2}{u|\psi''(c)|}} e^{-iu\psi(c)} \int_{-\infty}^{\infty} \exp\left[-i\alpha^2 \mathrm{sgn}\{\psi''(c)\}\right] d\alpha$$

$$= \lim_{u \to \infty} f(c)\sqrt{\frac{2}{u|\psi''(c)|}} e^{-iu\psi(c)} \left[\int_{-\infty}^{\infty} \cos\alpha^2 d\alpha - i\,\mathrm{sgn}\{\psi''(c)\} \int_{-\infty}^{\infty} \sin\alpha^2 d\alpha\right]$$

$$= \lim_{u \to \infty} f(c)\sqrt{\frac{\pi}{u|\psi''(c)|}} e^{-iu\psi(c)} \left[1 - i\,\mathrm{sgn}\{\psi''(c)\}\right],$$

$$(4.281)$$

since each of the two last integrals is known to have the value $\sqrt{\pi/2}$. Note that for $\psi(\varphi) = 1 - \cos\varphi$, we have $\psi'(\varphi) = 0$ for $\varphi = 0$ and $\varphi = \pi$. Moreover, observing that $\psi(0) = 0$, $\psi''(0) = 1$, $\psi(\pi) = 2$ and $\psi''(\pi) = -1$, we can easily see that Eq. (4.275) follows from Eqs. (4.277) and (4.281).

In the two-dimensional case ($\partial/\partial y \equiv 0$), the bodies and air chambers shown as cross sections in Figure 4.8 are assumed to have infinite extension in the $y$-direction. To make the integral (4.232) finite, we integrate over that part of the surface $S$ which is contained within the interval $-d/2 < y < d/2$, where we shall later let the width $d$ tend to infinity. The corresponding integral per unit width is $I' = I/d$, as in Eq. (4.248). The surface $S_\infty$ may be taken as the two planes $x = r^\pm$, where the constants $r^\pm$ are sufficiently large ($r^\pm \to \pm\infty$). The two-dimensional Kochin function $H'_j(\beta)$ is defined as in Eq. (4.268), with $I$ replaced by $I'$. In accordance with Eq. (4.233), we take the integral over $S_\infty$ instead of $S$, and we use the far-field approximation (4.246) to obtain

$$H_j(\beta) = -\frac{k}{D(kh)} \int_h^0 e^2(kz)\, dz \int_{-d/2}^{d/2} \exp(iky \sin \beta)\, dy \, (G^+ + G^-), \quad (4.282)$$

where

$$G^\pm = (1 \pm \cos \beta) \left(\frac{gk}{\omega}\right) A_j^\pm \exp(ikr^\pm \cos \beta) \exp\{\mp ikr^\pm\}. \quad (4.283)$$

The integral over $z$ is $D(kh)/2k$ in accordance with Eq. (4.108) and, moreover,

$$\lim_{d \to \infty} \frac{1}{d} \int_{-d/2}^{d/2} \exp(iky \sin \beta)\, dy = \begin{cases} 1 & \text{if } \sin \beta = 0 \\ 0 & \text{if } \sin \beta \neq 0. \end{cases} \quad (4.284)$$

Note that for a two-dimensional case, the angle of incidence is either $\beta = 0$ or $\beta = \pi$. In both cases, $\sin \beta = 0$. Thus the two-dimensional Kochin function per unit width, $H_j'(\beta) = H_j(\beta)/d$, is given by

$$H_j'(0) = -\frac{gk}{\omega} A_j^+, \quad (4.285)$$

$$H_j'(\pi) = -\frac{gk}{\omega} A_j^-. \quad (4.286)$$

Otherwise, if $\sin \beta \neq 0$, we have $H_j'(\beta) = 0$. Here we have derived the two-dimensional version of the relation (4.272) between the Kochin function and the far-field coefficient.

Let us finally consider some relations between the Kochin functions, the far-field coefficients and the useful integral (4.233). For outgoing wave source number $j$, we have the Kochin function as given by Eqs. (4.268) and (4.272). For the outgoing wave, we have the asymptotic approximation

$$\begin{aligned} \psi_j &\sim A_j(\theta)e(kz)(kr)^{-1/2}e^{-ikr} \\ &= H_j(\theta)e(kz)(2\pi kr)^{-1/2} \exp(-ikr - i\pi/4). \end{aligned} \quad (4.287)$$

For two waves $\psi_i$ and $\psi_j$ satisfying the same radiation condition, we have, from Eq. (4.244),

$$I(\psi_i, \psi_j^*) = i\frac{D(kh)}{k} \int_0^{2\pi} A_i(\theta)A_j^*(\theta)\, d\theta = \frac{iD(kh)}{2\pi k} \int_0^{2\pi} H_i(\theta)H_j^*(\theta)\, d\theta. \quad (4.288)$$

Using the relation (4.261) between the far-field coefficients referred to the global origin and the local origin for wave source number $j$, we have

$$\begin{aligned} I(\psi_i, \psi_j^*) &= i\frac{D(kh)}{k} \int_0^{2\pi} A_i(\theta)A_j^*(\theta)\, d\theta \\ &= i\frac{D(kh)}{k} \int_0^{2\pi} B_i(\theta)B_j^*(\theta) \exp\left[ik(r_j - r_i)\right]\, d\theta. \end{aligned} \quad (4.289)$$

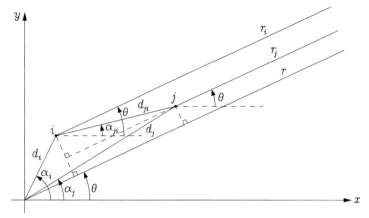

Figure 4.11: Definition sketch for horizontal distances and angles determining the position of the common origin and the local origins for outgoing wave sources number $i$ and number $j$.

The exponent in the latter integrand may be rewritten by using the geometrical relations

$$
\begin{aligned}
r_j - r_i &= (x_i - x_j)\cos\theta + (y_i - y_j)\sin\theta \\
&= d_i \cos(\alpha_i - \theta) - d_j \cos(\alpha_j - \theta) \\
&= d_{ij}\cos(\alpha_{ij} - \theta),
\end{aligned}
\tag{4.290}
$$

where distances and angles are defined in Figure 4.11 ($d_{ij} = d_{ji}$, $\alpha_{ij} = \alpha_{ji} + \pi$).

The two-dimensional version of Eq. (4.288) is

$$
I'(\psi_i, \psi_j^*) = \frac{iD(kh)}{k^2}\left[H_i'(0)H_j'^*(0) + H_i'(\pi)H_j'^*(\pi)\right],
\tag{4.291}
$$

which is obtained by combination of Eqs. (4.248), (4.285) and (4.286).

For waves not satisfying the radiation condition, such as $\phi_i$ and $\phi_j$ in Eq. (4.263), we shall now consider $I(\phi_i, \phi_j^*)$. Assuming that $\Phi_j^*$ commutes with $\psi_i$, we obtain

$$
I(\phi_i, \phi_j^*) = I(\Phi_i, \psi_j^*) - I(\Phi_j^*, \psi_i) + I(\psi_i, \psi_j^*)
\tag{4.292}
$$

in a similar way as we derived Eq. (4.265). Combining Eq. (4.292) with Eqs. (4.266) and (4.268), from which also follows

$$
\Phi_j^* = \frac{g}{i\omega}A_j^* e(kz)\exp\left[ikr(\beta_j)\right],
\tag{4.293}
$$

$$
H_j^*(\beta) = -\frac{k}{D(kh)}I\left\{e(kz)e^{-ikr(\beta)}, \psi_j^*\right\},
\tag{4.294}
$$

we find

$$
I(\phi_i, \phi_j^*) = \frac{gD(kh)}{i\omega k}\left[A_i H_j^*(\beta_i) + A_j^* H_i(\beta_j)\right] + I(\psi_i, \psi_j^*).
\tag{4.295}
$$

Here we may use Eq. (4.288) to express the last term in terms of Kochin functions. For the two-dimensional case, we instead use Eq. (4.291), and then we also replace $I$ and $H_{i,j}$ by $I'$ and $H'_{i,j}$, respectively.

If we compare Eqs. (4.271) and (4.295), we notice two differences apart from the conjugation of $\phi_j$. Firstly, because of Eq. (4.239), Eq. (4.271) contains no term corresponding to the last term in Eq. (4.295). Secondly, in contrast to Eq. (4.271), there is not a shift of an angle of $\pi$ in the argument of the Kochin functions in Eq. (4.295).

Equations (4.271) and (4.295) were used by Newman [34] as a base for deriving a set of reciprocity relations between parameters associated with various waves. In the next chapter, we shall derive and discuss some of these reciprocity relations.

## 4.9 Waves in the Time Domain

When analysing waves, we may choose between a frequency-domain or a time-domain approach. The frequency domain usually receives greater attention, as it also does here in this chapter, from Section 4.2 and onward. More recently, however, time-domain investigations have also been carried out [35, 36]. Such investigations may be based on a numerical solution of the time-domain Laplace equation (4.14) with boundary conditions (4.15), (4.16), (4.26) and (4.27). It is outside the scope of this chapter to discuss such investigations in detail, but let us here make a few comments based on applying inverse Fourier transform to some of the frequency-domain results. To be specific, let us discuss two linear systems: one which relates wave elevations at two locations and another which relates the hydrodynamic pressure to the wave elevation.

We start by assuming that an incident plane wave is propagating in the positive $x$-direction. In the frequency domain, the wave elevation may be expressed as

$$\eta(x,\omega) = A(\omega)e^{-ikx}. \tag{4.296}$$

Here $A(\omega)$ is the Fourier transform of the incident wave elevation at $x = 0$:

$$A(\omega) = \eta(0,\omega) = \int_{-\infty}^{\infty} \eta(0,t)e^{-i\omega t}\, dt. \tag{4.297}$$

To ensure wave propagation in the positive $x$-direction, we have to choose a solution of the dispersion equation, $k = k(\omega)$, for which the angular repetency $k$ has the same sign as the angular frequency $\omega$.

### 4.9.1 Relation between Wave Elevations at Two Locations

As real sea waves have a finite coherence time, it is possible to predict, with a certain probability, the wave elevation at a given point on the basis of wave measurement at the same point. However, a more deterministic type of prediction of the incident wave elevation may be expected if the wave is measured at some distance $l$ from the body's position $x_B$, in the 'upwave' direction (that is, opposite to the direction of wave propagation). We then assume that the measurement takes place at $x = x_A$, where

$$x_B - x_A = l > 0. \tag{4.298}$$

Choosing $\eta(x_A,\omega)$ and $\eta(x_B,\omega)$ as input and output, respectively, of a linear system, we find from Eq. (4.296) that the transfer function of the system is

$$H_l(\omega) = e^{-ik(\omega)l}. \tag{4.299}$$

The corresponding impulse response function is

$$h_l(t) = \frac{1}{2\pi} \int_{-\infty}^{\infty} H_l(\omega)e^{i\omega t}\, d\omega = \frac{1}{2\pi} \int_{-\infty}^{\infty} e^{-ik(\omega)l+i\omega t}\, d\omega. \tag{4.300}$$

It can be shown that this linear system is noncausal—that is, $h_l(t) \neq 0$ for $t < 0$ [cf. Eq. (2.129)]. We shall demonstrate this here for the case of waves on deep water ($h \to \infty$). Further details and generalisation to finite water depth can be found in [37].

We start by rewriting the impulse response (4.300) as

$$h_l(t) = \frac{1}{2\pi} \int_{-\infty}^{\infty} [\cos(\omega t - kl) + i\sin(\omega t - kl)]\, d\omega. \tag{4.301}$$

Since $k$ has the same sign as $\omega$ and since the sine function is odd, the imaginary part of the integral vanishes. Moreover, the cosine function is even, and for deep water, we have $k = \omega^2/g$ [cf. Eq. (4.55)]. Hence,

$$h_l(t) = \frac{1}{\pi} \int_0^{\infty} \cos(\omega t - \omega^2 l/g)\, d\omega, \tag{4.302}$$

which can further be decomposed into even and odd parts:

$$h_e(t) = \frac{1}{\pi} \int_0^{\infty} \cos\frac{\omega^2 l}{g} \cos(\omega t)\, d\omega, \tag{4.303}$$

$$h_o(t) = \frac{1}{\pi} \int_0^{\infty} \sin\frac{\omega^2 l}{g} \sin(\omega t)\, d\omega. \tag{4.304}$$

Using a table of Fourier transforms [38], we find, for $t > 0$,

$$h_e(t) = \frac{1}{4} \left(\frac{2g}{\pi l}\right)^{1/2} \left(\cos\frac{gt^2}{4l} + \sin\frac{gt^2}{4l}\right), \tag{4.305}$$

$$h_o(t) = \frac{1}{2} \left(\frac{2g}{\pi l}\right)^{1/2} \left(C_2\left\{t\left(\frac{g}{2\pi l}\right)^{1/2}\right\}\cos\frac{gt^2}{4l} + S_2\left\{t\left(\frac{g}{2\pi l}\right)^{1/2}\right\}\sin\frac{gt^2}{4l}\right). \tag{4.306}$$

Here, $C_2$ and $S_2$ are Fresnel integrals, which are defined as

$$C_2(x) = \frac{1}{\sqrt{2\pi}} \int_0^x t^{-1/2} \cos t \, dt \tag{4.307}$$

$$S_2(x) = \frac{1}{\sqrt{2\pi}} \int_0^x t^{-1/2} \sin t \, dt \tag{4.308}$$

for $x > 0$. As $|h_e(t)| \neq |h_o(t)|$, the impulse response $h(t) = h_e(t) + h_o(t)$ is not causal [cf. Eq. (2.176)].

That this linear system is noncausal may be explained by the fact that the wave elevation at one location (input) is not the actual cause of the wave elevation at another location (output). The real cause of the output, as well as the input, may be a wavemaker (in the laboratory) or a distant storm (on the ocean).

Nevertheless, although the impulse response function $h_l(t)$ is, strictly speaking, not causal [39], it is approximately causal and particularly so if $l$ is large. With $l = 400$ m, remarkably good prediction of computer-simulated sea swells may be obtained for at least half a minute into the future [40]. Previous experimental work [41] on real sea waves gave a less accurate prediction of the surface elevation but a more satisfactory prediction of the hydrodynamic pressure on the sea bottom at location $x_B$.

### 4.9.2 Relation between Hydrodynamic Pressure and Wave Elevation

Let us now consider a linear system in which the input is the wave elevation at some position $(x, y)$ on the average water surface and in which the output is the hydrodynamic pressure. Then according to Eq. (4.89), the transfer function is

$$H_p(\omega) = \frac{p(x, y, z; \omega)}{\eta(x, y; \omega)} = \rho g e(kz), \tag{4.309}$$

which for deep water becomes

$$H_p(\omega) = \rho g e^{\omega^2 z/g}. \tag{4.310}$$

(Note that $z \leq 0$.) Observe that $H_p(\omega)$ is real, and hence its inverse Fourier transform is an even function of $t$ (see Section 2.6.1), and for deep water is given by [38]

$$h_p(t) = h_p(-t) = \tfrac{1}{2}\rho g(-g/\pi z)^{1/2} e^{gt^2/4z}, \tag{4.311}$$

which is, evidently, not causal, except for the case of $z = 0$, which gives

$$h_p(t) = \rho g \delta(t). \tag{4.312}$$

Compared to the value at $t = 0$, $h_p(t)$ is reduced to 1% for $t = \pm 4.3\sqrt{-z/g}$, which increases with the submergence of the considered point. Thus a deeper point has a 'more noncausal' impulse response function for its hydrodynamic pressure.

Referring also to Eqs. (4.52) and (4.54), we see that $H_p(\omega)$ is an even function of $\omega$ also in the case of finite water depth. Moreover, $H_p(\omega)$ is real when $\omega$ is real. It follows that the corresponding inverse Fourier transform $h_p(t)$ is an even function of $t$, and hence, it is a noncausal impulse response function.

Physically we may interpret this as follows. There is an effect on the hydrodynamic pressure also from the previous and the following wave troughs at the instant when a wave crest is passing. The deeper the considered location is, the more this non-instantaneous effect contributes.

## Problems

### Problem 4.1: Deriving Dispersion Relation Including Capillarity

Consider an infinitesimal surface element of length $\Delta x$ in the $x$-direction and unit length in the $y$-direction. The net capillary attraction force is

$$p_k \Delta x = \gamma \sin(\theta) - \gamma \sin(\theta + \Delta\theta),$$

where $\gamma$ is the surface tension ($[\gamma] = \text{N/m}$). Further, $\theta$ and $\theta + \Delta\theta$ are the angles between the surface and the horizontal plane at the positions $x$ and $x + \Delta x$, respectively.

Derive a modified dispersion relation including capillarity, by inserting a linearisation of the previously mentioned surface tension force into the Bernoulli equation (4.10). Recall that a discussion of this dispersion relation, for the deepwater case, is the subject of Problem 3.2. [Hint: first show that the free-surface boundary condition (4.41) has to be replaced by

$$\left[ -\omega^2 \hat{\phi} + g\frac{\partial \hat{\phi}}{\partial z} - \frac{\gamma}{\rho}\frac{\partial^3 \hat{\phi}}{\partial x^2 \partial z} \right]_{z=0} = 0.$$

Then use this condition in combination with Eq. (4.80)—which already satisfies the Laplace equation as well as the sea-bed boundary condition (4.40)—to determine a new dispersion relation to replace Eq. (4.54).]

### Problem 4.2: Vertical Functions from Bottom Boundary Condition

Use the boundary condition (4.40) for $z = -h$ together with Eq. (4.42) in the determination of the integration constants $c_+$ and $c_-$ in Eq. (4.47) for $Z(z)$. One additional condition is required. Determine both integration constants, and write down an explicit expression for $Z(z)$ for both of the following two cases of a second condition:

(a) $Z(0) = 1$,
(b) $\int_{-h}^{0} |Z(z)|^2 \, dz = h$.

### Problem 4.3: Vertical Functions for Evanescent Solutions

(a) Show that the equation $Z_n''(z) = -m_n^2 Z_n(z)$ with the boundary condition $Z_n'(-h) = 0$ has a particular solution

$$Z_n(z) = C_n \cos(m_n z + m_n h).$$

(b) Further, show that the boundary condition $Z_n'(0) = \omega^2 Z_n(0)/g$ is satisfied if

$$\omega^2 = -g m_n \tan(m_n h). \tag{1}$$

(c) Discuss possible solutions of Eq. (1) as well as normalisation of the corresponding functions $Z_n(z)$.

### Problem 4.4: Distance to Wave Origin

Assume that the ocean is calm and that suddenly a storm develops. Waves are generated, and swells propagate away from the storm region. Far away, at a distance $l$ from the storm centre, swells of frequency $f = 1/T = \omega/2\pi$ are recorded a certain time $\tau$ after the start of the storm. We assume deep water between the storm centre and the place where the swells are recorded. Derive an expression for the 'waiting time' $\tau$ in terms of $l$, $f$ and $g$, where $g$ is the acceleration due to gravity. The starting point of the derivation should be the dispersion relationship $\omega^2 = gk$ for waves on deep water.

In the South Pacific, at Tuvalu, a long swell was registered as follows:

| Date | Time | $T$ |
|------|------|-----|
| 16 Nov 1991 | 0300 GMT | 22.2 s |
|  | 1400 GMT | 20.0 s |
| 17 Nov 1991 | 0100 GMT | 18.2 s |

Find the distance to the swell source. Approximately on which date did this storm occur? Assuming the swell was travelling southward, determine the latitude at which the swell was formed, given that Tuvalu is located at 8°30′ S (1′ corresponds to a nautical mile = 1852 m. Correspondingly, 90 latitude degrees— the distance from the equator to the globe's North or South Pole—is $10^7$ m).

### Problem 4.5: Reflection of Plane Wave at Vertical Wall

(a) Show that a gravity wave for which the elevation has a complex amplitude

$$\hat{\eta}_i = \eta_0 \exp(-ik_x x - ik_y y) \tag{2}$$

may propagate on an infinite lake of constant depth $h$. That is, show that this wave satisfies the partial differential equation

$$\left[\frac{\partial^2}{\partial x^2} + \frac{\partial^2}{\partial y^2} + k^2\right]\hat{\eta} = 0.$$

(b) Moreover, it is assumed that the boundary conditions at the free water surface $z = 0$ and at the bottom $z = -h$ are satisfied (although a proof of this statement is not required in the present problem). Express $k_x$ and $k_y$ by the angular repetency $k$ and the angle $\beta$ between the direction of propagation and the $x$-axis.

(c) Assume that the incident wave (2) in the region $x < 0$ is reflected at a fixed vertical wall in the plane $x = 0$. Use the additional boundary condition at the wall to determine the resultant wave

$$\hat{\eta} = \hat{\eta}_i + \hat{\eta}_r,$$

where $\hat{\eta}_r$ represents the reflected wave. Check that the reflection angle equals the angle of incidence.

## Problem 4.6: Cross Waves in a Wave Channel

An infinitely long wave channel in the region $-d/2 < x < d/2$ is bounded by fixed plane walls, the planes $x = -d/2$ and $x = d/2$. The channel has constant water depth $h$. Show that a harmonic wave of the type

$$\hat{\eta} = \left(Ae^{-ik_x x} + Be^{ik_x x}\right)e^{-ik_y y}$$

may propagate in the $y$-direction. Use the boundary conditions to express possible values of $k_x$ in terms of $d$. Find the limiting value of the angular frequency, below which no cross waves can propagate.

Derive expressions for the phase velocity $v_p = \omega/k_y$ and the group velocity $v_g = d\omega/dk_y$ for the cross waves which may propagate along the wave channel.

## Problem 4.7: Kinetic Energy for Progressive Wave

Derive an expression for the time-average of the stored kinetic energy per unit volume associated with a progressive wave

$$\hat{\phi} = -(g/i\omega)A\,e(kz)\,e^{-ikx}$$

in terms of $\rho, g, k, \omega, |A|$ and $z$.

## Problem 4.8: Kinetic Energy Density

Prove that a plane wave

$$\hat{\eta} = Ae^{-ikx} + Be^{ikx}$$

on water of constant depth $h$ is associated with a kinetic energy density per unit horizontal area equal to

$$\langle E_k \rangle = (\rho g/4)\left(|A|^2 + |B|^2\right)$$

(averaged over time and over the horizontal plane). Also prove that the time-averaged kinetic energy per unit volume is

$$e_k = \frac{\rho}{2}\left[\frac{e(kz)\,gk}{\omega}\right]^2\left(|A|^2 + |B|^2\right) - \frac{\rho g^2 k^2}{4\omega^2 \cosh^2(kh)}|\hat{\eta}|^2.$$

## Problem 4.9: Propagation Velocities on Intermediate Water Depth

Find numerical values for the group velocity and phase velocity for a wave of period $T = 9$ s on water of depth $h = 8$ m, and also for the case when $h = 6$ m. This problem involves numerical (or graphical) solution of the transcendental equation $\omega^2 = gk\tanh(kh)$ (the dispersion relation). The acceleration of gravity is $g = 9.81$ m/s$^2$.

## Problem 4.10: Evanescent Waves

The complex amplitude $\hat{\phi}$ of a velocity potential satisfying Laplace's equation may be written

$$\hat{\phi} = X(x)Z(z),$$

where $X(x)$ and $Z(z)$ have to satisfy the corresponding differential equations

$$X''(x) = -\lambda X(x),$$
$$Z''(z) = \lambda Z(z).$$

Further, $Z(z)$ has to satisfy the boundary conditions

$$Z'(-h) = 0, \quad -\omega^2 Z(0) + gZ'(0) = 0.$$

Solutions for $\hat{\phi}$ corresponding to negative values of $\lambda$—that is, $\lambda = -m^2$—represent the so-called evanescent waves.

Consider a propagating wave travelling in an infinite basin of water depth $h$, where some discontinuities at $x = x_1$ and $x = x_2$ ($x_2 > x_1$) introduce evanescent waves. Show that an evanescent wave in the $x$-region $-\infty < x < x_1$ or $x_2 < x < \infty$ does not transport any time-average power. Is it possible that real energy transport is associated with an evanescent wave in the region $x_1 < x < x_2$?

## Problem 4.11: Slowly Varying Water Depth and Channel Width

A regular wave $\eta = \eta(x,t) = A_0\cos(\omega t - k_0 x)$ is travelling in deep water. Its amplitude is $A_0 = 1$ m, and its period is $T = 9$ s.

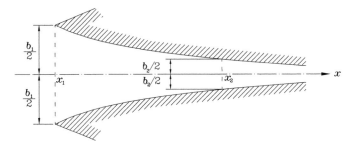

Figure 4.12: Tapered horizontal channel for wave concentration.

(a) Give mathematical expressions and numerical values for the angular frequency $\omega$ and the angular repetency (wave number) $k_0$. Also determine the wavelength $\lambda_0$. Give mathematical expressions and numerical values for the phase velocity $v_{p0}$, the group velocity $v_{g0}$ and the maximum water particle velocity $v_{0,max}$.

Now assume that this wave is moving toward a coast line which is normal to the $x$-axis. The water depth $h = h(x)$ varies slowly with $x$, which means that we may neglect reflection (or partial reflection) of the wave. If we also neglect energy losses (e.g., due to friction at the sea bed), the wave-energy transport $J$ (wave power per unit width of the wave front) stays constant as the wave travels into shallower water. (Note: for simplicity, we assume that the water depth does not vary with the $y$-coordinate.)

(b) At some position $x = x_1$, the water depth is $h = h(x_1) = 8$ m. Here the phase velocity is $v_{p1} = 8.27$ m/s, and the group velocity is $v_{g1} = 7.24$ m/s. Give relative numerical values of the wavelength $\lambda_1$ and the wave amplitude $A_1$ at this location; that is, state the numerical values of $(\lambda_0/\lambda_1)$ and $(A_1/A_0)$.

(c) Note that for a 9 s wave in 8 m depth, neither the deep-water nor the shallow-water approximations give accurate values for $v_{g1}$ and $v_{p1}$. If you had used those approximations, how many percent too small or too large would the computed values be for $v_{g1}$ and $v_{p1}$?

A tapered horizontal wave channel with vertical walls has its entrance (mouth) at the position $x_1$ (see Figure 4.12). The entrance width is $b_1$ (= 30 m). As an approximation, we assume that the wave continues as a plane wave into the channel, but the narrower width available to the wave results in an increased wave amplitude. If the channel width $b = b(x)$ changes slowly with $x$, we may neglect partial reflections of the wave. Neglecting also other energy losses:

(d) derive an expression for the wave amplitude $A_2$ at $x_2$ in terms of $A_1$, $b_1$ and $b_2$, where $b_2$ is the channel width at $x_2$ ($x_2 > x_1$). The water depth is still 8 m at $x_2$. The bottom of the vertical channel wall is at the same depth, of course. The top of the vertical wall is 3.5 m above the mean water level. Find the numerical value of $b_2$ when it is given that $A_2 = 3.5$ m. (Remember that in the deep sea, the wave amplitude is $A_0 = 1$ m.)

(e) If the water depth at $x_2$ had been 6 m instead of 8 m while the depth at the entrance remains 8 m, find for this case the wave amplitude $A_2'$ at $x_2'$ in terms of $A_1$, $b_1$, $b_2$, $v_{g1}$ and $v_{g2}$. In 6 m depth the 9 s wave has a group velocity $v_{g2} = 6.60$ m/s. With this channel (where the depth decreases slowly with increasing $x$), what would be the width $b_2'$ at $x_2'$ if $A_2' = 3.5$ m?

(f) If 5% of the energy was lost (in friction and other processes) when the wave travelled from deep water to the position $x_1$, how would this influence the answers under point (b)? (Will the numerical values be increased, decreased or unchanged?) If changed, find the new values.

(g) Moreover, if there is also energy loss when the wave travels from $x_1$ to $x_2$, how would that influence the answers under point (d)? Assuming that this energy loss is 20% (and that 5% of the deep-water wave energy was lost before the entrance of the channel), obtain new numerical values for the answers under (d).

## Problem 4.12: Depth Function $D(kh)$

(a) Show that the depth function $D(kh)$ defined by Eq. (4.77) has a maximum $x_0$ for $kh = x_0$, where $x_0$ is a root of the transcendental equation

$$x_0 \tanh(x_0) = 1,$$

which corresponds to $h = h_0 \equiv g/\omega^2$.

(b) Further, show that

$$2k \int_{-h}^{0} e^2(kz)\,dz = D(kh),$$

where

$$e(kz) = \frac{\cosh(kh + kz)}{\cosh(kh)}.$$

(See also Problem 4.2.)

(c) Show that the depth function may be written as

$$D(kh) = \left(1 + \frac{4khe^{-2kh}}{1 - e^{-4kh}}\right)\frac{1 - e^{-2kh}}{1 + e^{-2kh}}.$$

This expression is recommended for use in computer programmes when large values of $kh$ are used.

## Problem 4.13: Transmission and Reflection at a Barrier

We consider the following two-dimensional problem. In a wave channel which is infinitely long in the $x$-direction and which has a constant water depth $h$, there is at $x = 0$ a barrier, a stiff plate of negligible thickness, occupying the region $x = 0$, $-h_1 > z > -h_2$, where $h \geq h_2 > h_1 \geq 0$). A wave propagating in the positive $x$-direction originates from $x = -\infty$, and it is partially reflected and partially transmitted at the barrier.

In the region $x < 0$, the complex amplitude of the velocity potential may be written as

$$\hat{\phi}(x,z) = A_0 Z_0(z)e^{-ikx} + \sum_{n=0}^{\infty} b_n Z_n(z)e^{m_n x},$$

where the first term on the right-hand side represents the incident wave. The orthogonal and complete set of vertical eigenfunctions $\{Z_n(z)\}$ is defined by Eq. (4.70), and $m_n$ ($n \geq 1$) is a positive solution of Eq. (4.71), while $m_0 = ik$. See also Eqs. (4.74), (4.75) and (4.111). For the region $x > 0$, we correspondingly write

$$\hat{\phi}(x,z) = \sum_{n=0}^{\infty} a_n Z_n(z) \exp(-m_n x).$$

While the coefficients $\{a_n\}$ and $\{b_n\}$ are unknown, we consider $A_0$ to be known. We define the complex reflection and transmission coefficients by

$$\Gamma = b_0/A_0, \quad T = a_0/A_0,$$

respectively. Obviously, $\hat{\phi}(x,z)$ satisfies Laplace's equation and the homogeneous boundary conditions (4.40) at $z = -h$ and (4.41) at $z = 0$.

Next we have to apply continuity and boundary conditions at the plane $x = 0$. Firstly, $\partial\hat{\phi}/\partial x$, representing the horizontal component of the fluid velocity, has to be continuous there (for $0 > z > -h$), and moreover, it has to vanish on the barrier ($-h_1 > z > -h_2$). Finally, $\hat{\phi}$, representing the hydrodynamic pressure [see Eq. (4.37)], has to be continuous above and below the barrier ($0 > z > -h_1$ and $-h_2 > z > h$). On this basis, show that $a_n = -b_n$ for $n \geq 1$, whereas $a_0 = A_0 - b_0$. The latter result means that $\Gamma + T = 1$, and energy conservation means that $|\Gamma|^2 + |T|^2 = 1$. Using this, show that $\Gamma T^* + \Gamma^* T = 0$. Moreover, show that the set of coefficients $\{b_n\}$ has to satisfy an equation of the type

$$\sum_{m=0}^{\infty} B_{mn} b_m = c_n A_0 \quad \text{for } n = 0, 1, 2, 3, \ldots,$$

which may be solved approximately by truncating the infinite sum and then using a computer for a numerical solution. Derive an expression for $B_{mn}$ and for $c_n$. [Hint: use the orthogonality condition (4.67) and (4.73)—that is,

$$\int_{-h}^{0} Z_m(z)Z_m(z)dz = h\delta_{mn}.$$

If the integral is taken only over the interval $-h_1 > z > -h_2$, it takes another value, which we may denote by $hD_{mn}$. Likewise, we define $hE_{mn}$ as the integral taken over the remaining two parts of the interval $0 > z > -h$. Thus, $D_{mn} + E_{mn} = \delta_{mn}$. Observe that both $D_{mn}$ and $E_{mn}$ have to enter (explicitly or implicitly) into the expression for $B_{mn}$, in order for $B_{mn}$ to be influenced by the boundary condition on the barrier as well as by the condition of continuous hydrodynamic pressure above and below the barrier.]

## Problem 4.14: Energy-Absorbing Wall

Assume that a wave

$$\hat{\eta}_i = A \exp\left[-ik(x\cos\beta + y\sin\beta)\right], \quad |\beta| < \pi/2$$

is incident on a vertical wall $x = 0$ where there is a boundary condition

$$\hat{p} = Z_n \hat{v}_x = Z_n \frac{\partial\hat{\phi}}{\partial x} \quad \text{at } x = 0.$$

Here the constant parameter $Z_n$ is a distributed surface impedance (of dimension Pa s/m = N s/m$^3$). Show that the time-average absorbed power per unit surface area of the wall is $\frac{1}{2}\text{Re}\{Z_n\}|\hat{v}_x|^2$. Further, show that a wave

$$\hat{\eta}_r = \Gamma A \exp\left[ik(x\cos\beta - y\sin\beta)\right]$$

is reflected from the wall. Assume infinite water depth, and derive an expression for $\Gamma$ in terms of $\omega$, $\rho$, and $Z_n$. Show that $|\Gamma| = 1$ if $\text{Re}\{Z_n\} = 0$ and that $|\Gamma| < 1$ if $\text{Re}\{Z_n\} > 0$.

## Problem 4.15: Wave-Energy Transport with Two Waves

A harmonic wave on water of constant depth $h$ is composed of two plane waves propagating in different directions. Let the wave elevation be given by

$$\hat{\eta} = Ae^{-ikx} + Be^{-ik(x\cos\beta + y\sin\beta)}.$$

For the superposition of these two plane waves, derive an expression for the intensity

$$\vec{I} = \overline{p(t)\vec{v}(t)},$$

which, by definition, is a time-independent vector. Further, referring to Section 4.4.4, show that the wave-power-level vector may be expressed as

$$\vec{J} = (\rho g/2)v_g \left[\vec{e}_x\left(|A|^2 + |B|^2\cos\beta\right) + \vec{e}_y|B|^2\sin\beta\right] + \vec{s}(x, y),$$

where the spatially dependent vector $\vec{s}(x, y)$ is solenoidal. Find an expression for $\vec{s}(x, y)$, and show explicitly that $\nabla \cdot \vec{s} = 0$, which has the physical significance that there is nowhere, in the water, accumulation of any permanent wave energy (active energy—but possibly only of reactive energy). Discuss the cases $\beta = 0$, $\beta = \pi$ and $\beta = \pi/2$.

Derive expressions for the space-averaged wave-energy transport $\langle J_x \rangle$ and $\langle J_y \rangle$. Pay special attention to the case of $\cos \beta = 1$.

## Problem 4.16: One Plane Wave and One Circular Wave

Let us assume that a harmonic wave on water of constant depth $h$ is composed of one plane wave and one circular wave. The wave elevation $\eta(r,\theta,t)$ has a complex amplitude $\hat{\eta} = \hat{\eta}(r,\theta)$ given as

$$\hat{\eta} = \hat{\eta}_A + \hat{\eta}_B = Ae^{-ikr \cos \theta} + \frac{B}{\sqrt{kr}}e^{-ikr} + \cdots$$

$$= A\left(e^{-ikr \cos \theta} + \frac{b}{\sqrt{kr}}e^{-ikr} + \cdots\right),$$

where $b = B/A$ is dimensionless. Cartesian coordinates $(x,y,z)$ have been replaced by cylindrical ones $(r,\theta,z)$; thus, $x = r \cos \theta$ and $y = r \sin \theta$. In relation to $\hat{\eta}_B$, additional terms of order $(kr)^{-3/2}$, which may be neglected for $kr \gg 1$, are denoted by three dots $(\cdots)$. The wave's velocity potential is given by its complex amplitude

$$\hat{\phi}(r,\theta,z) = -\hat{\eta}(r,\theta)\, e(kz)g/i\omega,$$

where $e(kz) = \cosh(kz + kh)/\cosh kh$.

Our aim is to derive and discuss an expression for the wave power $P_{\text{cyl}}$, which is propagated outwards through an envisaged vertical, cylindrical, axisymmetric surface of radius $r \gg 1/k$.

(a) As a first step, derive complex-amplitude expressions for the hydrodynamic pressure $\hat{p} = \hat{p}_A + \hat{p}_B$ and for the radial component of the water-particle velocity $\hat{v}_r = \hat{v}_{A,r} + \hat{v}_{B,r}$. Here we may neglect terms that, for $kr \gg 1$, decrease faster towards zero than $1/\sqrt{kr}$. Consequently, we shall, in the following, neglect terms that decrease faster towards zero than $1/kr$.

(b) Secondly, derive an expression for the resulting radial component of the intensity

$$I_r(r,\theta,z) = \frac{1}{2}\text{Re}\{\hat{p}\hat{v}_r^*\} = \frac{1}{4}(\hat{p}\hat{v}_r^* + \hat{p}^*\hat{v}_r).$$

(c) Further, show that the radial component $J_r$ of the wave-power level (wave-energy transport) is

$$J_r(r,\theta) = J_A\left(\cos \theta + \frac{|b|^2}{kr} + \frac{\text{Re}\{be^{-ikr(1-\cos \theta)}\cos \theta + b^*e^{-ikr(1-\cos \theta)}\}}{\sqrt{kr}} + \cdots\right),$$

where

$$J_A = \frac{\rho g^2 D(kh)|A|^2}{4\omega} = v_g(\rho g/2)|A|^2 = v_g(\rho g/2)|\hat{\eta}_A|^2$$

is the wave-power level for the plane wave alone. Note that, in the preceding expression for $J_r(r,\theta)$, the first and second terms correspond to the direct contributions from the plane wave alone and the circular wave alone, respectively.

(d) Next, assuming that $kr \to \infty$, show that the time-average wave-power

$$P_{cyl} = P_{cyl,\text{ outwards}} = \int_0^{2\pi} J_r(r,\theta)rd\theta = r \int_{-\beta}^{2\pi-\beta} J_r(r,\theta)d\theta$$

which passes outwards through the envisaged cylindrical surface of radius $r$ is given by

$$P_{cyl} = \frac{J_A}{k}(|1 + \zeta|^2 - 1), \quad \text{where} \quad \zeta = b\sqrt{2\pi}e^{-i\pi/4} = (1 - i)b\sqrt{\pi}.$$

(e) Is it reasonable that this power is independent of $r$, the radius of the envisaged cylindrical surface? Explain—for instance, by physical arguments—why this expression for $P_{cyl}$ may be applicable/valid even for values of $r$ that are too small to satisfy the condition $kr \gg 1$.

(f) Let us assume that the complex plane-wave elevation amplitude $A$ has a fixed value, and then discuss how the real-valued (positive or negative) $P_{cyl}$ varies with the complex ratio $B/A = b$, or with the dimensionless complex variable $\zeta$. Which kind of a closed curve, in the complex $B/A$ (or $\zeta$) plane, separates regions for which $P_{cyl}$ has opposite signs?

(g) Show that $P_{cyl,\text{ min}} = -J_A/k = -(\lambda/2\pi)J_A$.

[Hint: when solving this Problem 4.16, it may be helpful to apply, e.g., Eqs. (4.36)–(4.37), (4.52), (4.85)–(4.90), (4.126), (4.129)–(4.130), (4.214)–(4.219), and, finally, (4.274)–(4.276).]

# CHAPTER FIVE

# Wave–Body Interactions

The subject of this chapter is, as with Chapter 3, a discussion on the interaction between waves and oscillating bodies. Now, however, the discussion is limited to the case of interaction with water waves, which we discussed in some detail in Chapter 4. We start by studying body oscillations and wave forces on bodies. Next, we consider the phenomenon of wave generation by oscillating bodies, and we discuss some general relationships between the wave forces on bodies when they are held fixed and the waves they generate when they oscillate in otherwise calm water. Some particular consideration is given to two-dimensional cases and to axisymmetric cases. While most of the analysis is carried out in the frequency domain, some studies in the time domain are also made.

## 5.1 Six Modes of Body Motion: Wave Forces and Moments

To describe the motion of a body immersed in water, we need a coordinate system, for which we choose to have the $z$-axis passing through the centre of gravity of the body. The plane $z = 0$ defines the mean free surface. Further, we choose a reference point $(0, 0, z_0)$ on the $z$-axis—for instance, the centre of gravity or the origin (as in Figure 5.1). The body has a wet surface $S$, which separates it from the water. Consider a surface element $dS$ at position $\vec{s}$ (Figure 5.1). The vector $\vec{s}$ originates in the chosen reference point. Let $\vec{U}$ be the velocity of the reference point. The surface element $dS$ has a velocity

$$\vec{u} = \vec{U} + \vec{\Omega} \times \vec{s}, \tag{5.1}$$

where $\vec{\Omega}$ is the angular velocity vector corresponding to rotation about the reference point. Thus, with no rotation, $\vec{u} = \vec{U}$, and with no translation, $\vec{u} = \vec{\Omega} \times \vec{s}$. For a given body, the motion of each point $\vec{s}$ (of the wet body surface $S$) is characterised by the time-dependent vectors $\vec{U}$ and $\vec{\Omega}$.

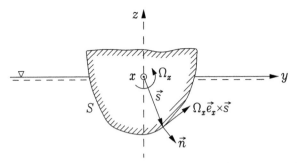

Figure 5.1: Body oscillating in water. Vector $\vec{s}$ gives the position of a point on the wet surface $S$, where the unit normal is $\vec{n}$.

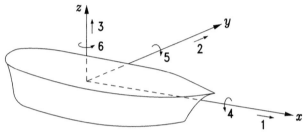

Figure 5.2: A rigid body has six modes of motion: surge (1), sway (2), heave (3), roll (4), pitch (5) and yaw (6).

The velocity potential $\phi$ must satisfy the boundary condition (4.15) every-where on $S$:

$$\frac{\partial \phi}{\partial n} = \vec{n} \cdot \nabla \phi = u_n = \vec{U} \cdot \vec{n} + (\vec{\Omega} \times \vec{s}) \cdot \vec{n} = \vec{U} \cdot \vec{n} + \vec{\Omega} \cdot (\vec{s} \times \vec{n}). \tag{5.2}$$

Here, $u_n$ is the normal component of the velocity $\vec{u}$ of the wet-surface element $dS$ of the body.

### 5.1.1 Six Modes of Motion

The motion of the rigid body is characterised by six components, corresponding to six degrees of freedom or modes of (oscillatory) motion, as shown in Figure 5.2. The three translational modes are conventionally named surge, sway and heave, whereas the three rotational modes are named roll, pitch and yaw. Note that in the two-dimensional case (see Section 5.8)—that is, when there is no variation in the $y$-direction—there are only three modes of motion, namely surge, heave and pitch (two translations and one rotation).

We introduce six-dimensional generalised vectors: a velocity vector **u** with components

$$(u_1, u_2, u_3) \equiv (U_x, U_y, U_z) = \vec{U}, \tag{5.3}$$

$$(u_4, u_5, u_6) \equiv (\Omega_x, \Omega_y, \Omega_z) = \vec{\Omega}; \tag{5.4}$$

and a normal vector **n** with components

$$(n_1, n_2, n_3) \equiv (n_x, n_y, n_z) = \vec{n}, \tag{5.5}$$

$$(n_4, n_5, n_6) \equiv \vec{s} \times \vec{n}. \tag{5.6}$$

Thus, $n_4 = (\vec{s} \times \vec{n})_x = s_y n_z - s_z n_y$, and so on. Note that the components numbered 1 to 3 have different dimensions than the remaining components numbered 4 to 6. Thus, $u_1$, $u_2$ and $u_3$ have SI units of m/s, whereas $u_4$, $u_5$ and $u_6$ have SI units of rad/s.

Inserting Eqs. (5.3)–(5.6) into the boundary condition (5.2), we can write it as

$$\frac{\partial \phi}{\partial n} = \sum_{j=1}^{6} u_j n_j = \mathbf{u}^{\mathsf{T}} \mathbf{n} \quad \text{on } S. \tag{5.7}$$

Here the superscript T denotes transpose. Thus, $\mathbf{u}^{\mathsf{T}} = (u_1, u_2, u_3, u_4, u_5, u_6)$ is a row vector, while $\mathbf{u}$ is the corresponding column vector. Moreover, $\mathbf{n} = (n_1, n_2, n_3, n_4, n_5, n_6)^{\mathsf{T}}$. These generalised vectors are six-dimensional.

Note that in the preceding equations, the variables $\vec{U}$, $\vec{\Omega}$, $u_j(t)$ and $\phi(x, y, z, t)$ may, in the case of harmonic oscillations and waves, be replaced by their complex amplitudes $\hat{\vec{U}}$, $\hat{\vec{\Omega}}$, $\hat{u}_j$ and $\hat{\phi}(x, y, z)$. Thus,

$$\frac{\partial \hat{\phi}}{\partial n} = \sum_{j=1}^{6} \hat{u}_j n_j = \hat{\mathbf{u}}^{\mathsf{T}} \mathbf{n} \quad \text{on } S. \tag{5.8}$$

Let us, for a while, consider the radiation problem. When the body oscillates, a radiated wave $\phi_r$ is generated which is a superposition (linear combination) of radiated waves due to each of the six oscillation modes:

$$\hat{\phi}_r = \sum_{j=1}^{6} \varphi_j \hat{u}_j, \tag{5.9}$$

where $\varphi_j = \varphi_j(x, y, z)$ is a complex coefficient of proportionality. This frequency-dependent coefficient $\varphi_j$ must satisfy the body-boundary condition

$$\frac{\partial \varphi_j}{\partial n} = n_j \quad \text{on } S, \tag{5.10}$$

since the complex amplitude $\hat{\phi}_r$ of the radiated velocity potential must satisfy the boundary condition (5.8). Whereas $\phi_r$ has SI unit m²/s, $\varphi_1$, $\varphi_2$ and $\varphi_3$ have unit m, and $\varphi_4$, $\varphi_5$ and $\varphi_6$ have unit m² (or, more precisely, m²/rad). The coefficient $\varphi_j$ may be interpreted as the complex amplitude of the radiated velocity potential due to body oscillation in mode $j$ with unit velocity amplitude ($\hat{u}_j = 1$). Coefficients $\varphi_j$ are independent of the oscillation amplitude because boundary condition (5.2), as well as Laplace's equation (4.33), is linear. Finally, let us mention that, in the present linear theory, viscosity is neglected.

Coefficients $\varphi_j$ must satisfy the same homogeneous equations as $\hat{\phi}_r$, namely the Laplace equation

$$\nabla^2 \varphi_j = 0, \tag{5.11}$$

the sea-bed boundary condition

$$\left[ \frac{\partial \varphi_j}{\partial z} \right]_{z=-h} = 0 \tag{5.12}$$

and the free-surface boundary condition

$$\left[ -\omega^2 \varphi_j + g \frac{\partial \varphi_j}{\partial z} \right]_{z=0} = 0. \tag{5.13}$$

[See Eqs. (4.33), (4.40) and (4.41).] Moreover, $\varphi_j$ (and $\hat{\phi}_r$) must satisfy a radiation condition at infinite distance. This will be discussed later (also see Section 4.6).

Remember that in this linear theory it is consistent to take the boundary condition (5.13) at the mean free surface $z = 0$ instead of the instantaneous free surface $z = \eta(t)$. Similarly, when we use the boundary condition (5.10), we may consider $S$ to be the mean wet surface of the body instead of the instantaneous wet surface.

### 5.1.2 Hydrodynamic Force Acting on a Body

Next, let us derive expressions for the force and the moment which act on the body, in terms of a given velocity potential $\phi$.

Firstly, let us consider the vertical (or heave) force component $F_z$. The vertical force on the element $dS$ (Figure 5.3) is $p(-n_z)dS = -pn_3 dS$. Here $p$ is the hydrodynamic pressure. Integrating over the wet surface $S$ gives the total heave force

$$F_3 \equiv F_z = - \iint_S pn_3 \, dS. \tag{5.14}$$

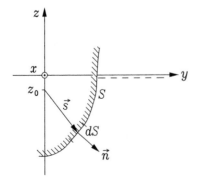

Figure 5.3: Surface element $dS$ of the wet surface $S$ of a rigid body with unit normal $\vec{n}$ and with position $\vec{s}$ relative to the chosen reference point $(x, y, z) = (0, 0, z_0)$ of the body.

In terms of complex amplitudes, we have, using Eq. (4.37),

$$\hat{F}_3 = \hat{F}_z = i\omega\rho \iint_S \hat{\phi} n_3 \, dS. \tag{5.15}$$

Analogous expressions apply for the horizontal force components—that is, surge force $F_1$ and sway force $F_2$.

Next, let us consider the moment about the $x$-axis. The moment acting on surface element $dS$ is

$$dM_x = s_y dF_z - s_z dF_y = (-p n_z s_y + p n_y s_z) dS. \tag{5.16}$$

Integrating over the whole wet surface $S$ gives

$$M_x = -\iint_S p(s_y n_z - s_z n_y) dS = -\iint_S p(\vec{s} \times \vec{n})_x \, dS. \tag{5.17}$$

Thus, using Eq. (5.6)—that is, $n_4 = (\vec{s} \times \vec{n})_x$—we find that the roll moment is

$$M_x = -\iint_S p n_4 \, dS. \tag{5.18}$$

In terms of complex amplitudes,

$$\hat{M}_x = i\omega\rho \iint_S \hat{\phi} n_4 \, dS. \tag{5.19}$$

Similar expressions may be obtained for pitch moment $M_y$ and yaw moment $M_z$.

We now define a generalised force vector having six components:

$$\mathbf{F} \equiv (F_1, F_2, F_3, F_4, F_5, F_6) \equiv (F_x, F_y, F_z, M_x, M_y, M_z) = (\vec{F}, \vec{M}). \tag{5.20}$$

The first three components have SI unit N (newton), and the three remaining ones have SI unit N m. For component $j$ ($j = 1, 2, \ldots, 6$) we have

$$F_j = -\iint_S p n_j \, dS, \tag{5.21}$$

$$\hat{F}_j = i\omega\rho \iint_S \hat{\phi} n_j \, dS \tag{5.22}$$

for an arbitrary, given potential $\hat{\phi}$.

If the body is oscillating, surface element $dS$ receives an instantaneous power

$$\begin{aligned}
dP(t) &= -\vec{u} \cdot \vec{n} p \, dS \\
&= -(\vec{U} \cdot \vec{n} + \vec{\Omega} \times \vec{s} \cdot \vec{n}) p \, dS = -[\vec{U} \cdot \vec{n} + \vec{\Omega} \cdot (\vec{s} \times \vec{n})] p \, dS \\
&= -(u_1 n_1 + u_2 n_2 + u_3 n_3 + u_4 n_4 + u_5 n_5 + u_6 n_6) p \, dS.
\end{aligned} \tag{5.23}$$

Integrating over the wet surface, we get the power received by the oscillating body:

$$P(t) = \vec{F}(t) \cdot \vec{U}(t) + \vec{M}(t) \cdot \vec{\Omega}(t) = \sum_{j=1}^{6} F_j u_j, \qquad (5.24)$$

where we have used Eqs. (5.1), (5.3)–(5.6), (5.20) and (5.21).

### 5.1.3 Excitation Force

If the body is fixed, a non-vanishing potential $\hat{\phi}$ is the result of an incident wave only. In such a case, $F_j$ is called the 'excitation force' (or, for $j = 4, 5, 6$, the 'excitation moment'). With no body motion, no radiated wave is generated. The generalised vector of the *excitation force* is written

$$\mathbf{F}_e \equiv (F_{e,1}, F_{e,2}, \ldots, F_{e,6}) = (\vec{F}_e, \vec{M}_e), \qquad (5.25)$$

and using Eq. (5.22), we have

$$\hat{F}_{ej} = i\omega\rho \iint_S (\hat{\phi}_0 + \hat{\phi}_d) n_j \, dS, \qquad (5.26)$$

where $\hat{\phi}_0$ represents the undisturbed incident wave and $\hat{\phi}_d$ the diffracted wave. The latter is induced when the former does not satisfy the homogeneous boundary condition (4.16) on the fixed wet surface $S$. The boundary condition

$$-\frac{\partial \hat{\phi}_d}{\partial n} = \frac{\partial \hat{\phi}_0}{\partial n} \quad \text{on } S \qquad (5.27)$$

has to be satisfied for the diffraction problem. Note that $\hat{\phi}_d$ and $\hat{\phi}_0$ satisfy the homogeneous boundary conditions (4.40) on the sea bed, $z = -h$, and (4.41) on the free water surface, $z = 0$.

If the diffraction term $\hat{\phi}_d$ is neglected in Eq. (5.26), the resulting force is termed the Froude–Krylov force, which is considered in more detail in Section 5.6. It may represent a reasonable approximation to the excitation force, in particular if the horizontal extension of the immersed body is very small compared to the wavelength. It may be computationally convenient to use such an approximation since it is not then required to solve the boundary-value problem for finding the diffraction potential $\hat{\phi}_d$.

Except for very simple geometries, it is not possible to find solutions to the diffraction problem in terms of elementary functions. However, for the case of a wave

$$\hat{\eta}_0 = A e^{-ikx} \qquad (5.28)$$

incident upon a vertical wall at $x = 0$, the diffracted wave is simply the totally reflected wave

$$\hat{\eta}_d = A e^{ikx} \qquad (5.29)$$

(cf. Problem 4.5). Then, from Eq. (4.88), the velocity potential is given by

$$\hat{\phi}_0 + \hat{\phi}_d = -\frac{g}{i\omega}e(kz)(\hat{\eta}_0 + \hat{\eta}_d) = -\frac{2g}{i\omega}e(kz)A\cos(kx). \tag{5.30}$$

Using Eq. (5.26) and noting that $x = 0$ and $n_1 = -1$ on $S$, we now find the surge component of the excitation force (per unit width in the $y$-direction):

$$\hat{F}'_{e,1} = -i\omega\rho \int_{-a_2}^{-a_1} (\hat{\phi}_0 + \hat{\phi}_d)\, dz = 2\rho g A \int_{-a_2}^{-a_1} e(kz)\, dz \tag{5.31}$$

for the striplike piston shown in Figure 5.4. (The prime is used to denote a quantity per unit width. $F'_{e,1}$ has SI unit N/m.) Inserting the integrand from Eq. (4.52) and performing the integration give (cf. Problem 5.2)

$$\hat{F}'_{e,1} = 2\rho g A\,\frac{\sinh(kh - ka_1) - \sinh(kh - ka_2)}{k\cosh(kh)}. \tag{5.32}$$

Setting $a_2 = h$ and $a_1 = 0$, we find the surge excitation force per unit width of the entire vertical wall:

$$\hat{F}'_{e,1} = 2\rho g A\,\frac{\sinh(kh)}{k\cosh(kh)} = 2\rho A\left(\frac{\omega}{k}\right)^2, \tag{5.33}$$

where we have made use of the dispersion relationship (4.54).

If the incident wave is as given by Eq. (5.28), then, according to Eqs. (4.88) and (5.27), $\hat{\phi}_0$ and $\hat{\phi}_d$ are proportional to the complex elevation amplitude $A$ at the origin $(x, y) = (0, 0)$. Hence, also all excitation force components $\hat{F}_{e,j}$ are proportional to $A$. It should be remembered that the excitation force on a body is the wave force when the body is not moving. Thus, for instance, the surge velocity $u_1$ (see Figure 5.4) was zero for the situation analysed earlier. In Section 5.2.3, however, we shall consider the case when $u_1 \neq 0$.

Although the excitation force in the preceding example is in phase with the incident wave elevation at the origin, there may, in the general case, be a

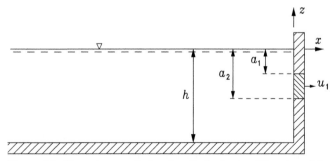

Figure 5.4: Piston in a vertical end wall of a basin with water of depth $h$. The piston, which is able to oscillate in surge (horizontal motion), occupies, in this two-dimensional problem, the strip $-a_2 < z < -a_1$.

nonzero phase difference $\theta_j$ between these two variables such that the *excitation-force coefficient*

$$f_j = \frac{\hat{F}_{e,j}}{A} = \frac{|\hat{F}_{e,j}|}{|A|} e^{i\theta_j} \qquad (5.34)$$

is, in general, a complex coefficient of proportionality. As we shall see later (in Section 5.6), for a floating body which is very much smaller than the wavelength, we have $\theta_1 \approx \pi$ and $\theta_3 \approx 0$ for the surge and heave modes, respectively. For a floating semisubmerged sphere on deep water, numerically computed values for $\theta_3$ are given by the graph in Figure 5.5 (provided the sphere is located such that its vertical axis coincides with the $z$-axis). It can be seen from the graph that $\theta_3 < 0.1$ for $ka < 0.25$. Additional numerical information relating to a semisubmerged sphere is presented in Section 5.2.4.

## 5.2 Radiation from an Oscillating Body

### 5.2.1 The Radiation Impedance Matrix

In the following, let us consider a case when a body is oscillating in the absence of an incident wave. We shall study the forces acting on the body due to the wave which is radiated as a result of the body's oscillation. Let the body oscillate in a single mode $j$ with complex velocity amplitude $\hat{u}_j$. The radiated wave is associated with a velocity potential $\phi_r$ given by

$$\hat{\phi}_r = \varphi_j \hat{u}_j, \qquad (5.35)$$

where $\varphi_j = \varphi_j(x,y,z)$ is a coefficient of proportionality, as introduced in Eq. (5.9). The radiated wave reacts with a force on the body. Component $j'$ of the force is

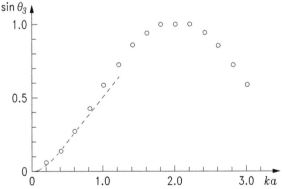

Figure 5.5: Phase angle $\theta_3$ between the heave excitation force and the undisturbed incident wave elevation at the body centre for a semisubmerged floating sphere of radius $a$. The circle points are numerically computed by Greenhow [42]. The curve is computed by Kyllingstad [43] using a second-order scattering method.

$$\hat{F}_{r,j'} = i\omega\rho \iint_S \varphi_j \hat{u}_j n_{j'} dS, \tag{5.36}$$

according to Eqs. (5.22) and (5.35).

Whether $j$ denotes a translation mode ($j = 1, 2, 3$) or a rotation mode ($j = 4, 5, 6$), $\hat{u}_j$ is a constant under the integration. Hence, we may write

$$\hat{F}_{r,j'} = -Z_{j'j}\hat{u}_j, \tag{5.37}$$

where

$$Z_{j'j} = -i\omega\rho \iint_S \varphi_j n_{j'} \, dS \tag{5.38}$$

is an element of the so-called *radiation impedance* matrix. The SI units are N s/m = kg/s for $Z_{q'q}$ ($q', q = 1, 2, 3$), N s m (or N s m/rad) for $Z_{p'p}$ ($p', p = 4, 5, 6$) and N s for $Z_{qp}$ and $Z_{pq}$.

Using the wet-surface boundary condition (5.10), we have

$$Z_{j'j} = -i\omega\rho \iint_S \varphi_j \frac{\partial \varphi_{j'}}{\partial n} \, dS. \tag{5.39}$$

Whereas $\varphi_j$ is complex, $\partial \varphi_j / \partial n$ is real on $S$, since $n_j$ is real [cf. Eq. (5.10)]. Hence, we may in the integrand replace $\partial \varphi_{j'} / \partial n$ by $\partial \varphi_{j'}^* / \partial n$ if we so wish. Thus, we have the alternative formula

$$Z_{j'j} = -i\omega\rho \iint_S \varphi_j \frac{\partial \varphi_{j'}^*}{\partial n} \, dS. \tag{5.40}$$

Note that $\varphi_j$ has to satisfy boundary conditions elsewhere [see Eqs. (5.12)–(5.13)]. This means, for instance, that the radiation impedance matrix of a body in a wave channel does not have the same value as that of the same body in open sea [44, 45]. Note that a homogeneous boundary condition such as (4.16), namely $\partial \varphi_j / \partial n = 0$, has to be satisfied on the vertical walls of the wave channel.

We may interpret $-Z_{j'j}$ as the $j'$ component of the reaction force due to the wave radiated by the body oscillating with unit amplitude in mode $j$ ($\hat{u}_j = 1$). In fact, it is equal to the $j$ component of the reaction force due to wave radiation from unit-amplitude oscillation in mode $j'$ ($\hat{u}_{j'} = 1$). This is true because of the reciprocity relation

$$Z_{jj'} = Z_{j'j}, \tag{5.41}$$

which follows from Eqs. (5.39), (4.230) and (4.239), when we utilise the fact that $\varphi_j$ and $\varphi_{j'}$ satisfy the same radiation condition. Thus, the radiation impedance matrix is symmetric.

For certain body geometries, some of the elements of the radiation impedance matrix vanish. If $y = 0$ is a plane of symmetry (which is typical for a ship hull), then $n_2$, $n_4$ and $n_6$ in Eq. (5.38) are odd functions of $y$, while $\varphi_1$, $\varphi_3$ and $\varphi_5$ are even functions. Hence, $Z_{21} = Z_{23} = Z_{25} = Z_{41} = Z_{43} = Z_{45} = Z_{61} = Z_{63} = Z_{65} = 0$. On the other hand, if $x = 0$ is a plane of symmetry, then $n_1$, $n_5$ and $n_6$ are odd functions of $x$, while $\varphi_2$, $\varphi_3$ and $\varphi_4$ are even functions. Hence, $Z_{12} = Z_{13} = Z_{14} = Z_{52} = Z_{53} = Z_{54} = Z_{62} = Z_{63} = Z_{64} = 0$. From this observation, while noting Eq. (5.41), it follows that if both $y = 0$ and $x = 0$ are planes of symmetry, then the only non-vanishing off-diagonal elements of the radiation impedance matrix are $Z_{15} = Z_{51}$ and $Z_{24} = Z_{42}$.

Since $\omega$ is real, it is convenient to split $Z_{j'j}$ into real and imaginary parts:

$$Z_{j'j} = R_{j'j} + iX_{j'j} = R_{j'j} + i\omega m_{j'j}, \tag{5.42}$$

where we call $\mathbf{R}$ the *radiation resistance* matrix, $\mathbf{X}$ the *radiation reactance* matrix and $\mathbf{m}$ the *added-mass* matrix. Note that some authors call $\mathbf{R}$ the 'added damping coefficient' matrix. Eq. (5.42) is a generalisation of the scalar equations (3.33) and (3.37).

### 5.2.2 Energy Interpretation of the Radiation Impedance

In Eq. (3.29), we have defined a radiation resistance by considering the radiated power. We shall later, in a rather general way, relate radiation resistance and reactance to power and energy associated with wave radiation. At present, however, let us just consider a diagonal element of the radiation impedance matrix. Using Eq. (5.40), we have

$$\frac{1}{2}Z_{jj}|\hat{u}_j|^2 = \frac{1}{2}Z_{jj}\hat{u}_j\hat{u}_j^* = \frac{1}{2}\iint_S (-i\omega\rho\varphi_j\hat{u}_j)\frac{\partial}{\partial n}\left(\varphi_j^*\hat{u}_j^*\right)dS. \tag{5.43}$$

Further, using Eqs. (5.35), (4.7) and (4.37), this becomes

$$\frac{1}{2}Z_{jj}|\hat{u}_j|^2 = \frac{1}{2}\iint_S (-i\omega\rho\hat{\phi}_r)\frac{\partial\hat{\phi}_r^*}{\partial n}dS = \frac{1}{2}\iint_S \hat{p}\hat{v}_n^* \, dS = \mathcal{P}_r. \tag{5.44}$$

Referring to Eqs. (3.17) and (3.20), we see that the real part of $\mathcal{P}_r$ is the (time-average) power $P_r$ delivered to the fluid from the body oscillating in mode $j$:

$$P_r = \mathrm{Re}\{\mathcal{P}_r\} = \tfrac{1}{2}\mathrm{Re}\{Z_{jj}\}|\hat{u}_j|^2 = \tfrac{1}{2}R_{jj}|\hat{u}_j|^2. \tag{5.45}$$

Here $R_{jj}$ is the diagonal element of the radiation resistance matrix. The real part $P_r$ of the radiated 'complex power' $\mathcal{P}_r$ [see Eq. (2.90)] represents the radiated 'active power'. The imaginary part

$$\mathrm{Im}\{\mathcal{P}_r\} = \tfrac{1}{2}\mathrm{Im}\{Z_{jj}\}|\hat{u}_j|^2 = \tfrac{1}{2}X_{jj}|\hat{u}_j|^2 \tag{5.46}$$

represents the radiated 'reactive power'. We may note that (see Problem 5.7 or Section 6.5)

$$\mathcal{P}_r = \sum_{j'=1}^{6} \sum_{j=1}^{6} Z_{j'j} \hat{u}_{j'} \hat{u}_j^*$$

(5.47)

is a generalisation of Eq. (5.45), which is valid when all except one of the oscillation modes have a vanishing amplitude.

### 5.2.3 Wavemaker in a Wave Channel

Let us consider an example referring to Figure 5.4, when the piston has a surge velocity with complex amplitude $\hat{u}_1 \neq 0$. To be slightly more general, we shall, for this example, write the inhomogeneous boundary condition (5.10) as

$$\frac{\partial \varphi_1}{\partial x} = c(z) \quad \text{for } x = 0,$$

(5.48)

where $c(z)$ is a given function. For the case shown in Figure 5.4, we must set $c(z) = 1$ for $-a_2 < z < -a_1$ and $c(z) = 0$ elsewhere. If we set $c(z) = 1$ for $-h < z < 0$, this corresponds to the entire vertical wall oscillating as a surging piston. If $c(z) = 1 + z/h$ for $-h < z < 0$, this corresponds to an oscillating flap, hinged at $z = -h$. Both are typical as wavemakers in wave channels.

The given oscillation at $x = 0$ generates a wave which propagates in the negative $x$-direction towards $x = -\infty$ (Figure 5.4). Hence, it is a requirement that our solution of the boundary-value problem is finite in the region $x < 0$ and that it satisfies the radiation condition at $x = -\infty$. Referring to Eq. (4.111), which is a solution to the boundary-value problem given by Eqs. (4.33), (4.40) and (4.41), we can immediately write down the following solution:

$$\varphi_1 = c_0 e^{ikx} Z_0(z) + \sum_{n=1}^{\infty} c_n e^{m_n x} Z_n(z) = \sum_{n=0}^{\infty} X_n(x) Z_n(z),$$

(5.49)

which (for $j = 1$) satisfies the Laplace equation (5.11), the homogeneous boundary conditions (5.12)–(5.13), the finiteness condition for $-\infty < x < 0$ and the radiation condition for $x = -\infty$. Concerning the orthogonal functions $Z_n(z)$ used in Eq. (5.49), we may refer to Eqs. (4.70)–(4.76). For convenience, we have introduced $m_0 = ik$ and

$$X_n(x) = c_n e^{m_n x}.$$

(5.50)

In Eq. (5.49), the terms corresponding to $n \geq 1$ represent evanescent waves which are non-negligible only near the wave generator. Thus, in the far-field region $-x \gg 1/m_1$, we have, asymptotically,

$$\varphi_1 \approx c_0 e^{ikx} Z_0(z).$$

(5.51)

In order to determine the unknown constants $c_n$ in Eq. (5.49), we use the boundary condition (5.48) at $x = 0$:

$$c(z) = \left[\frac{\partial \varphi_1}{\partial x}\right]_{x=0} = \sum_{n=0}^{\infty} X'_n(0) Z_n(z). \tag{5.52}$$

We multiply by $Z^*_m(z)$, integrate from $z = -h$ to $z = 0$ and use the orthogonality and normalisation conditions (4.67) and (4.73). This gives

$$\int_{-h}^{0} c(z) Z^*_m(z)\,dz = \sum_{n=0}^{\infty} X'_n(0) \int_{-h}^{0} Z^*_m(z) Z_n(z)\,dz = X'_m(0) h. \tag{5.53}$$

That is,

$$X'_n(0) = \frac{1}{h}\int_{-h}^{0} c(z) Z^*_n(z)\,dz. \tag{5.54}$$

According to Eq. (5.50), we have

$$X'_0(0) = ikc_0, \quad X'_n(0) = m_n c_n, \tag{5.55}$$

and hence,

$$c_0 = \frac{1}{ikh}\int_{-h}^{0} c(z) Z^*_0(z)\,dz, \tag{5.56}$$

$$c_n = \frac{1}{m_n h}\int_{-h}^{0} c(z) Z^*_n(z)\,dz. \tag{5.57}$$

Thus, $c_n$ is a kind of 'Fourier' coefficient for the expansion of the velocity-distribution function $c(z)$ in the function 'space' spanned by the complete function set $\{Z_n(z)\}$.

The velocity potential of the radiated wave is

$$\hat{\phi}_r = \varphi_1 \hat{u}_1, \tag{5.58}$$

where $\varphi_1$ is given by Eq. (5.49). The wave elevation of the radiated wave is [cf. Eq. (4.39)]

$$\hat{\eta}_r = -\frac{i\omega}{g}\left[\hat{\phi}_r\right]_{z=0} = -\frac{i\omega}{g}\hat{u}_1 \sum_{n=0}^{\infty} c_n Z_n(0) e^{m_n x}. \tag{5.59}$$

In the far-field region—that is, for $-x \gg 2h/\pi > 1/m_1$—the evanescent waves are negligible, and we have the asymptotic solution

$$\hat{\eta}_r \sim A^-_r e^{ikx} = a^-_1 \hat{u}_1 e^{ikx}, \tag{5.60}$$

where

$$a^-_1 = \frac{A^-_r}{\hat{u}_1} = -\frac{i\omega}{g} c_0 Z_0(0) = -\frac{i\omega}{g} c_0 \sqrt{\frac{2kh}{D(kh)}}. \tag{5.61}$$

Eqs. (4.52) and (4.76) have been used in the last step. Note that $a_1^-$ is real, since $c(z)$ and $Z_0(z)$ are real, and hence [cf. Eq. (5.56)], $ic_0$ is also real.

If Figure 5.4 represents a wave channel of width $d$, the radiation impedance of the wavemaker is, according to Eq. (5.40), with $\partial/\partial n = -\partial/\partial x$,

$$Z_{11} = i\omega\rho d \int_{-h}^{0} \left[ \varphi_1 \frac{\partial \varphi_1^*}{\partial x} \right]_{x=0} dz. \tag{5.62}$$

Considering Figure 5.4 as representing a two-dimensional problem, we find that the radiation impedance per unit width is

$$
\begin{aligned}
Z_{11}' = \frac{Z_{11}}{d} &= i\omega\rho \int_{-h}^{0} \left[ c_0 Z_0(z) + \sum_{n=1}^{\infty} c_n Z_n(z) \right] c^*(z)\, dz \\
&= i\omega\rho \left( -ikhc_0 c_0^* + \sum_{n=1}^{\infty} m_n h c_n c_n^* \right),
\end{aligned} \tag{5.63}
$$

where we have used Eqs. (5.48), (5.49), (5.56) and (5.57) in Eq. (5.62). Hence,

$$Z_{11}' = \frac{Z_{11}}{d} = \omega k\rho h |c_0|^2 + i\omega\rho h \sum_{n=1}^{\infty} m_n |c_n|^2. \tag{5.64}$$

The radiation resistance is

$$R_{11} = \text{Re}\{Z_{11}\} = \omega k\rho |c_0|^2 hd. \tag{5.65}$$

The added mass is

$$m_{11} = \frac{1}{\omega} \text{Im}\{Z_{11}\} = \rho hd \sum_{n=1}^{\infty} m_n |c_n|^2 \tag{5.66}$$

[cf. Eq. (5.42)]. Because all $m_n$ are positive, we see that all contributions to the radiation reactance $\omega m_{11}$ are positive. We also see that the radiation resistance is associated with the far field (corresponding to the propagating wave), whereas the radiation reactance and, hence, the added mass are associated with the near field (corresponding to the evanescent waves).

It might be instructive to consider the 'complex energy transport' $\mathcal{J}(x)$ associated with the radiated wave. Referring to Eqs. (2.90) and (4.126), we obtain the complex wave-energy transport (which, in this case, is in the negative $x$-direction):

$$
\begin{aligned}
\mathcal{J}(x) &= \frac{1}{2} \int_{-h}^{0} \hat{p}(-\hat{v}_x)^* dz = -\frac{1}{2} i\omega\rho \int_{-h}^{0} \hat{\phi}_r \left( -\frac{\partial\hat{\phi}_r^*}{\partial x} \right) dz \\
&= \frac{i\omega\rho}{2} \sum_{n=0}^{\infty} \sum_{l=0}^{\infty} X_n(x) X_l'^*(x) \int_{-h}^{0} Z_n(z) Z_l^*(z)\, dz\, \hat{u}_1 \hat{u}_1^* \\
&= \frac{1}{2} i\omega\rho h \sum_{n=0}^{\infty} X_n(x) X_n'^*(x) |\hat{u}_1|^2,
\end{aligned} \tag{5.67}
$$

where we have utilised Eqs. (5.49), (5.58) and (4.67). Using Eq. (5.50) gives

$$J(x) = \left[ \frac{1}{2}\omega kh\rho|c_0|^2 + \frac{1}{2}i\omega\rho h \sum_{n=1}^{\infty} m_n|c_n|^2 \exp(2m_n x) \right] |\hat{u}_1|^2. \tag{5.68}$$

We see that, because of orthogonality condition (4.67), there are no cross terms from different vertical eigenfunctions in the product.

The (active) energy transport per unit width of wave frontage is

$$J = \mathrm{Re}\{J(x)\} = \tfrac{1}{2}\omega kh\rho|c_0|^2|\hat{u}_1|^2. \tag{5.69}$$

Moreover, we see that there is a reactive power $\mathrm{Im}\{J(x)\}$ which exists only in the near field. This demonstrates that the radiation reactance and, hence, the added mass are related to reactive energy transport in the near-field region of the wavemaker. The active energy transport $J$ in a plane wave in an ideal (loss-free) liquid is, of course, independent of $x$. We see that the radiated power, in accordance with Eq. (3.29), is

$$\tfrac{1}{2}R_{11}|\hat{u}_1|^2 = P_r = Jd = \mathrm{Re}\{J(x)\}d = J(-\infty)\,d. \tag{5.70}$$

Correspondingly, the radiation impedance $Z_{11}$ may be expressed as

$$\tfrac{1}{2}Z_{11}|\hat{u}_1|^2 = J(0)d, \tag{5.71}$$

in accordance with Eq. (5.44).

Imagine now that we have a slightly flexible wavemaker such that, for a particular value of $k$ (and, hence, of $\omega$), a horizontal oscillation can be realised where

$$c(z) = e(kz). \tag{5.72}$$

Using Eq. (4.76) in Eq. (5.56), we then find that

$$c_0 = -\frac{i}{k}\sqrt{\frac{D(kh)}{2kh}}, \tag{5.73}$$

and, from Eq. (4.67), that $c_n = 0$ for $n \geq 1$. From Eq. (5.65), we then find that

$$R_{11} = \frac{\omega\rho D(kh)d}{2k^2} = \frac{\rho d}{g}v_p^2 v_g \tag{5.74}$$

and, from Eq. (5.66), that $m_{11} = 0$. In such a case, we have wave generation without added mass and without any evanescent wave. The reason is that the far-field solution alone [cf. Eq. (5.51)] matches the wavemaker boundary condition (5.48) on the wavemaker's wet surface, when $c(z)$ is given as in Eq. (5.72).

In constrast, if we are able to realise

$$c(z) = \frac{\cos(m_1 z + m_1 h)}{\cos(m_1 h)}, \tag{5.75}$$

we generate only one evanescent mode and no progressive wave. (Note that, in this case, the lower and upper parts of the vertical wall have to oscillate in opposite phases.) For this example, the radiation resistance vanishes at the particular frequency for which Eq. (5.75) is satisfied. This shows that it is possible, at least in principle, to make oscillations in the water without producing a propagating wave. This knowledge may be of use if it is desired that a body oscillating in the sea with a particular frequency shall not generate undesirable waves.

Finally, let us consider the case of a stiff surging vertical plate. That is, we choose $c(z) = 1$ for $-h < z < 0$. Then, in general, $c_n \neq 0$ for all $n$. In particular, we have

$$c_0 = -\frac{i\omega^2}{gk^3h}\sqrt{\frac{2kh}{D(kh)}}, \tag{5.76}$$

as shown in Problem 5.3. From Eqs. (5.61) and (5.65), we then find that the far-field coefficient is

$$a_1^- = -\frac{2\omega^3}{g^2k^2D(kh)} = -\frac{v_p^2}{gv_g}, \tag{5.77}$$

and the radiation resistance is

$$R_{11} = \frac{2\omega^5\rho d}{g^2k^4D(kh)} = \frac{\rho d}{g}\frac{v_p^4}{v_g}. \tag{5.78}$$

Let us now, as a prelude to sections on reciprocity relations (Sections 5.4 and, in particular, 5.8), establish that the following relations hold in the present example, with respect to the interaction of a surging vertical plate with a wave on one of its sides (the left-hand side, as shown in Figure 5.4). The radiation resistance (5.78) and the excitation force (5.33) may be expressed in terms of the far-field coefficient (5.77) as

$$R_{11} = \frac{\rho g^2 D(kh)d}{2\omega}|a_1^-|^2 = \rho g d v_g|a_1^-|^2, \tag{5.79}$$

$$\hat{F}_{e,1} = -A\frac{\rho g^2 D(kh)d}{\omega}a_1^- = -2\rho g d v_g A a_1^-, \tag{5.80}$$

respectively. Using these relations in Eq. (3.45), we see that at optimum oscillation, the maximum power absorbed by the plate is

$$P_{max} = \frac{|\hat{F}_{e,1}|^2}{8R_{11}} = \frac{\rho g^2 D(kh)d}{4\omega}|A|^2 = \frac{1}{2}\rho g d v_g|A|^2 = Jd, \tag{5.81}$$

which, according to Eq. (4.130), is the complete energy transport of the incident wave in the wave channel of width $d$. In this case, the absorbing plate generates a wave which, in the far-field region, cancels the reflected wave. Thus, $A_r^- = a_1^-\hat{u}_1 = -A$, or

$$\hat{u}_1 = \hat{u}_{1,opt} = -A/a_1^- = -Aa_1^-/|a_1^-|^2 = \hat{F}_{e,1}/2R_{11}, \tag{5.82}$$

in accordance with Eq. (3.46). Note that $a_1^-$ is real in the present example. The results of Eqs. (5.81) and (5.82) are examples of matter discussed in a more general and systematic way in Section 8.3.

### 5.2.4 Other Body Geometries

For a general geometry, it is complicated to solve the radiation problem for an oscillating body; that is, it is complicated to solve the boundary-value problem represented by Eqs. (5.10)–(5.13). A simple example which is amenable to analytical solution was discussed in Section 5.2.3. For almost all other cases, a numerical solution is necessary. In the present subsection, some numerically obtained results will be presented for the radiation impedance and the excitation force of some particular bodies. These quantities are examples of the so-called hydrodynamic parameters.

Let us first consider a sphere of radius $a$ semisubmerged on water of infinite depth. With this body geometry, it is obvious that rotary modes cannot generate any wave in an ideal fluid. This, together with the fact that both $x = 0$ and $y = 0$ are planes of symmetry, means that all off-diagonal elements of the radiation impedance vanish. Thus, the only non-vanishing elements are $Z_{11} = Z_{22}$ and $Z_{33}$, which we may write as

$$Z_{jj} = \tfrac{2}{3}\pi a^3 \rho\omega(\epsilon_{jj} + i\mu_{jj}),\tag{5.83}$$

where $\epsilon$ and $\mu$ are non-dimensionalised radiation resistance and added mass, respectively. These parameters are shown by curves in Figure 5.6, which is based on previous numerical results [46, 47]. We may note that the radiation resistance tends to zero as $ka$ approaches infinity or zero. The added mass is finite in both of these limits. Observe that

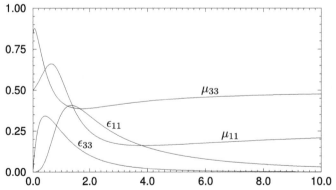

Figure 5.6: Non-dimensionalised surge and heave coefficients of radiation resistance and of added mass versus $ka$ for a semisubmerged sphere of radius $a$ on deep water. Here, $k = \omega^2/g$ is the angular repetency. The numerical values are from Havelock [46] and Hulme [47]. The radiation impedance for surge ($j = 1$) and heave ($j = 3$) is given by $Z_{jj} = \tfrac{2}{3}\pi a^3 \rho\omega(\epsilon_{jj} + i\mu_{jj})$.

$$\frac{R_{jj}}{\omega m_{jj}} = \frac{\epsilon_{jj}}{\mu_{jj}} = \mathcal{O}\{(ka)^n\} \quad \text{as } ka \to 0, \tag{5.84}$$

where $n = 2$ for the surge mode and $n = 1$ for the heave mode. It can be shown (cf. Problem 5.10) that

$$\epsilon_{33} \to (3\pi/4)ka \quad \text{as } ka \to 0. \tag{5.85}$$

When $ka \ll 1$, the diameter is much shorter than the wavelength. Also, for body geometries other than a sphere, it is generally true that the radiation reactance dominates over the radiation resistance when the extension of the body is small compared to the wavelength.

As the next example, we present in Figure 5.7 the excitation-force coefficient

$$f_3 \equiv F_{e,3}/A = \kappa_3 \rho g \pi a^2 \tag{5.86}$$

and the radiation impedance

$$Z_{33} = \tfrac{2}{3}\pi a^3 \rho\omega(\epsilon_{33} + i\mu_{33}) \tag{5.87}$$

for the heave mode of a floating truncated vertical cylinder of radius $a$ and draft $b$. The cylinder axis coincides with the $z$-axis ($x = y = 0$). The numerical values for these hydrodynamic parameters have been computed using a method described by Eidsmoen [48].

We observe from Figure 5.7 that, for low frequencies ($ka \to 0$), the heave excitation force has a magnitude as we could expect from simply applying Archimedes' law, which neglects effects of wave interference and wave diffraction. For high frequencies ($ka \to \infty$), the force tends to zero, which is to be expected because of the mentioned effects, but also because of the decrease of hydrodynamic pressure with increasing submergence below the free water surface. From the lower-left-hand graph, we see that for large frequencies, the curve for the phase approaches a straight line which (in radians) has a steepness equal to 1. This means that the heave excitation force is, for $ka \to \infty$, in phase with the incident wave elevation at $x = -a$ (that is, where the wave first hits the cylinder), whereas for $ka \to 0$, it is in phase with the (undisturbed) incident wave elevation at $x = 0$.

The curves for $\epsilon_{33}$ and $\mu_{33}$ in Figure 5.7 are qualitatively similar to the corresponding curves in Figure 5.6 for the semisubmerged sphere. Note, however, that $\epsilon_{33}$ is, at larger frequencies, significantly smaller for the vertical cylinder. The limiting values for $ka \to 0$ are also different. It can be shown (cf. Problem 5.10) that for finite water depth $h$, $\epsilon_{33} \to 3\pi a/8h$, while for infinite water depth, $\epsilon_{33} \to 3\pi ka/4$, as $ka \to 0$. The reason for the different appearance of the two $\epsilon_{33}$ curves for $ka \to 0$ is that infinite water depth was assumed with the semisubmerged sphere results in Figure 5.6.

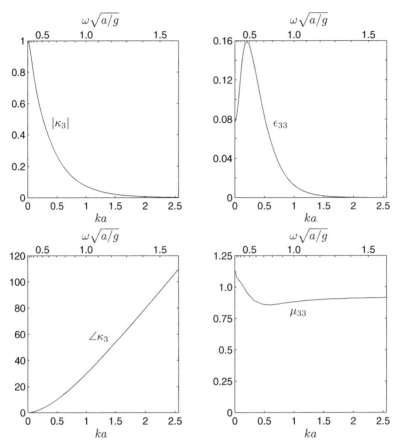

Figure 5.7: Non-dimensional hydrodynamic parameters for the heave mode of a floating truncated vertical cylinder of radius $a$ and draft $b = 1.88a$ on water of depth $h = 15a$. In each graph, the horizontal scales are $ka$ (lower scale) and $\omega\sqrt{a/g}$ (upper scale). The four graphs are, in dimensionless values, the amplitude $|\kappa_3|$ and phase $\angle\kappa_3$ (in degrees) of the excitation force, and the real and imaginary parts of the radiation impedance $\epsilon_{33} + i\mu_{33}$.

The final example for which numerical results are presented in this subsection is the International Ship and Offshore Structures Congress (ISSC) tension-leg platform (TLP) [49] in water of depth $h = 450$ m. The platform consists of four cylindrical columns (each with a radius of 8.435 m) and four pontoons of rectangular cross section (7.5 m wide and 10.5 m high) connecting the columns. The distance between the centres of adjacent columns is $2L = 86.25$ m. The draft of the TLP is 35.0 m, corresponding to the lowermost surface of the columns as well as of the pontoons. This is a large structure with a non-simple geometry. To numerically solve the boundary-value problems as given in Section 5.1.1 or in the first part of Section 4.2, we have applied the computer programme WAMIT [50]. The wet surface of the immersed TLP is approximated to 4,048 plane panels

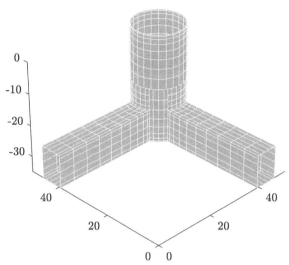

Figure 5.8: One quadrant of the ISSC TLP as approximated by 1,012 panels [50]. The scales on the indicated coordinate system are in metres.

(see Figure 5.8). The structure is symmetric with respect to the planes $x = 0$ and $y = 0$, and the width and length are equal. The computer programme WAMIT provides values for the fluid velocity, hydrodynamic pressure, body parameters (e.g., immersed volume, centre of gravity and hydrostatic coefficients), drift forces and motion response, in addition to the hydrodynamic parameters (i.e., excitation forces and moments, radiation resistance, added masses and inertia moments).

Dimensionless graphs based on the computed results for the hydrodynamic parameters of the TLP body are shown in Figure 5.9. The dimensionless hydrodynamic parameters $A_{jj'}$, $B_{jj'}$ and $X_j$ are defined as follows from the radiation impedance and the excitation force:

$$Z_{jj'} = \omega \rho L^k (B_{jj'} + iA_{jj'}), \tag{5.88}$$
$$\hat{F}_{e,j} = f_j A = \rho g A L^m X_j, \tag{5.89}$$

where $m = 2$ for $j = 1, 3$, while $m = 3$ for $j = 5$, and $k = 3$ for $(j, j') = (1, 1)$ and $(j, j') = (3, 3)$, while $k = 4$ for $(j, j') = (1, 5)$ and $k = 5$ for $(j, j') = (5, 5)$. The excitation force computed here applies for the case when the incident wave propagates in the $x$-direction, and $A$ is the wave elevation at the origin $(x, y) = (0, 0)$.

We observe that the curves in Figure 5.9 have a somewhat wavy nature. This is due to constructive and destructive wave interference associated with different parts of the TLP body. Let us consider, for instance, the surge mode. When the distance $2L$ between adjacent columns is an odd multiple of half wavelengths,

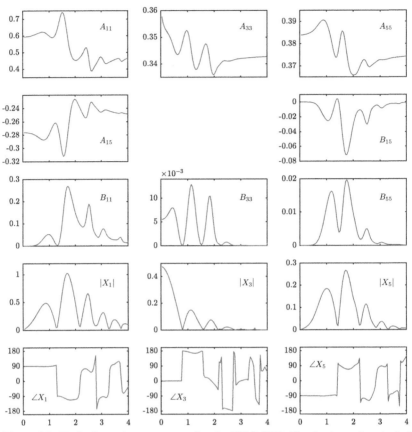

Figure 5.9: Hydrodynamic parameters for the ISSC TLP. The horizontal scale is the non-dimensional frequency $\Omega = \omega\sqrt{L/g}$, where $L = 43.125$ m is half of the distance between centres of adjacent TLP columns and $g = 9.81$ m/s². The eight graphs in the three upper rows give the radiation impedance [cf. Eq. (5.88)], and the six graphs in the two lower rows give the excitation force [cf. Eq. (5.89)] in dimensionless quantities for the modes surge ($j = 1$), heave ($j = 3$) and pitch ($j = 5$). In the lowest row, the phase is given in degrees.

corresponding to values of dimensionless frequency $\Omega = 1.25, 2.17, 2.80, \ldots,$ the surge excitation force contribution to one pair of columns is cancelled by the contribution from the remaining pair. Similarly, generated waves from the two pairs cancel each other. This explains why $|X_1|$ and $B_{11}$ are so small at these frequencies. If, however, the distance $2L$ corresponds to a multiple of wavelengths, which happens for $\Omega = 1.77, 2.51, \ldots,$ then we expect constructive interference, which explains why $|X_1|$ and $B_{11}$ are large at these frequencies. We see similar behaviour with the pitch mode. However, with the heave mode, maxima and minima for $|X_3|$ and $B_{33}$ occur at other frequencies. Graphs similar to those shown in Figure 5.9 have been published for a TLP body with six columns [51]. They show, as we would expect, even more interference effects than disclosed by the wavy nature of the curves in Figure 5.9 for a four-column TLP.

A submerged body of the pontoon's cross section and of total length $8L$ displaces a water mass of 0.339 (in normalised units). It is interesting to note that this value is rather close to the added-mass value $A_{33}$ for heave, as given in Figure 5.9. A deeply submerged horizontal cylinder of infinite length has an added mass for heave, as well as for surge, that equals the mass of the displaced water. If, however, the cylinder is only slightly below the water surface, the added mass may depend strongly on frequency, and in some cases, it may even be negative. Numerical results for the added mass and radiation resistance of a submerged horizontal cylinder, as well as a submerged vertical cylinder, have been published by McIver and Evans [23].

## 5.3 Impulse Response Functions in Hydrodynamics

We have discussed two kinds of wave–body interaction in the frequency domain. First (in Section 5.1), we introduced the excitation force—that is, the wave force on the body when the body is held fixed ($u_j = 0$ for $j = 1, \ldots, 6$); see Eqs. (5.26) and (5.27). Then, in Section 5.2.1, we introduced the hydrodynamic radiation parameters: the radiation impedance $Z_{jj'}$—see Eq. (5.39)—and its real and imaginary parts—that is, the radiation resistance $R_{jj'}$ and the radiation reactance $X_{jj'}$, respectively. Also, the added mass $m_{jj'} = X_{jj'}/\omega$ was introduced; see Eq. (5.42).

To the two kinds of interaction there correspond two kinds of boundary-value problems, which we call the excitation problem and the radiation problem. They have different inhomogeneous boundary conditions on the wet body surface $S$, namely as given by Eqs. (5.27) and (5.10), respectively.

Following the introduction of generalised six-dimensional vectors (in Section 5.1), we may define a six-dimensional column vector $\hat{\mathbf{F}}_e$ for the excitation force's complex amplitude. Similarly, we have a vector $\hat{\mathbf{u}}$, which represents the complex amplitudes of the six components of the velocity of the oscillating body. Correspondingly, we denote the radiation impedance matrix by $\mathbf{Z} = \mathbf{Z}(\omega)$, which is of dimension $6 \times 6$. The reaction force due to the oscillating body's radiation is given by

$$\hat{\mathbf{F}}_r = -\mathbf{Z}\hat{\mathbf{u}} = -\mathbf{Z}(\omega)\hat{\mathbf{u}}, \tag{5.90}$$

which is an alternative way of writing Eq. (5.37). We may interpret $-\mathbf{Z}$ as the transfer function of a linear system, where $\hat{\mathbf{u}}$ is the input and $\hat{\mathbf{F}}_r$ the output. Similar to this definition of a linear system for the radiation problem, we may define the following linear system for the excitation problem:

$$\hat{\mathbf{F}}_e = \mathbf{f}(\omega)A. \tag{5.91}$$

Here the system's input is $A = \hat{\eta}_0(0,0)$, which is the complex elevation amplitude of the (undisturbed) incident wave at the origin $(x, y) = (0, 0)$. Further, the transfer function is the six-dimensional column vector $\mathbf{f}$, which we shall call the excitation-force coefficient vector.

In the following, let us generalise this concept to situations in which the oscillation need not be sinusoidal. Then, referring to Eqs. (2.133), (2.158) and (2.163), we have the Fourier transforms of the reaction force $\mathbf{F}_{r,t}(t)$, due to radiation, and the excitation force $\mathbf{F}_{e,t}(t)$:

$$\mathbf{F}_r(\omega) = -\mathbf{Z}(\omega)\mathbf{u}(\omega), \tag{5.92}$$

$$\mathbf{F}_e(\omega) = \mathbf{f}(\omega)A(\omega). \tag{5.93}$$

Here $\mathbf{u}(\omega)$ is the Fourier transform of $\mathbf{u}_t(t)$, and $A(\omega)$ is the Fourier transform of $a(t) \equiv \eta_0(0,0,t)$, the wave elevation of the undisturbed incident wave at the origin $(x,y) = (0,0)$. (A subscript $t$ is used to denote the inverse Fourier transforms, which are functions of time.)

The inverse Fourier transforms of the transfer functions $\mathbf{f}(\omega)$ and $\mathbf{Z}(\omega)$ correspond to time-domain impulse response functions, introduced into ship hydrodynamics by Cummins [52]. Let us first discuss the causal linear system given by Eq. (5.92). Afterwards, we shall discuss the system given by Eq. (5.93); we shall see that this system may be noncausal.

### 5.3.1 The Kramers–Kronig Relations in Hydrodynamic Radiation

We consider a body which oscillates and thereby generates a radiated wave on otherwise calm water. Let the body's oscillation velocity (in the time domain) be given by $\mathbf{u}_t(t)$. The reaction force $\mathbf{F}_{r,t}(t)$ is given by the convolution product

$$\mathbf{F}_{r,t}(t) = -\mathbf{z}(t) * \mathbf{u}_t(t) \tag{5.94}$$

in the time domain. [See Eq. (2.126), and remember that convolution is commutative. Note, however, that the matrix multiplication implied here is not commutative.] The impulse response matrix $-\mathbf{z}(t)$ is the negative of the inverse Fourier transform of the radiation impedance

$$\mathbf{z}(t) = \mathcal{F}^{-1}\{\mathbf{Z}(\omega)\}. \tag{5.95}$$

Note that this system is causal; that is,

$$\mathbf{z}(t) = 0 \quad \text{for } t < 0, \tag{5.96}$$

which means that there is no output $\mathbf{F}_{r,t}(t)$ before a non-vanishing $\mathbf{u}_t(t)$ is applied as input. For this reason, when we compute the convolution integral in Eq. (5.94), we may apply Eq. (2.172) instead of the more general Eq. (2.126), where the range of integration is from $t = -\infty$ to $t = +\infty$. The fact that $\mathbf{z}(t)$ is a causal function also has some consequence for its Fourier transform

$$\mathbf{Z}(\omega) = \mathbf{R}(\omega) + i\omega\mathbf{m}(\omega), \tag{5.97}$$

namely that the radiation-resistance matrix $\mathbf{R}(\omega)$ and the added-mass matrix $\mathbf{m}(\omega)$ are related by Kramers–Kronig relations. Apparently, Kotik and Mangulis

[53] were the first to apply these relations to a hydrodynamic problem. Later, some consequences of the relations were discussed by, for example, Greenhow [54]. Here let us derive the relations in the following way. First we note that $\mathbf{m}(\omega)$ does not, in general, vanish in the limit $\omega \rightarrow \infty$. We wish to remove this singularity by considering the following related transfer function:

$$\mathbf{H}(\omega) = \mathbf{m}(\omega) - \mathbf{m}(\infty) + \mathbf{R}(\omega)/i\omega. \tag{5.98}$$

Note that $\mathbf{R}(\omega) \rightarrow 0$ as $\omega \rightarrow \infty$. We may assume that $\mathbf{R}(\omega)$ tends sufficiently fast to zero as $\omega \rightarrow 0$ to make $\mathbf{H}(\omega)$ non-singular at $\omega = 0$. The Kramers–Kronig relations (2.197) and (2.199) give for the real and imaginary parts of $\mathbf{H}(\omega)$

$$\mathbf{m}(\omega) - \mathbf{m}(\infty) = -\frac{2}{\pi} \int_0^\infty \frac{\mathbf{R}(y)}{\omega^2 - y^2} \, dy, \tag{5.99}$$

$$\mathbf{R}(\omega) = \frac{2\omega^2}{\pi} \int_0^\infty \frac{\mathbf{m}(y) - \mathbf{m}(\infty)}{\omega^2 - y^2} \, dy. \tag{5.100}$$

By using Eqs. (5.92), (5.97) and (5.98), we find that the reaction force due to radiation is given by

$$\mathbf{F}_r(\omega) = -\mathbf{Z}(\omega)\mathbf{u}(\omega) = \mathbf{F}'_r(\omega) - i\omega\mathbf{m}(\infty)\mathbf{u}(\omega), \tag{5.101}$$

where

$$\mathbf{F}'_r(\omega) = -i\omega\mathbf{H}(\omega)\mathbf{u}(\omega) = -\mathbf{K}(\omega)\mathbf{u}(\omega). \tag{5.102}$$

Here we have introduced an alternative transfer function

$$\begin{aligned}\mathbf{K}(\omega) = i\omega\mathbf{H}(\omega) = \mathbf{Z}(\omega) - i\omega\mathbf{m}(\infty) \\ = \mathbf{R}(\omega) + i\omega[\mathbf{m}(\omega) - \mathbf{m}(\infty)].\end{aligned} \tag{5.103}$$

The corresponding inverse Fourier transforms are

$$\mathbf{F}_{r,t}(t) = \mathbf{F}'_{r,t}(t) - \mathbf{m}(\infty)\dot{\mathbf{u}}_t(t), \tag{5.104}$$

with

$$\mathbf{F}'_{r,t}(t) = -\mathbf{h}(t) * \dot{\mathbf{u}}_t(t) = -\mathbf{k}(t) * \mathbf{u}_t(t), \tag{5.105}$$

where

$$\mathbf{h}(t) = \frac{1}{2\pi} \int_{-\infty}^\infty \mathbf{H}(\omega)e^{i\omega t} \, dt, \tag{5.106}$$

$$\mathbf{k}(t) = \frac{1}{2\pi} \int_{-\infty}^\infty \mathbf{K}(\omega)e^{i\omega t} \, dt. \tag{5.107}$$

Note that

$$\mathbf{k}(t) = \frac{d}{dt}\mathbf{h}(t). \tag{5.108}$$

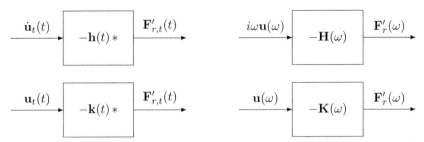

Figure 5.10: Block diagrams of linear systems in the time (left) and frequency (right) domains, where the acceleration (upper) or the velocity (lower) vectors of the oscillating body are considered as input and where a certain part [cf. Eqs. (5.98) and (5.102)] of the radiation reaction force vector is, in both cases, considered as the output.

The linear systems corresponding to the two alternative transfer functions $\mathbf{H}(\omega)$ and $\mathbf{K}(\omega)$—or, equivalently, to the convolutions in Eq. (5.105)—are represented by block diagrams in Figure 5.10.

Because of the condition of causality, we have $\mathbf{h}(t) = 0$ and $\mathbf{k}(t) = 0$ for $t < 0$. Moreover, using Eqs. (2.178), (2.179), (5.98) and (5.103), we find for $t > 0$

$$
\begin{aligned}
\mathbf{h}(t) &= \frac{2}{\pi} \int_0^\infty \left[ \mathbf{m}(\omega) - \mathbf{m}(\infty) \right] \cos(\omega t) \, d\omega \\
&= \frac{2}{\pi} \int_0^\infty \frac{\mathbf{R}(\omega)}{\omega} \sin(\omega t) \, d\omega,
\end{aligned}
\tag{5.109}
$$

$$
\begin{aligned}
\mathbf{k}(t) &= \frac{2}{\pi} \int_0^\infty \mathbf{R}(\omega) \cos(\omega t) \, d\omega = 2\mathcal{F}^{-1}\{\mathbf{R}(\omega)\} \\
&= -\frac{2}{\pi} \int_0^\infty \omega[\mathbf{m}(\omega) - \mathbf{m}(\infty)] \sin(\omega t) \, d\omega.
\end{aligned}
\tag{5.110}
$$

Note that as a consequence of the principle of causality, all information contained in the two real matrix functions $\mathbf{R}(\omega)$ and $\mathbf{m}(\omega)$ is contained alternatively in the single real matrix function $\mathbf{k}(t)$ together with the constant matrix $\mathbf{m}(\infty)$. A function of the type $\mathbf{k}(t)$ has been applied in a time-domain analysis of a heaving body—for instance, by Count and Jefferys [55].

As an example, let us consider the heave mode of the floating truncated vertical cylinder for which the radiation resistance was given in Figure 5.7. The corresponding impulse response function $k_3(t)$ for radiation is given by the curve in the left-hand graph of Figure 5.11.

### 5.3.2 Noncausal Impulse Response for the Excitation Force

The impulse responses corresponding to the radiation impedance $\mathbf{Z}(\omega)$ and the related transfer functions $\mathbf{H}(\omega)$ and $\mathbf{K}(\omega)$, as described earlier, are causal because their inputs (velocity or acceleration of the oscillating body) are the actual causes of their responses (reaction forces from the radiated wave which is generated by the oscillating body).

In contrast, the impulse response functions related to the excitation forces are not necessarily causal, since the output excitation force, as well as the input

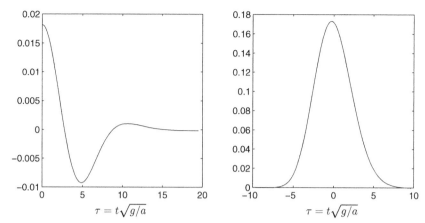

Figure 5.11: Impulse-response functions for the heave mode of a floating truncated vertical cylinder of radius $a$ and draft $b = 1.88a$ on water of depth $h = 15a$. The scales are dimensionless. The dimensionless time on the horizontal scale is $t\sqrt{g/a}$. The curve in the left graph gives the dimensionless impulse-response function $k_3/(\pi\rho g a^2)$ for the radiation problem. The curve in the right graph gives the dimensionless impulse-response function $f_{t,3}/[\pi\rho(ga)^{3/2}]$ for the excitation force [cf. Eqs. (5.110) and (5.112)].

incident wave elevation at the origin (or at some other specified reference point), has a distant primary cause such as a storm or an oscillating wavemaker. The linear system, whose output is the excitation force, has the undisturbed incident wave elevation at the origin of the body as input. However, this wave may hit part of the body and exert a force before the arrival of the wave at the origin [56]. Note, however, that this is not sufficient [37] for a complete explanation of the noncausality of the impulse response function $\mathbf{f}_t(t)$ defined as follows.

The linear system associated with the excitation problem is represented by Eq. (5.93) or its inverse Fourier transform

$$\mathbf{F}_{e,t}(t) = \mathbf{f}_t(t) * a(t)$$
$$= \int_{-\infty}^{\infty} \mathbf{f}_t(t - \tau)a(\tau)\, d\tau = \int_{-\infty}^{\infty} \mathbf{f}_t(\tau)a(t - \tau)\, d\tau, \tag{5.111}$$

where

$$\mathbf{f}_t(t) = \frac{1}{2\pi}\int_{-\infty}^{\infty} \mathbf{f}(\omega)e^{i\omega t}\, d\omega \tag{5.112}$$

is the inverse Fourier transform of the excitation-force-coefficient vector $\mathbf{f}(\omega)$, and where $a(t)$ is the wave elevation due to the (undisturbed) incident wave at the reference point $(x, y) = (x_0, y_0)$ of the body. We may choose this to be the origin—that is, $(x_0, y_0) = (0, 0)$. Note that, as a result of the noncausality of $\mathbf{f}_t(t)$, we could apply neither Eq. (2.178) nor Eq. (2.179) to compute the inverse Fourier transform in Eq. (5.112). Rather, we had to apply Eq. (2.167). When $\mathbf{f}_t(t) \neq 0$ for $t < 0$, it appears from Eq. (5.111) that in order to compute $\mathbf{F}_{e,t}(t)$, future information is required on the wave elevation $a(t)$. For this reason, it may be desirable to predict the incident wave a certain time length $t_1$ into the future, where $t_1$ is sufficiently large to make $\mathbf{f}_t(t)$ negligible for $t < -t_1$.

The statement that $\mathbf{f}_t(t)$ is not, in general, causal is easy to demonstrate by considering the example of Figure 5.4, where the surge excitation-force coefficient per unit width is

$$f_1'(\omega) = \hat{F}_{e,1}'/A = 2\rho(\omega/k)^2,$$

$(5.113)$

according to Eq. (5.33). Note that $k$ is a function of $\omega$ as determined through the dispersion relationship (4.54). In this case, $f_1'(\omega)$ is real and an even function of $\omega$. Hence, the inverse Fourier transform $f_{t,1}(t)$ is an even, and hence noncausal, function of time. Let us, for mathematical simplicity, consider the surge excitation force on an infinitesimal horizontal strip, corresponding to $z = -a_1$ and $z + \Delta z = -a_2$ in Figure 5.4, when the water is deep. Then with $k = \omega^2/g$ and $e(kz) = e^{kz}$ [see Eqs. (4.52) and (4.55)], we have, from Eq. (5.31),

$$\Delta f_1'(\omega) = \Delta F_{e,1}'/A = 2\rho g e^{\omega^2 z/g} \Delta z.$$

$(5.114)$

By taking the inverse Fourier transform [38, formula (1.4.11)], we have the corresponding impulse response

$$\Delta f_{t,1}'(t) = \rho g(-g/\pi z)^{1/2} \exp(gt^2/4z)\Delta z,$$

$(5.115)$

which is evidently noncausal, except for the case of $z = 0$, for which

$$\Delta f_{t,1}'(t) = 2\rho g\delta(t)\Delta z$$

$(5.116)$

(see Section 4.9.2 and Problem 5.8).

Taking the inverse Fourier transform of the heave excitation force given in Figure 5.7 for a floating truncated vertical cylinder, we find for this example the excitation impulse-response function $f_{3,t}$ as shown in the right-hand graph of Figure 5.11. Looking at the curves in this figure, we may say that the causal radiation impulse response 'remembers' roughly 15 to 20 $\sqrt{a/g}$ time units back into the past. In contrast, the noncausal excitation impulse 'remembers' only approximately eight $\sqrt{a/g}$ time units back into the past, but it also 'requires' information on the incident wave elevation approximately seven $\sqrt{a/g}$ time units into the future.

## 5.4 Reciprocity Relations

In this section, let us state and prove some relations (so-called reciprocity relations) which may be very useful when we study the interaction between a wave and an oscillating body.

We have already proven Eq. (5.41), which states that the radiation impedance is symmetric, $Z_{jj'} = Z_{j'j}$ or, in the matrix notation,

$$\mathbf{Z} = \mathbf{Z}^{\mathrm{T}},$$

$(5.117)$

where $\mathbf{Z}^T$ denotes the transpose of matrix $\mathbf{Z}$. It follows that also the radiation resistance matrix and the added mass matrix are symmetric:

$$\mathbf{R} = \mathbf{R}^T, \quad \mathbf{m} = \mathbf{m}^T. \tag{5.118}$$

Reciprocity relations of this type are well known in several branches of physics. An analogy is, for instance, the symmetry of the electric impedance matrix for a linear electric network (see, e.g., section 2.13 in Goldman [57]).

Another analogy, from the subject of statics, is the relationship between a set of forces applied to a body of a linear elastic material and the resulting linear displacements of the points of force attack, where each displacement is in the same direction as the corresponding force. In this case, a reciprocity relation may be derived by considering a force $F$ applied statically to two different points $a$ and $b$. Applied to point $a$, the displacement is $\delta_{aa}$ at point $a$ and $\delta_{ba}$ at point $b$, and the work done is $(F/2)\delta_{aa}$. If applied to point $b$, the displacement is $\delta_{ab}$ at point $a$ and $\delta_{bb}$ at point $b$, and the work done is $(F/2)\delta_{bb}$. Further, if, in addition, an equally large force $F$ is afterwards applied to point $a$, the displacements are $\delta_{ab} + \delta_{aa}$ at point $a$ and $\delta_{bb} + \delta_{ba}$ at point $b$, since the principle of superposition is applicable for the linear system. The work done is

$$W_{ba} = (F/2)\delta_{bb} + (F/2)\delta_{aa} + F\delta_{ba}. \tag{5.119}$$

However, if a force $F$ is first applied to point $a$ and then a second force $F$ applied to point $b$, the work would be

$$W_{ab} = (F/2)\delta_{aa} + (F/2)\delta_{bb} + F\delta_{ab}. \tag{5.120}$$

The work done is stored as elastic energy in the material, and this energy must be independent of the order of applying the two forces—that is, $W_{ba} = W_{ab}$. Hence,

$$\delta_{ab} = \delta_{ba}, \tag{5.121}$$

which proves the following reciprocity relation from the subject of statics: The displacement at point $b$ due to a force at point $a$ equals the displacement at point $a$ if the same force is applied to point $b$ (see, e.g., p. 423 in Timoshenko and Gere [58]).

Several kinds of reciprocity relations relating to hydrodynamics have been formulated and proven by Newman [34]. Some of these relations will be considered in the following.

### 5.4.1 Radiation Resistance in Terms of Far-Field Coefficients

The complex amplitude of the velocity potential of the wave radiated from an oscillating body may, according to Eq. (5.9), be written as

$$\hat{\phi}_r(x, y, z) = \boldsymbol{\varphi}^T(x, y, z)\hat{\mathbf{u}} = \hat{\mathbf{u}}^T \boldsymbol{\varphi}(x, y, z), \tag{5.122}$$

where we have introduced the column vectors $\boldsymbol{\varphi}(x, y, z)$ and $\hat{\mathbf{u}}$, which are six-dimensional if the wave is radiated from a single body oscillating in all its six degrees of freedom [see Eqs. (5.3)–(5.4)]. Note that, because $\hat{\phi}_r$ satisfies the radiation condition of an outgoing wave at infinity, we have [see Eq. (4.212) or Eq. (4.256)] the far-field approximations

$$\hat{\phi}_r \sim A_r(\theta)e(kz)(kr)^{-1/2}e^{-ikr},\tag{5.123}$$

$$\boldsymbol{\varphi} \sim \mathbf{a}(\theta)e(kz)(kr)^{-1/2}e^{-ikr},\tag{5.124}$$

where

$$A_r(\theta) = \mathbf{a}^{\mathrm{T}}(\theta)\hat{\mathbf{u}} = \hat{\mathbf{u}}^{\mathrm{T}}\mathbf{a}(\theta)\tag{5.125}$$

is the far-field coefficient for the radiated wave. Note that each component $\varphi_j$ of the column vector satisfies the radiation condition and that $a_j(\theta)$ is the far-field coefficient for this component. Using relation (4.272), we have the following Kochin functions for the radiated wave:

$$\begin{bmatrix} H_r(\theta) \\ \mathbf{h}(\theta) \end{bmatrix} = \sqrt{2\pi} \begin{bmatrix} A_r(\theta) \\ \mathbf{a}(\theta) \end{bmatrix} e^{i\pi/4}.\tag{5.126}$$

Here $\mathbf{h}(\theta)$ and $\mathbf{a}(\theta)$ are column vectors composed of the components $h_j(\theta)$ and $a_j(\theta)$, respectively.

As a further preparation for proof of reciprocity relations involving radiation parameters, let us introduce the following matrix version of Eq. (4.230):

$$\mathbf{I}(\boldsymbol{\varphi}, \boldsymbol{\varphi}^{\mathrm{T}}) = \iint_S \left( \boldsymbol{\varphi} \frac{\partial \boldsymbol{\varphi}^{\mathrm{T}}}{\partial n} - \frac{\partial \boldsymbol{\varphi}}{\partial n} \boldsymbol{\varphi}^{\mathrm{T}} \right) dS.\tag{5.127}$$

Because $\boldsymbol{\varphi}$ satisfies the radiation condition, we may apply Eq. (4.239), which means that

$$\mathbf{I}(\boldsymbol{\varphi}, \boldsymbol{\varphi}^{\mathrm{T}}) = 0.\tag{5.128}$$

However, the integral does not vanish when we take the complex conjugate of one of the two functions entering into the integral. Thus, the complex conjugate of Eqs. (4.244) and (4.245) may in matrix notation be rewritten as

$$\begin{aligned} \mathbf{I}(\boldsymbol{\varphi}^*, \boldsymbol{\varphi}^{\mathrm{T}}) &= -i\frac{D(kh)}{k} \int_0^{2\pi} \mathbf{a}^*(\theta)\mathbf{a}^{\mathrm{T}}(\theta)\, d\theta \\ &= -\lim_{r\to\infty} 2 \iint_{S_\infty} \frac{\partial \boldsymbol{\varphi}^*}{\partial r} \boldsymbol{\varphi}^{\mathrm{T}}\, dS. \end{aligned}\tag{5.129}$$

Note that $\mathbf{I}(\boldsymbol{\varphi}^*, \boldsymbol{\varphi}^{\mathrm{T}})$, which should not be confused with the identity matrix $\mathbf{I}$, is a square matrix, which is of dimension 6×6 for the case of one (three-dimensional) single body oscillating in all its degrees of freedom.

As Eq. (5.35) has been generalised to Eq. (5.122), it follows that expression (5.40) for the radiation impedance may be rewritten in matrix notation as

$$\mathbf{Z} = -i\omega\rho \iint_S \frac{\partial \boldsymbol{\varphi}^*}{\partial n} \boldsymbol{\varphi}^T \, dS. \tag{5.130}$$

The complex conjugate of this matrix is

$$\mathbf{Z}^* = i\omega\rho \iint_S \frac{\partial \boldsymbol{\varphi}}{\partial n} \boldsymbol{\varphi}^\dagger \, dS, \tag{5.131}$$

where $\boldsymbol{\varphi}^\dagger = (\boldsymbol{\varphi}^T)^*$ is the complex-conjugate transpose of $\boldsymbol{\varphi}$. Note, however, that $\mathbf{Z}^T = \mathbf{Z}$ [see Eq. (5.117)], and hence,

$$\mathbf{Z}^* = \mathbf{Z}^\dagger = i\omega\rho \iint_S \boldsymbol{\varphi}^* \frac{\partial \boldsymbol{\varphi}^T}{\partial n} \, dS. \tag{5.132}$$

The real part of $\mathbf{Z}$ is the radiation resistance matrix

$$
\begin{aligned}
\mathbf{R} &= \frac{1}{2}(\mathbf{Z} + \mathbf{Z}^*) = \frac{1}{2}(\mathbf{Z} + \mathbf{Z}^\dagger) \\
&= -\frac{i}{2}\omega\rho \iint_S \left( \frac{\partial \boldsymbol{\varphi}^*}{\partial n} \boldsymbol{\varphi}^T - \boldsymbol{\varphi}^* \frac{\partial \boldsymbol{\varphi}^T}{\partial n} \right) dS = \frac{i}{2}\omega\rho \, \mathbf{I}(\boldsymbol{\varphi}^*, \boldsymbol{\varphi}^T),
\end{aligned}
\tag{5.133}
$$

where Eq. (5.127) has been used. (It has been assumed that $\omega$ is real.) Now, using Eq. (5.129) in Eq. (5.133), we have

$$\mathbf{R} = -i\omega\rho \lim_{r \to \infty} \iint_{S_\infty} \frac{\partial \boldsymbol{\varphi}^*}{\partial r} \boldsymbol{\varphi}^T \, dS. \tag{5.134}$$

It is interesting to compare this with Eq. (5.130). If the integral is taken over $S_\infty$ in the far field instead of over the wet surface $S$ of the body, we arrive at the radiation resistance matrix instead of the radiation impedance matrix. This indicates that the radiation reactance and hence the added mass are somehow related to the near-field region of the wave-generating oscillating body.

The radiation resistance matrix may be expressed in terms of the far-field coefficient vector $\mathbf{a}(\theta)$ of the radiated wave. From Eqs. (5.126), (5.129) and (5.133), we obtain

$$\mathbf{R} = \frac{\omega\rho D(kh)}{2k} \int_0^{2\pi} \mathbf{a}^*(\theta)\, \mathbf{a}^T(\theta) \, d\theta = \frac{\omega\rho D(kh)}{4\pi k} \int_0^{2\pi} \mathbf{h}^*(\beta)\, \mathbf{h}^T(\beta) \, d\beta. \tag{5.135}$$

We have here been able to express the radiation resistance matrix in terms of far-field quantities of the radiated wave. The physical explanation of this is that the energy radiated from the oscillating body must be retrieved without loss in the far-field region of the assumed ideal fluid. It is possible to derive Eq. (5.135) simply by using this energy argument as a basis for the derivation (see Problem 5.7).

Whereas the radiation resistance is related to energy in the far-field region, the radiation reactance and, hence, the added mass are somehow related to the near-field region, as we remarked earlier when comparing Eqs. (5.130) and (5.134). We shall return to this matter in a more quantitative fashion in Section 5.5.4.

### 5.4.2 The Excitation Force: The Haskind Relation

Some important relations which express the excitation force in terms of radiation parameters are derived in the following. Using the wet-surface conditions (5.10) and (5.27), we may rewrite expression (5.26) for the excitation force as

$$
\begin{aligned}
\hat{F}_{e,j} &= i\omega\rho \iint_S \left[ (\hat{\phi}_0 + \hat{\phi}_d)\frac{\partial\varphi_j}{\partial n} - \varphi_j\frac{\partial}{\partial n}(\hat{\phi}_0 + \hat{\phi}_d) \right] dS \\
&= i\omega\rho I(\hat{\phi}_0 + \hat{\phi}_d, \varphi_j) \\
&= i\omega\rho I(\hat{\phi}_0, \varphi_j) + i\omega\rho I(\hat{\phi}_d, \varphi_j),
\end{aligned}
\tag{5.136}
$$

where integral (4.230) has been used. Further, since $\hat{\phi}_d$ and $\varphi_j$ both satisfy the radiation condition, which means that $I(\hat{\phi}_d, \varphi_j) = 0$ according to Eq. (4.239), we obtain

$$
\hat{F}_{e,j} = i\omega\rho I(\hat{\phi}_0, \varphi_j).
\tag{5.137}
$$

In order to calculate the excitation force from the formula (5.26), we need to know the diffraction potential at the wet surface $S$. Alternatively, the excitation force may be calculated from Eq. (5.137), where we need to know not the diffraction potential $\phi_d$ but the radiation potential $\varphi_j$, either at the wet surface $S$ or at the control surface $S_\infty$ in the far-field region [see Eq. (4.233)]. This latter method of calculating the excitation force was first pointed out by Haskind [59, 60] for the case of a single body. Equation (5.137) is one way of formulating the so-called Haskind relation.

Using the column vector $\varphi$, which we introduced in Eq. (5.122), we may, as in Eq. (5.91), write the excitation force as a column vector

$$
\hat{\mathbf{F}}_e = \mathbf{f}A = i\omega\rho\mathbf{I}(\hat{\phi}_0, \varphi)
\tag{5.138}
$$

as an alternative to Eq. (5.137). We have also here introduced the excitation-force coefficient vector $\mathbf{f}$.

In another version of the Haskind relation, the excitation force may also be expressed in terms of the radiated wave's far-field coefficients or its Kochin functions. We assume now that the incident plane wave is given as

$$
\hat{\eta}_0 = A \exp[-ik(x\cos\beta + y\sin\beta)] = Ae^{-ikr(\beta)}.
\tag{5.139}
$$

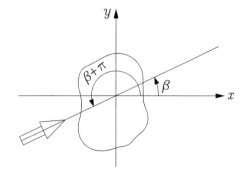

Figure 5.12: Excitation force experienced by an oscillating body due to a wave with angle of incidence $\beta$ is related to the body's ability to radiate a wave in the opposite direction, $\theta = \beta + \pi$.

Here, $\beta$ is the angle of wave incidence. See Eqs. (4.98) and (4.267). Note that the frequency-dependent excitation-force coefficient vector $\mathbf{f} = \hat{\mathbf{F}}_e/A$ is a function also of $\beta$—that is, $\mathbf{f} = \mathbf{f}(\beta)$. Applying now Eq. (4.271) with $\phi_i = \hat{\phi}_0$ (that is, $A_i = A$ and $\beta_i = \beta$) and with $\phi_j = \hat{\phi}_r$ (that is, $A_j = 0$), we have

$$I(\hat{\phi}_0, \hat{\phi}_r) = \frac{gD(kh)}{i\omega k} H_r(\beta \pm \pi)A. \tag{5.140}$$

Further, using this together with Eqs. (5.122), (5.125) and (5.126) in Eq. (5.138), we may write the excitation-force vector as

$$\hat{\mathbf{F}}_e = \mathbf{f}(\beta)A = i\omega\rho\mathbf{I}(\hat{\phi}_0, \boldsymbol{\varphi}) = \frac{\rho g D(kh)}{k}\mathbf{h}(\beta \pm \pi)A, \tag{5.141}$$

or for the excitation-force coefficient vector,

$$\mathbf{f}(\beta) = \frac{\rho g D(kh)}{k}\mathbf{h}(\beta \pm \pi) = \frac{\rho g D(kh)}{k}\sqrt{2\pi}\,\mathbf{a}(\beta \pm \pi)e^{i\pi/4}. \tag{5.142}$$

Reciprocity relation (5.142), which is another form of the Haskind relation, means that a body's ability to radiate a wave *into* a certain direction is related to the excitation force which the body experiences when a plane wave is incident *from* just that direction. See Figure 5.12. If we, in an experiment, measure the excitation-force coefficients for various directions of wave incidence $\beta$, we can also, by application of the reciprocity relation (5.142), obtain the corresponding far-field coefficients for waves radiated in the opposite directions.

### 5.4.3 Reciprocity Relation between Radiation Resistance and Excitation Force

By using the Haskind relation in the form (5.142) together with reciprocity relation (5.135) between the radiation resistance and the far-field coefficients, we find

$$\mathbf{R} = \frac{\omega k}{4\pi\rho g^2 D(kh)}\int_{-\pi}^{\pi}\mathbf{f}(\beta)\mathbf{f}^\dagger(\beta)\,d\beta = \frac{\omega k}{4\pi\rho g^2 D(kh)}\int_{-\pi}^{\pi}\mathbf{f}^*(\beta)\mathbf{f}^\mathrm{T}(\beta)\,d\beta. \tag{5.143}$$

Here we have also utilised the fact that $\mathbf{R}$ is a symmetrical matrix; that is, it equals its own transpose. Remembering that $\mathbf{f}(\beta) = \hat{\mathbf{F}}_e(\beta)/A$, we may rewrite Eq. (5.143) as

$$\mathbf{R} = \frac{\omega k}{4\pi\rho g^2 D(kh)|A|^2} \int_{-\pi}^{\pi} \hat{\mathbf{F}}_e(\beta)\,\hat{\mathbf{F}}_e^\dagger(\beta)\,d\beta. \tag{5.144}$$

Introducing the wave-energy transport $J$ per unit frontage [see Eq. (4.130)], we may also write the radiation resistance matrix as

$$\mathbf{R} = \frac{k}{16\pi J} \int_{-\pi}^{\pi} \hat{\mathbf{F}}_e(\beta)\,\hat{\mathbf{F}}_e^\dagger(\beta)\,d\beta. \tag{5.145}$$

If the body is axisymmetric, the heave component $F_{e,3}$ of the excitation force is independent of $\beta$ (the angle of wave incidence), and we have

$$R_{33} = \frac{k}{8J}|\hat{F}_{e,3}|^2. \tag{5.146}$$

The coefficient in front of the integral in Eqs. (5.144) and (5.145) may also be written as $k/(8\pi\rho g v_g |A|^2)$ if Eq. (4.106) for the group velocity $v_g$ is used.

### 5.4.4 Reciprocity Relation between Diffraction and Radiation Kochin Functions

A relation which may prove useful when considering wave interactions in the far field relates the diffraction and radiation Kochin functions. The starting point is the integral $I(\hat{\phi}_0 + \hat{\phi}_d, \varphi_j^*)$, which is a special case of integral (4.292):

$$I(\hat{\phi}_0 + \hat{\phi}_d, \varphi_j^*) = \iint_S \left[ (\hat{\phi}_0 + \hat{\phi}_d)\frac{\partial \varphi_j^*}{\partial n} - \varphi_j^* \frac{\partial}{\partial n}(\hat{\phi}_0 + \hat{\phi}_d) \right] dS. \tag{5.147}$$

Before we proceed to evaluate this integral, observe that the body boundary condition (5.10) requires that $\partial \varphi_j/\partial n = n_j$ on $S$, where $n_j$ is the $j$-component of $\mathbf{n}$. Since $n_j$ is real, we have $\partial \varphi_j^*/\partial n = \partial \varphi_j/\partial n$ on $S$. Also, $\partial(\hat{\phi}_0 + \hat{\phi}_d)/\partial n = 0$ on $S$, due to boundary condition (5.27). Thus, we have the following identity:

$$I(\hat{\phi}_0 + \hat{\phi}_d, \varphi_j^*) = I(\hat{\phi}_0 + \hat{\phi}_d, \varphi_j) = I(\hat{\phi}_0, \varphi_j), \tag{5.148}$$

where the last equality follows from the fact that $I(\hat{\phi}_d, \varphi_j) = 0$ because $\hat{\phi}_d$ and $\varphi_j$ satisfy the same radiation condition [cf. Eq. (4.239)].

We may express both sides of the identity in terms of Kochin functions. For the right-hand side, we have, from Eqs. (4.265)–(4.271),

$$I(\hat{\phi}_0, \varphi_j) = \frac{2v_p v_g A}{i\omega} h_j(\beta \pm \pi). \tag{5.149}$$

Likewise, for the left-hand side, we have, from Eqs. (4.288) and (4.292)–(4.295),

$$I(\hat{\phi}_0 + \hat{\phi}_d, \varphi_j^*) = I(\hat{\phi}_0, \varphi_j^*) + I(\hat{\phi}_d, \varphi_j^*)$$

$$= \frac{2v_p v_g A}{i\omega} h_j^*(\beta) + \frac{iv_p v_g}{\pi g} \int_0^{2\pi} H_d(\theta) h_j^*(\theta) \, d\theta. \tag{5.150}$$

Equating the two sides yields

$$h_j(\beta \pm \pi) = h_j^*(\beta) - \frac{\omega}{2\pi g} \int_0^{2\pi} \frac{H_d(\theta)}{A} h_j^*(\theta) \, d\theta. \tag{5.151}$$

Multiplying by $u_j^*$ and taking the sum, we have

$$\bar{H}_r(\beta \pm \pi) = H_r^*(\beta) - \frac{\omega}{2\pi g} \int_0^{2\pi} \frac{H_d(\theta)}{A} H_r^*(\theta) \, d\theta, \tag{5.152}$$

where

$$\bar{H}_r(\theta) \equiv \sum_j h_j(\theta) u_j^*. \tag{5.153}$$

## 5.5 Several Bodies Interacting with Waves

So far in this chapter, we have studied the interaction between waves and a body which may oscillate in six independent modes, as introduced in Section 5.1. There are two different kinds of interaction, as represented by two hydrodynamic parameters: the excitation force and the radiation impedance.

Let us now consider the interaction between waves and an arbitrary number of oscillating bodies, partly or totally submerged in water (see Figure 5.13). A systematic study of many hydrodynamically interacting bodies was made independently by Evans [61] and Falnes [25], following Budal's analysis [62] of wave-energy absorption by such a system of several bodies.

First, in the following subsection, we shall introduce the excitation force and the radiation impedance in a phenomenological way. Later, we shall relate these phenomenologically defined parameters with hydrodynamic theory.

In general, each oscillating body has six degrees of freedom: three translatory modes and three rotary modes. We chose a coordinate system with the $z$-axis pointing upwards and with the plane $z = 0$ coinciding with the mean free water surface. We consider each body as corresponding to six *oscillators*. Thus, if the number of oscillating bodies is $N$, there are $6N$ oscillators. Parameters pertaining to a particular oscillator are denoted by a subscript

$$i = 6(p - 1) + j. \tag{5.154}$$

Here, $p$ is the number of the body, and $j = 1, 2, \ldots, 6$ is its mode number.

With an assumed angular frequency $\omega$, the state of an oscillator is given by amplitude and phase, two real quantities which are conveniently incorporated

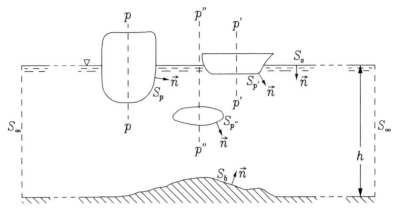

Figure 5.13: System of interacting oscillating bodies contained within an imaginary cylindrical control surface $S_\infty$ at water depth $h$. Vertical lines through (the average position of) the centre of gravity of the bodies are indicated by $p–p$, $p'–p'$ and $p''–p''$. Wet surfaces of oscillating bodies are indicated by $S_p$, $S_{p'}$ and $S_{p''}$. Fixed surfaces, including the sea bed, are given by $S_b$, while $S_0$ denotes the external free water surface. The arrows indicate unit normals pointing into the fluid region.

into a single complex quantity, the complex   amplitude. We specify the state of oscillator number $i$ by a complex velocity amplitude $\hat{u}_i$ or, alternatively, by $\hat{u}_{pj}$, which is a translatory velocity for the modes surge, sway and heave ($j = 1, 2, 3$), or an angular velocity for the modes roll, pitch and yaw ($j = 4, 5, 6$). See Section 5.1.1.

### 5.5.1 Phenomenological Discussion

When no oscillators are moving, an excitation force, represented by its complex amplitude $\hat{F}_{e,i}$, acts on oscillator number $i$. This excitation force is caused by an incoming wave and includes diffraction effects due to the fixed bodies. When oscillator number $i'$ oscillates with a complex velocity amplitude $\hat{u}_{i'}$, it radiates a wave acting on oscillator $i$ with an additional force, which has a complex amplitude $-Z_{ii'}\hat{u}_{i'}$. The complex coefficient $-Z_{ii'}$ is a factor of proportionality, and it depends on $\omega$ and on the geometry of the problem. Based on the principle of superposition, the total force acting on oscillator number $i$ is given as

$$\hat{F}_{t,i} = \hat{F}_{e,i} - \sum_{i'} Z_{ii'}\hat{u}_{i'} = f_i A - \sum_{i'} Z_{ii'}\hat{u}_{i'}, \qquad (5.155)$$

where the sum is taken over all oscillators, including oscillator number $i$.

The set of complex amplitudes of the total force, the excitation force and the oscillator velocity may be assembled into column vectors, $\hat{\mathbf{F}}_t$, $\hat{\mathbf{F}}_e$ and $\hat{\mathbf{u}}$, respectively. For all oscillators, the set of equations (5.155) may be written in matrix form as

$$\hat{\mathbf{F}}_t = \hat{\mathbf{F}}_e - \mathbf{Z}\hat{\mathbf{u}} = \mathbf{f}A - \mathbf{Z}\hat{\mathbf{u}}. \qquad (5.156)$$

Here $\mathbf{Z}$ is a square matrix composed of elements $Z_{ii'}$.

This complex matrix $\mathbf{Z}$ may be decomposed into its real part $\mathbf{R}$, the radiation resistance matrix, and its imaginary part $\mathbf{X}$, the radiation reactance matrix. Thus,

$$\mathbf{Z} = \mathbf{R} + i\mathbf{X} = \mathbf{R} + i\omega\mathbf{m}. \tag{5.157}$$

Here $\omega$ is the angular frequency, and $\mathbf{m}$ is the so-called hydrodynamic added-mass matrix.

In general, $\mathbf{Z}$ is a $6N \times 6N$ matrix. It may be partitioned into $6 \times 6$ matrices $\mathbf{Z}_{pp'}$, defined as follows. Let us assume that body number $p'$ is oscillating with complex velocity amplitudes, as given by the generalised six-dimensional vector $\hat{\mathbf{u}}_{p'}$. This results in a contribution $-\mathbf{Z}_{pp'}\hat{\mathbf{u}}_{p'}$ to the generalised six-dimensional vector, which represents the complex amplitudes of the radiation force on body number $p$. For the particular case of $p' = p$, $\mathbf{Z}_{pp'} = \mathbf{Z}_{pp}$ represents the ordinary radiation impedance matrix for body number $p$, but modified due to the presence of all the other bodies. The radiation impedance matrix for the whole system of the $N$ oscillating bodies is, thus, a partitioned matrix of the type

$$\mathbf{Z} = \begin{bmatrix} \ddots & \vdots & & \vdots & \\ \cdots & \mathbf{Z}_{pp} & \cdots & \mathbf{Z}_{pp'} & \cdots \\ & \vdots & \ddots & \vdots & \\ \cdots & \mathbf{Z}_{p'p} & \cdots & \mathbf{Z}_{p'p'} & \cdots \\ & \vdots & & \vdots & \ddots \end{bmatrix}. \tag{5.158}$$

In this way, the matrix $\mathbf{Z}$ is composed of $N^2$ matrices.

If $p' \neq p$, and if the distance $d_{pp'}$ between the two bodies' vertical reference axes $p$–$p$ and $p'$–$p'$ (see Figure 5.13) is sufficiently large, we expect that matrix $\mathbf{Z}_{pp'}$ depends on the distance $d_{pp'}$ approximately as

$$\mathbf{Z}_{pp'} \sim \left(\mathbf{Z}_{pp'}\right)_0 \sqrt{d_0/d_{pp'}} \, \exp(-ikd_{pp'}), \tag{5.159}$$

where $\left(\mathbf{Z}_{pp'}\right)_0$ is independent of $d_{pp'}$. Compare this with approximation (4.212) or (4.236), and remember that in linear theory, the wave force contribution on a body is proportional to the corresponding wave hitting the body. We expect that approximation (5.159) might be useful when the spacing $d_{pp'}$ is large in comparison with the wavelength and with the largest horizontal extension (diameter) of bodies $p$ and $p'$.

### 5.5.2 Hydrodynamic Formulation

In general, the velocity potential is composed of three main contributions:

$$\phi = \phi_0 + \phi_d + \phi_r. \tag{5.160}$$

Here $\phi_0$ represents the given incident wave, which results in a diffracted wave $\phi_d$ when all bodies are fixed. If the bodies are oscillating, then, additionally, a radiated wave $\phi_r$ is set up.

Note that, in general, $\phi_0$ is the given velocity potential in case all the $N$ bodies are absent. It could, for instance, represent a plane wave with complex amplitude, as given by Eqs. (4.88) and (4.98), or alternatively by Eq. (4.80). In the former case, it is a propagating incident wave. In the latter case, it is a plane wave which includes also a wave which is partly reflected from, for example, a straight coast line. If the coast topography is more irregular, the mathematical description of $\phi_0$ would be more complicated. The term $\phi_d$ in Eq. (5.160) thus represents diffraction due to the immersed $N$ bodies only.

The velocity potential associated with the radiated wave has a complex amplitude

$$\hat{\phi}_r = \sum_i \varphi_i \hat{u}_i, \tag{5.161}$$

where the sum is taken over all oscillators. The complex coefficient $\varphi_i$ represents the velocity potential resulting from a unit velocity amplitude of oscillator number $i$, when all other oscillators are not moving. For one single body, we introduced the six-dimensional column vector $\boldsymbol{\varphi}$ and its transpose $\boldsymbol{\varphi}^T$ in connection with Eqs. (5.122) and (5.127). If we extend the dimension to $6N$, the vector includes all $\varphi_i$ as its components. Then we may write Eqs. (5.161) as

$$\hat{\phi}_r = \boldsymbol{\varphi}^T \hat{\mathbf{u}} = \hat{\mathbf{u}}^T \boldsymbol{\varphi}. \tag{5.162}$$

For each of the bodies, we may define corresponding six-dimensional column vectors $\boldsymbol{\varphi}_p$ and $\mathbf{u}_p$. Then we may write

$$\hat{\phi}_r = \sum_{p=1}^N \boldsymbol{\varphi}_p^T \hat{\mathbf{u}}_p = \sum_{p=1}^N \hat{\mathbf{u}}_p^T \boldsymbol{\varphi}_p. \tag{5.163}$$

At large horizontal distance $r_p$ from the reference axis $p$–$p$ of body number $p$, we have, according to Eq. (4.255), an asymptotic expression

$$\boldsymbol{\varphi}_p \sim \mathbf{b}_p(\theta_p) e(kz) (kr_p)^{-1/2} \exp\{-ikr_p\}. \tag{5.164}$$

Here $\mathbf{b}_p(\theta_p)$ is a six-dimensional vector composed of body number $p$'s six far-field coefficients referred to the local origin (the point where the axis $p$–$p$ crosses the plane of the mean water surface). See also Figure 4.9 (with subscript $p$ instead of $i$) for a definition of the local coordinates $r_p$ and $\theta_p$.

All terms in Eq. (5.160) satisfy the Laplace equation (4.14) in the fluid domain and the usual homogeneous boundary conditions (4.27) and (4.16) or (4.17) on the free surface, $z = 0$, and on the sea bed. The radiation condition of outgoing waves at infinity has to be satisfied for the velocity potentials $\phi_d$, $\phi_r$

and all $\varphi_i$. The inhomogeneous boundary condition (4.15) on the wet surfaces $S_p$ of the oscillating bodies is satisfied, if

$$\frac{\partial}{\partial n}(\hat{\phi}_0 + \hat{\phi}_d) = 0 \quad \text{on all } S_p, \tag{5.165}$$

$$\frac{\partial \varphi_i}{\partial n} = \begin{cases} n_{pj} & \text{on } S_p \\ 0 & \text{on } S_{p'}\,(p' \neq p) \end{cases}, \tag{5.166}$$

where $\partial/\partial n$ is the normal derivative in the direction of the outward unit normal $\vec{n}$ to the surface of the body. In accordance with Eq. (5.154), $n_{pj}$ is defined as the $x$, $y$ or $z$ component of $\vec{n}$ when $j = 1, 2$ or 3, respectively. Furthermore, $n_{pj}$ is the $x$, $y$ or $z$ component of vector $\vec{s}_p \times \vec{n}$ when $j = 4, 5$ or 6, respectively. Here $\vec{s}_p$ is the position vector referred to a selected point on the reference axis $p$–$p$ of body number $p$—for instance, the centre of gravity or the centre of buoyancy. Note that Eqs. (5.165) and (5.166) represent a generalisation of the corresponding boundary conditions (5.10) and (5.27) for the single body.

### 5.5.3 Radiation Impedance and Radiation Resistance Matrices

According to Eq. (5.21), the force on oscillator number $i$ due to the potential $\varphi_{i'}$ (that is, due to a unit velocity amplitude of oscillator number $i'$) is

$$-Z_{ii'} = -\iint_{S_p} p_{i'} n_{pj}\, dS_p = i\omega\rho \iint_S \varphi_{i'}\frac{\partial \varphi_i}{\partial n}\, dS, \tag{5.167}$$

where $p_{i'}$ represents the hydrodynamic pressure corresponding to $\varphi_{i'}$. The last integral, which represents a generalisation of Eq. (5.39), is, by virtue of Eq. (5.166), to be taken over $S_p$ or over the totality of wet body surfaces

$$S = \sum_{p=1}^{N} S_p. \tag{5.168}$$

Since $\partial \varphi_i/\partial n$ is real everywhere on S according to boundary condition (5.166), we may use $\partial \varphi_i/\partial n$ or $\partial \varphi_i^*/\partial n$ as we wish in the preceding integrand. Note that Eq. (5.167) is valid for $i' = i$ as well as for $i' \neq i$. When $i' \neq i$, oscillator $i'$ may pertain to any of the oscillating bodies, including body number $p$.

Matrix $\mathbf{Z}_{pp'}$, representing the radiation interaction between bodies $p$ and $p'$ [see Eq. (5.158)], may be written in terms of the (generally six-dimensional) column vectors $\boldsymbol{\varphi}_p$ and $\boldsymbol{\varphi}_{p'}$ as

$$\mathbf{Z}_{pp'} = -i\omega\rho \iint_S \frac{\partial \boldsymbol{\varphi}_p}{\partial n}\boldsymbol{\varphi}_{p'}^{\mathrm{T}}\, dS. \tag{5.169}$$

In the integrand here, we may replace $\boldsymbol{\varphi}_p$ by $\boldsymbol{\varphi}_p^*$.

The radiation impedance matrix for the total system of $N$ oscillating bodies may be written as

$$\mathbf{Z} = -i\omega\rho \iint_S \frac{\partial\varphi}{\partial n} \varphi^T \, dS. \tag{5.170}$$

If in the integrand we replace $\partial\varphi/\partial n$ with $\partial\varphi^*/\partial n$, we get the alternative expression

$$\mathbf{Z} = -i\omega\rho \iint_S \frac{\partial\varphi^*}{\partial n} \varphi^T \, dS, \tag{5.171}$$

which is a generalisation of Eq. (5.40).

Next, we show that the radiation impedance matrix $\mathbf{Z}$ is symmetric; that is,

$$\mathbf{Z} = \mathbf{Z}^T. \tag{5.172}$$

From Eqs. (4.230) and (5.170) [see also Eq. (5.127)], it follows that

$$\mathbf{Z} - \mathbf{Z}^T = -i\omega\rho \iint_S \left( \frac{\partial\varphi}{\partial n} \varphi^T - \varphi \frac{\partial\varphi^T}{\partial n} \right) dS = i\omega\rho \, \mathbf{I}(\varphi, \varphi^T) = 0, \tag{5.173}$$

where the last equality follows from Eq. (4.239) [see also Eq. (5.128)], since all components of vector $\varphi$ satisfy the same radiation condition at infinity. We may note that reciprocity relation (5.172) is a generalisation of Eq. (5.41) or (5.117). Taking a look at Eq. (5.158), we see that Eq. (5.172) means

$$\mathbf{Z}_{p'p} = \mathbf{Z}_{pp'}^T. \tag{5.174}$$

Note that matrix $\mathbf{Z}_{p'p}$ is not necessarily symmetric if $p' \neq p$. Physically, this means that the $i$ component of the force on body $p'$ due to body $p$ oscillating with a unit amplitude in mode $j$ is the same as the $j$ component of the force on body $p$ due to body $p'$ oscillating with a unit amplitude in mode $i$ but not the same as the $j$ component of the force on body $p'$ due to body $p$ oscillating with a unit amplitude in mode $i$. Moreover, from Eq. (5.172), it follows that

$$\mathbf{Z}^* = (\mathbf{Z}^T)^* = \mathbf{Z}^\dagger. \tag{5.175}$$

Splitting the radiation impedance matrix into real and imaginary parts, as in Eq. (5.157), we may write the radiation resistance matrix as

$$\begin{aligned}
\mathbf{R} &= \frac{1}{2}(\mathbf{Z} + \mathbf{Z}^*) = \frac{1}{2}(\mathbf{Z} + \mathbf{Z}^\dagger) = -\frac{1}{2}i\omega\rho \iint_S \left( \frac{\partial\varphi^*}{\partial n} \varphi^T - \varphi^* \frac{\partial\varphi^T}{\partial n} \right) dS \\
&= \frac{i}{2}\omega\rho \, \mathbf{I}(\varphi^*, \varphi^T) = -\frac{i}{2}\omega\rho \, \mathbf{I}(\varphi, \varphi^\dagger).
\end{aligned} \tag{5.176}$$

Here we have used Eqs. (5.171), (5.172) and (5.175). Note that Eq. (5.176) is an extension of Eq. (5.133) from the case of $6 \times 6$ matrix to the case of $6N \times 6N$ matrix. In an analogous way, we may extend Eqs. (5.134) and (5.135). Thus, we may also write the $6N \times 6N$ radiation resistance matrix as

$$\mathbf{R} = -i\omega\rho \lim_{r \to \infty} \iint_{S_\infty} \frac{\partial \varphi^*}{\partial r} \varphi^{\mathrm{T}} \, dS \tag{5.177}$$

or as

$$\mathbf{R} = \frac{\omega\rho D(kh)}{2k} \int_0^{2\pi} \mathbf{a}^*(\theta)\mathbf{a}^{\mathrm{T}}(\theta) \, d\theta = \frac{\omega\rho D(kh)}{4\pi k} \int_0^{2\pi} \mathbf{h}^*(\theta)\mathbf{h}^{\mathrm{T}}(\theta) \, d\theta, \tag{5.178}$$

where $\mathbf{a}(\theta)$ and $\mathbf{h}(\theta)$ are $6N$-dimensional column vectors consisting of all far-field coefficients $a_i(\theta)$ for the radiated wave and of the corresponding Kochin functions $h_i(\theta)$, respectively. Note that the far-field coefficients $a_i(\theta)$ are referred to the common (global) origin. We may observe that since $\mathbf{R}$ is real, we may take the complex conjugate of the integrands without changing the resulting integral in Eq. (5.178).

We shall now prove the following relationship between the radiation resistance matrix $\mathbf{R}$ and the radiated power $P_r$:

$$P_r = \tfrac{1}{2}\hat{\mathbf{u}}^{\mathrm{T}}\mathbf{R}\hat{\mathbf{u}}^* = \tfrac{1}{2}\hat{\mathbf{u}}^{\dagger}\mathbf{R}\hat{\mathbf{u}}. \tag{5.179}$$

The last equality follows from the fact that $P_r$ is real ($P_r^* = P_r$) and that $\mathbf{R}$ is a real matrix. Since $P_r \geq 0$, $\mathbf{R}$ is a positive semidefinite matrix (see Section 6.5).

In the far-field region ($r \to \infty$), the curvature of the wave front is negligible. The radiated power transport per unit width of the wave front is then

$$J(r, \theta) = \frac{\rho g^2 D(kh)}{4\omega} |\hat{\eta}_r(r, \theta)|^2 = \frac{\omega\rho D(kh)}{4} |\hat{\phi}_r(r, \theta, 0)|^2, \tag{5.180}$$

where we have used Eqs. (4.39) and (4.130). In the far-field region ($r \to \infty$), we may use the asymptotic approximation (4.212)

$$\hat{\phi}_r \sim A_r(\theta)e(kz)(kr)^{-1/2}e^{-ikr}, \tag{5.181}$$

where

$$A_r(\theta) = \hat{\mathbf{u}}^{\mathrm{T}}\mathbf{a}(\theta) = \mathbf{a}^{\mathrm{T}}(\theta)\hat{\mathbf{u}} \tag{5.182}$$

is the far-field coefficient for the radiated wave, in accordance with Eq. (5.162). Thus, the power radiated through the cylindrical control surface $S_\infty$ of radius $r$, ($r \to \infty$) is

$$P_r = \int_0^{2\pi} J_r(r, \theta)r \, d\theta = \frac{\omega\rho D(kh)}{4k} \int_0^{2\pi} A_r^*(\theta)A_r(\theta) \, d\theta$$

$$= \frac{\omega\rho D(kh)}{4k} \int_0^{2\pi} \hat{\mathbf{u}}^{\dagger}\mathbf{a}^*(\theta)\mathbf{a}^{\mathrm{T}}(\theta)\hat{\mathbf{u}} \, d\theta. \tag{5.183}$$

[Also cf. Eq. (4.217).] Remembering that $\hat{\mathbf{u}}$ may be taken outside the last integral, we obtain Eq. (5.179) if we make use of Eq. (5.178).

Components $a_i(\theta)$ of the vector $\mathbf{a}(\theta)$ may be arranged in groups $\mathbf{a}_p(\theta)$ pertaining to body $p$, $(p = 1, 2, \ldots, N)$. The real part of matrices $\mathbf{Z}_{pp'}$ on the right-hand side of Eq. (5.158) may, thus, according to Eqs. (5.169), (5.176) and (5.178), be written as

$$\mathbf{R}_{pp'} = \frac{i}{2}\omega\rho\mathbf{I}(\boldsymbol{\varphi}_p^*, \boldsymbol{\varphi}_{p'}^{\mathrm{T}}) = \frac{\omega\rho D(kh)}{2k}\int_0^{2\pi} \mathbf{a}_p^*(\theta)\mathbf{a}_{p'}^{\mathrm{T}}(\theta)\,d\theta. \tag{5.184}$$

In analogy with Eqs. (4.255) and (4.256), we may use far-field coefficient vectors $\mathbf{b}_p(\theta_p)$ or $\mathbf{a}_p(\theta)$ referred to the local origin $(x_p, y_p, 0)$ or the global origin $(0, 0, 0)$, respectively. See Figure 4.9, which—with subscript $i$ replaced by $p$—defines the angles $\theta_p$ and $\alpha_p$ and the distances $x_p$, $y_p$ and $d_p$. The vertical axis $p$–$p$ in Figure 5.13 is given by $(x, y) = (x_p, y_p)$. Note that vector $\mathbf{b}_p(\theta_p)$ appears also in Eq. (5.164). According to Eq. (4.261), we have

$$\mathbf{a}_p(\theta) = \mathbf{b}_p(\theta)\exp\left[ikd_p\cos(\alpha_p - \theta)\right] = \mathbf{b}_p(\theta)\exp\left[ik(x_p\cos\theta + y_p\sin\theta)\right]. \tag{5.185}$$

If we use this, or Eq. (4.289), in Eq. (5.184), we get

$$\mathbf{R}_{pp'} = \frac{\omega\rho D(kh)}{2k}\int_0^{2\pi} \mathbf{b}_p^*(\theta)\mathbf{b}_{p'}^{\mathrm{T}}(\theta)\exp\left[ik(r_p - r_{p'})\right]d\theta. \tag{5.186}$$

According to Eq. (4.290), the distance difference $r_p - r_{p'}$ entering in the exponent may be expressed in various ways, such as

$$\begin{aligned} r_p - r_{p'} &= (x_{p'} - x_p)\cos\theta + (y_{p'} - y_p)\sin\theta \\ &= d_{p'}\cos(\alpha_{p'} - \theta) - d_p\cos(\alpha_p - \theta) = d_{p'p}\cos(\alpha_{p'p} - \theta). \end{aligned} \tag{5.187}$$

Using the last expression, we have

$$\mathbf{R}_{pp'} = \frac{\omega\rho D(kh)}{2k}\int_0^{2\pi} \mathbf{b}_p^*(\theta)\mathbf{b}_{p'}^{\mathrm{T}}(\theta)\exp\left[ikd_{p'p}\cos(\alpha_{p'p} - \theta)\right]d\theta. \tag{5.188}$$

### 5.5.4 Radiation Reactance Relation to Near-Field Energy Mismatch

We note that, according to Eq. (5.157), the radiation reactance matrix may be written as

$$\mathbf{X} = \omega\mathbf{m} = -i(\mathbf{Z} - \mathbf{R}) = \omega\rho\left(\lim_{r\to\infty}\iint_{S_\infty}\frac{\partial\boldsymbol{\varphi}^*}{\partial r}\boldsymbol{\varphi}^{\mathrm{T}}dS - \iint_S\frac{\partial\boldsymbol{\varphi}^*}{\partial n}\boldsymbol{\varphi}^{\mathrm{T}}dS\right), \tag{5.189}$$

where we have used Eqs. (5.171) and (5.177). This indicates that the reactance matrix $\mathbf{X}$ and, hence, the added-mass matrix $\mathbf{m}$ are somehow related to the radiated wave in the near-field region. Equation (2.88) shows how the reactance of a simple mechanical oscillator is related to the difference between the kinetic and potential energy. In the following paragraphs, let us relate the radiation reactance matrix to the difference between kinetic and potential energy in the near-field region. We note that in the far-field region, as also in a plane wave, this difference vanishes [cf. Eqs. (4.117) and (4.123)]. To be explicit, we shall show that

$$\frac{1}{4}\hat{\mathbf{u}}^T\mathbf{m}\hat{\mathbf{u}}^* = \frac{1}{4\omega}\hat{\mathbf{u}}^T\mathbf{X}\hat{\mathbf{u}}^* = W_k - W_p, \tag{5.190}$$

where $W_k - W_p$ is the difference between the kinetic energy and the potential energy associated with the radiated wave $\hat{\phi}_r$, as given by Eq. (5.162). Negative added mass is possible if $W_p > W_k$ [23].

The (time-average) kinetic energy is [see Eq. (2.83)]

$$W_k = \frac{\rho}{4}\iiint_V \hat{\vec{v}} \cdot \hat{\vec{v}}^* dV = \frac{\rho}{4}\iiint_V \nabla\hat{\phi}_r^* \cdot \nabla\hat{\phi}_r\, dV, \tag{5.191}$$

where the integral is taken over the fluid volume shown in Figure 5.13. This volume is bounded by a closed surface consisting of the sea bed $S_b$, the control surface $S_\infty$, the free water surface $S_0$ and the wave-generating surface $S$, defined by Eq. (5.168). Since $\hat{\phi}_r$ obeys Laplace's equation (4.33), we have

$$\nabla \cdot (\hat{\phi}_r^*\nabla\hat{\phi}_r) \equiv \hat{\phi}_r^*\nabla^2\hat{\phi}_r + \nabla\hat{\phi}_r^* \cdot \nabla\hat{\phi}_r = \nabla\hat{\phi}_r^* \cdot \nabla\hat{\phi}_r. \tag{5.192}$$

Hence, since the integrand is the divergence of the vector $\hat{\phi}_r^*\nabla\hat{\phi}_r$, we may, by using Gauss's divergence theorem, replace the volume integral (5.191) by the following closed-surface integral

$$W_k = \frac{\rho}{4}\oiint \hat{\phi}_r^*\nabla\hat{\phi}_r \cdot (-\vec{n})\, dS = -\frac{\rho}{4}\oiint \hat{\phi}_r^*\frac{\partial\hat{\phi}_r}{\partial n}\, dS, \tag{5.193}$$

where the unit normal $\vec{n}$ is pointing into the fluid (see Figure 5.13). Note that $\partial/\partial n = -\partial/\partial z$ on $S_0$ and $\partial/\partial n = -\partial/\partial r$ on $S_\infty$. Further, we observe that the integrand vanishes on the sea bed $S_b$ as a result of boundary condition (4.16).

By using Eqs. (4.39), (4.41) and (4.115), we obtain the (time-average) potential energy

$$W_p = \iint_{S_0} E_p dS = \frac{\rho g}{4}\iint_{S_0} \hat{\eta}\hat{\eta}^* dS = \frac{\omega^2\rho}{4g}\iint_{S_0} \hat{\phi}_r^*\hat{\phi}_r dS$$

$$= \frac{\rho}{4}\iint_{S_0} \hat{\phi}_r^*\frac{\partial\hat{\phi}_r}{\partial z}dS = -\frac{\rho}{4}\iint_{S_0} \hat{\phi}_r^*\frac{\partial\hat{\phi}_r}{\partial n}dS. \tag{5.194}$$

Subtraction between Eqs. (5.193) and (5.194) gives

$$W_k - W_p = \frac{\rho}{4}\left(\iint_{S_\infty} \hat{\phi}_r^*\frac{\partial\hat{\phi}_r}{\partial r}dS - \iint_S \hat{\phi}_r^*\frac{\partial\hat{\phi}_r}{\partial n}dS\right). \tag{5.195}$$

Observing from Eq. (5.162) that we may write $\hat{\phi}_r^* = \hat{\mathbf{u}}^\dagger \boldsymbol{\varphi}^*$ and $\hat{\phi}_r = \boldsymbol{\varphi}^T \hat{\mathbf{u}}$ and, further, that $\hat{\mathbf{u}}^\dagger$ and $\hat{\mathbf{u}}$ may be taken outside the integral, we find from Eqs. (5.189) and (5.195) that

$$W_k - W_p = \frac{1}{4\omega}\hat{\mathbf{u}}^\dagger \mathbf{X} \hat{\mathbf{u}} = \frac{1}{4}\hat{\mathbf{u}}^\dagger \mathbf{m}\hat{\mathbf{u}}. \tag{5.196}$$

Since $W_k - W_p$ is a real scalar and $\mathbf{X}$ (and $\mathbf{m}$) is a real symmetric matrix, the right-hand sides of Eqs. (5.189) and (5.196) must equal their own conjugates as well as their own transposes. Hence, we obtain Eq. (5.190) if we take the complex conjugate of Eq. (5.196). Note that the control cylinder $S_\infty$ has to be sufficiently large in order to contain the complete near-field region inside $S_\infty$.

An alternative expression for the added mass matrix is obtained as follows. Using Eqs. (5.157), (5.171) and (5.175) gives

$$
\begin{aligned}
\mathbf{m} &= \frac{\mathbf{X}}{\omega} = \frac{1}{2i\omega}(\mathbf{Z} - \mathbf{Z}^*) = \frac{1}{2i\omega}(\mathbf{Z} - \mathbf{Z}^\dagger) \\
&= -\frac{\rho}{2} \iint_S \left( \frac{\partial \boldsymbol{\varphi}^*}{\partial n} \boldsymbol{\varphi}^T + \boldsymbol{\varphi}^* \frac{\partial \boldsymbol{\varphi}^T}{\partial n} \right) dS = -\frac{\rho}{2} \iint_S \frac{\partial}{\partial n}(\boldsymbol{\varphi}^* \boldsymbol{\varphi}^T)\, dS.
\end{aligned}
\tag{5.197}
$$

In this way, the added mass matrix is expressed as an integral over the wave-generating surface $S$ (the totality of wet body surfaces) only.

It is emphasised that $W_k - W_p$ in identity (5.190) or (5.196) is the kinetic–potential energy difference when there is no other wave than a radiated wave resulting from a forced body motion on otherwise still water. Thus, only $\phi_r$ is accounted for in Eqs. (5.191)–(5.195). If, in addition, an incident wave is present, then there is an additional contribution to $W_k - W_p$. Let us now consider this more general case, where we shall follow the derivation given in [63, appendix B].

Since the total velocity potential $\phi = \phi_0 + \phi_d + \phi_r$ satisfies Laplace's equation and the homogeneous boundary condition on the sea bed, Eq. (5.195) applies also to $\phi$ as to $\phi_r$, where it is now understood that $W_k - W_p$ in the equation is the kinetic–potential energy difference associated with $\phi$. Denoting $\phi_s \equiv \phi_0 + \phi_d$, we thus have

$$W_k - W_p = \frac{\rho}{4}\left[ \iint_{S_\infty} (\hat{\phi}_s + \hat{\phi}_r)^* \frac{\partial(\hat{\phi}_s + \hat{\phi}_r)}{\partial r} dS - \iint_S (\hat{\phi}_s + \hat{\phi}_r)^* \frac{\partial(\hat{\phi}_s + \hat{\phi}_r)}{\partial n} dS \right]. \tag{5.198}$$

We see that the integrands in this equation contain products of the type $(\hat{\phi}_s + \hat{\phi}_r)^*(\hat{\phi}_s + \hat{\phi}_r) = \hat{\phi}_s^* \hat{\phi}_s + \hat{\phi}_s^* \hat{\phi}_r + \hat{\phi}_r^* \hat{\phi}_s + \hat{\phi}_r^* \hat{\phi}_r$. The fourth term corresponds to Eq. (5.195). The first term corresponds to pure scattering—that is, the case in which the bodies are not oscillating. The remaining terms, $\hat{\phi}_s^* \hat{\phi}_r + \hat{\phi}_r^* \hat{\phi}_s$, represent the interaction between the radiated and scattered waves.

Let us first consider the pure scattering problem. In this case,

$$W_k - W_p = \frac{\rho}{4} \left( \iint_{S_\infty} \hat{\phi}_s^* \frac{\partial \hat{\phi}_s}{\partial r} dS - \iint_S \hat{\phi}_s^* \frac{\partial \hat{\phi}_s}{\partial n} dS \right). \tag{5.199}$$

Because of boundary condition (5.165), the second integral vanishes. The first integral may be split into

$$\iint_{S_\infty} \hat{\phi}_s^* \frac{\partial \hat{\phi}_s}{\partial r} dS \equiv I_{\infty,ss} = I_{\infty,00} + I_{\infty,0d} + I_{\infty,dd}, \tag{5.200}$$

where

$$I_{\infty,00} = \iint_{S_\infty} \hat{\phi}_0^* \frac{\partial \hat{\phi}_0}{\partial r} dS, \tag{5.201}$$

$$I_{\infty,0d} = \iint_{S_\infty} \left( \hat{\phi}_0^* \frac{\partial \hat{\phi}_d}{\partial r} + \hat{\phi}_d^* \frac{\partial \hat{\phi}_0}{\partial r} \right) dS, \tag{5.202}$$

$$I_{\infty,dd} = \iint_{S_\infty} \hat{\phi}_d^* \frac{\partial \hat{\phi}_d}{\partial r} dS. \tag{5.203}$$

Since $W_k - W_p$ is real, we may write

$$W_k - W_p = \frac{\rho}{4} I_{\infty,ss} = \frac{\rho}{8}(I_{\infty,ss} + I_{\infty,ss}^*). \tag{5.204}$$

The incident wave potential is given by [cf. Eq. (4.262)]

$$\hat{\phi}_0 = -\frac{g}{i\omega} A e(kz) \exp[-ikr\cos(\theta - \beta)], \tag{5.205}$$

while the diffracted wave potential in the far-field region has an asymptotic expression [cf. Eq. (4.212)]

$$\hat{\phi}_d \sim A_d(\theta) e(kz)(kr)^{-1/2} e^{-ikr}. \tag{5.206}$$

Inserting these expressions into Eqs. (5.201) and (5.203), we see that $I_{\infty,00} + I_{\infty,00}^*$ as well as $I_{\infty,dd} + I_{\infty,dd}^*$ vanish, and hence,

$$I_{\infty,ss} + I_{\infty,ss}^* = I_{\infty,0d} + I_{\infty,0d}^*. \tag{5.207}$$

Using the product rule, we find from Eq. (5.202) that

$$I_{\infty,0d} + I_{\infty,0d}^* = \iint_{S_\infty} \frac{\partial}{\partial r}(\hat{\phi}_0^* \hat{\phi}_d) dS + \text{c. c.} \equiv I'_{\infty,0d} + \text{c. c.} \tag{5.208}$$

Inserting Eqs. (5.205)–(5.206) and denoting $\varphi = \theta - \beta$, we then have

$$
\begin{aligned}
I'_{\infty,0d} &= -\iint_{S_\infty} ik(1 - \cos\varphi)\hat{\phi}_0^*\hat{\phi}_d \, dS \\
&= -\int_{-h}^{0} \lim_{kr\to\infty} \int_0^{2\pi} ikr(1 - \cos\varphi)\hat{\phi}_0^*\hat{\phi}_d \, d\varphi \, dz \\
&= -\frac{v_p v_g}{\omega} A^* \lim_{kr\to\infty} \int_0^{2\pi} \sqrt{kr}(1 - \cos\varphi) A_d(\varphi + \beta) e^{-ikr(1-\cos\varphi)} \, d\varphi,
\end{aligned}
$$

$$(5.209)$$

where we have used Eqs. (4.107)–(4.108) to evaluate the integral over $z$. The integral over $\varphi$ may be evaluated using the method of stationary phase, as described in Section 4.8. Application of Eq. (4.275) gives, after further manipulation,

$$
I'_{\infty,0d} = \frac{v_p v_g}{\omega} |A|\sqrt{8\pi} \lim_{kr\to\infty} \exp\left[-i(2kr + 3\pi/4 + \alpha_0 - \alpha_d)\right] |A_d(\pi + \beta)|, \quad (5.210)
$$

where $\alpha_0 = \arg(A)$ and $\alpha_d = \arg[A_d(\pi + \beta)]$. Putting this back into Eq. (5.204), we finally arrive at an expression for the kinetic–potential energy difference in the pure scattering case:

$$
W_k - W_p = \frac{\rho}{4}\frac{v_p v_g}{\omega} |A|\sqrt{8\pi} \lim_{kr\to\infty} \cos(2kr + 3\pi/4 + \alpha_0 - \alpha_d)|A_d(\pi + \beta)|. \quad (5.211)
$$

We may observe from this expression that, in the pure scattering case, $W_k - W_p = 0$ for discrete values of the distance $r$ from origin:

$$
kr = \frac{n\pi}{2} - \frac{\pi}{8} - \frac{\alpha_0}{2} + \frac{\alpha_d}{2}, \quad n = 0, 1, 2, \ldots. \quad (5.212)
$$

Further, $W_k - W_p$ averages to zero over every half-wavelength increment of $r$.

Next, let us consider the kinetic–potential energy difference associated with the interaction between the radiated and scattered waves. According to Eq. (5.198),

$$
W_k - W_p = \frac{\rho}{4}\left[\iint_{S_\infty}\left(\hat{\phi}_s^*\frac{\partial\hat{\phi}_r}{\partial r} + \hat{\phi}_r^*\frac{\partial\hat{\phi}_s}{\partial r}\right) dS - \iint_S\left(\hat{\phi}_s^*\frac{\partial\hat{\phi}_r}{\partial n} + \hat{\phi}_r^*\frac{\partial\hat{\phi}_s}{\partial n}\right) dS\right]. \quad (5.213)
$$

Let us first discuss the integral over $S_\infty$, which may be split into

$$
\iint_{S_\infty}\left(\hat{\phi}_s^*\frac{\partial\hat{\phi}_r}{\partial r} + \hat{\phi}_r^*\frac{\partial\hat{\phi}_s}{\partial r}\right) dS \equiv I_{\infty,sr} = I_{\infty,0r} + I_{\infty,dr}, \quad (5.214)
$$

where

$$
I_{\infty,0r} = \iint_{S_\infty}\left(\hat{\phi}_0^*\frac{\partial\hat{\phi}_r}{\partial r} + \hat{\phi}_r^*\frac{\partial\hat{\phi}_0}{\partial r}\right) dS, \quad (5.215)
$$

$$I_{\infty,dr} = \iint_{S_\infty} \left( \hat{\phi}_d^* \frac{\partial \hat{\phi}_r}{\partial r} + \hat{\phi}_r^* \frac{\partial \hat{\phi}_d}{\partial r} \right) dS. \tag{5.216}$$

Since $\hat{\phi}_r$ has the same asymptotic form as $\hat{\phi}_d$ in the far field, the preceding discussion pertaining to $I_{\infty,0d}$ is also applicable to $I_{\infty,0r}$, namely that the integral $I_{\infty,0r}$ contributes essentially nothing to the kinetic–potential energy difference. As for the integral $I_{\infty,dr}$, we may write it as $(I_{\infty,dr} + I_{\infty,dr}^*)/2$ because $W_k - W_p$ is real. Then, because Eqs. (4.237) and (4.241) are applicable to both $\hat{\phi}_r$ and $\hat{\phi}_d$, it follows that $I_{\infty,dr} + I_{\infty,dr}^* = 0$. What remains then is to evaluate the integral over $S$ in Eq. (5.213). We observe that the second term of the integrand vanishes because of boundary condition (5.165), whereas

$$\iint_S \hat{\phi}_s^* \frac{\partial \hat{\phi}_r}{\partial n} dS = \iint_S \hat{\phi}_s^* \hat{\mathbf{u}}^{\mathrm{T}} \mathbf{n}\, dS = \hat{\mathbf{u}}^{\mathrm{T}} \iint_S \hat{\phi}_s^* \mathbf{n}\, dS = -\frac{1}{i\omega\rho} \hat{\mathbf{u}}^{\mathrm{T}} \hat{\mathbf{F}}_e^* \tag{5.217}$$

because of boundary condition (5.166) and Eq. (5.26) [also see Eq. (5.221)]. Here, $\hat{\mathbf{F}}_e$ is the excitation force vector [see Eq. (5.220)]. Therefore, the kinetic–potential energy difference due to the interaction between the radiated and scattered waves is

$$W_k - W_p = -\frac{\rho}{8} \left( \iint_S \hat{\phi}_s^* \frac{\partial \hat{\phi}_r}{\partial n} dS + \text{c. c.} \right) = -\frac{\rho}{8} \left( -\frac{1}{i\omega\rho} \hat{\mathbf{u}}^{\mathrm{T}} \hat{\mathbf{F}}_e^* + \frac{1}{i\omega\rho} \hat{\mathbf{u}}^\dagger \hat{\mathbf{F}}_e \right)$$
$$= -\frac{1}{4\omega} \mathrm{Im}\{\hat{\mathbf{u}}^\dagger \hat{\mathbf{F}}_e\} = -\frac{1}{4\omega} \mathrm{Im}\{\hat{\mathbf{F}}_e^{\mathrm{T}} \hat{\mathbf{u}}^*\}. \tag{5.218}$$

Returning to the expression (5.198), we find that the kinetic–potential energy difference in the general case is given as

$$W_k - W_p = \frac{1}{4\omega} \left( \hat{\mathbf{u}}^{\mathrm{T}} \mathbf{X} \hat{\mathbf{u}}^* - \mathrm{Im}\{\hat{\mathbf{F}}_e^{\mathrm{T}} \hat{\mathbf{u}}^*\} \right). \tag{5.219}$$

### 5.5.5 Excitation Force Vector: The Haskind Relation

For each of the $N$ bodies, we may define an excitation force vector $\mathbf{F}_{e,p}$, which has, in general, six components, as in Eq. (5.26) or (5.138). Note that boundary condition (5.27) or (5.165) for $\phi_d$ has to be satisfied on the wet surface, not only of body $p$ but also of all other bodies. Thus, $\mathbf{F}_{e,p}$ represents the wave force on body $p$ when *all* bodies are in a non-oscillating state.

For the total system of $N$ bodies, we define a column vector $\mathbf{F}_e$ (in general, of dimension $6N$) for the excitation forces:

$$\mathbf{F}_e = (\mathbf{F}_{e,1}^{\mathrm{T}} \cdots \mathbf{F}_{e,p}^{\mathrm{T}} \cdots \mathbf{F}_{e,N}^{\mathrm{T}})^{\mathrm{T}}. \tag{5.220}$$

Here $\mathbf{F}_{e,p}$ is the (generally six-dimensional) excitation force vector for body $p$. In accordance with Eq. (5.26), we may write

$$\hat{\mathbf{F}}_{e,p} = i\omega\rho \iint_{S_p} (\hat{\phi}_0 + \hat{\phi}_d)\mathbf{n}_p \, dS, \tag{5.221}$$

where $\mathbf{n}_p$ is the (six-dimensional) unit normal vector, as defined by Eqs. (5.5) and (5.6), but now for body $p$.

Since all components $\varphi_i$ of vector $\boldsymbol{\varphi}$ satisfy boundary condition (5.166) and also the same radiation condition at infinity as $\hat{\phi}_d$, we may generalise Eqs. (5.136) and (5.138) to the case with N bodies:

$$\hat{\mathbf{F}}_e = i\omega\rho\mathbf{I}[(\hat{\phi}_0 + \hat{\phi}_d), \boldsymbol{\varphi}] = i\omega\rho\mathbf{I}(\hat{\phi}_0, \boldsymbol{\varphi}). \tag{5.222}$$

In particular, we have for body $p$

$$\hat{\mathbf{F}}_{e,p} = i\omega\rho\mathbf{I}(\hat{\phi}_0, \boldsymbol{\varphi}_p). \tag{5.223}$$

If the incident wave is plane in accordance with Eq. (5.139), we may also generalise Eqs. (5.141) and (5.142) to the case of $N$ bodies. Thus, the excitation force vector is

$$\hat{\mathbf{F}}_e(\beta) = \mathbf{f}_g(\beta)A = \frac{\rho g D(kh)}{k}\mathbf{h}(\beta \pm \pi)A, \tag{5.224}$$

where

$$\mathbf{f}_g(\beta) = \frac{\rho g D(kh)}{k}\mathbf{h}(\beta \pm \pi) = \frac{\rho g D(kh)}{k}\sqrt{2\pi}\,\mathbf{a}(\beta \pm \pi)e^{i\pi/4} \tag{5.225}$$

is, in general, a $6N$-dimensional column vector consisting of the excitation force coefficients referred to the global origin $(x, y) = (0, 0)$. Note that the excitation force depends on the angle of incidence $\beta$ of the incident wave.

Similarly, reciprocity relations (5.143)–(5.145), which express the radiation resistance matrix in terms of excitation force vectors, may be directly generalised to the $6N$-dimensional case.

Note that $A$ in Eq. (5.224) is the elevation of the undisturbed incident wave at the (global) origin $(x, y) = (0, 0)$. Considering body $p$'s excitation force vector $\mathbf{F}_{e,p}$, we have from Eqs. (5.224) and (5.225) that

$$\hat{\mathbf{F}}_{e,p}(\beta) = \mathbf{f}_{g,p}(\beta)A = \frac{\rho g D(kh)}{k}\sqrt{2\pi}\,\mathbf{a}_p(\beta \pm \pi)e^{i\pi/4}A. \tag{5.226}$$

Using relation (5.185) between the local and global far-field coefficients, we find

$$\hat{\mathbf{F}}_{e,p}(\beta) = \mathbf{f}_p(\beta)A_p = \frac{\rho g D(kh)}{k}\sqrt{2\pi}\,\mathbf{b}_p(\beta \pm \pi)e^{i\pi/4}A_p, \tag{5.227}$$

where

$$A_p = A\exp[-ik(x_p \cos\beta + y_p \sin\beta)] \tag{5.228}$$

is [see Eq. (5.139)] the wave elevation corresponding to the undisturbed incident wave at body $p$'s local origin $(x, y) = (x_p, y_p)$. In Eqs. (5.226) and (5.227), the excitation-force coefficients $\mathbf{f}_{g,p}(\beta)$ and $\mathbf{f}_p(\beta)$ for body $p$ are referred to the global origin $(x, y) = (0, 0)$ and to the local origin $(x, y) = (x_p, y_p)$, respectively.

By using Eqs. (4.82) and (4.106) for the phase and group velocities, we may write the coefficient $\rho g D(kh)/h$, appearing in Eqs. (5.224)–(5.227), as $2v_g v_p \rho$.

### 5.5.6 Wide-Spacing Approximation

If the distance between two different bodies ($p$ and $p'$, for example) is sufficiently large, the diffraction and radiation interactions between the two bodies may be well represented by the wide-spacing approximation. The requirement is that one body is not influenced by the other body's near-field contribution to the diffracted or radiated waves. Here, let us illustrate the wide-spacing approximation for the radiation problem. We shall apply the asymptotic approximation (5.164) for both bodies. That means we have to assume equal water depth $h$ for both bodies.

The interaction between the two bodies is then represented by the $6 \times 6$ matrix $\mathbf{Z}_{pp'}$. We consider the force

$$\Delta' \hat{\mathbf{F}}_p = -\mathbf{Z}_{pp'} \hat{\mathbf{u}}_{p'} \tag{5.229}$$

on body $p$ due to the oscillation of body $p'$ with velocity $\mathbf{u}_{p'}$. Using far-field approximation (5.164) with $p'$ instead of $p$ and also Eqs. (4.39) and (5.122), we find the corresponding wave elevation

$$\Delta' \hat{\eta}_r(r_{p'}, \theta_{p'}) = -\frac{i\omega}{g} \boldsymbol{\varphi}_{p'}^{\mathrm{T}}(r_{p'}, \theta_{p'}, 0) \hat{\mathbf{u}}_{p'} \tag{5.230}$$

$$\sim -\frac{i\omega}{g} \mathbf{b}_{p'}^{\mathrm{T}}(\theta_{p'})(kr_{p'})^{-1/2} \exp\{-ikr_{p'}\} \hat{\mathbf{u}}_{p'}. \tag{5.231}$$

If $kd_{pp'} \to \infty$, this wave may be considered as a plane wave when it arrives at body $p$, where $(r_{p'}, \theta_{p'}) = (d_{pp'}, \alpha_{pp'})$; see Figure 4.11. Using also Eq. (5.227) yields

$$\Delta' \hat{\mathbf{F}}_p \sim \frac{\rho g D(kh)}{k} \sqrt{2\pi} \, \mathbf{b}_p(\alpha_{pp'} \pm \pi) e^{i\pi/4} \Delta' \hat{\eta}_r(d_{pp'}, \alpha_{pp'}). \tag{5.232}$$

Inserting here from the approximation (5.231) and comparing with Eq. (5.229) result in the following asymptotic approximation for $\mathbf{Z}_{pp'}$ as $kd_{pp'} \to \infty$:

$$\mathbf{Z}_{pp'} \sim \frac{i\omega \rho D(kh)}{k} \sqrt{2\pi} \mathbf{b}_p(\alpha_{p'p}) \mathbf{b}_{p'}^{\mathrm{T}}(\alpha_{pp'})(kd_{pp'})^{-1/2} \exp\left(-ikd_{pp'} + \frac{\pi}{4}\right). \tag{5.233}$$

Note that $\mathbf{b}_p(\alpha_{p'p})$ represents the action on body $p'$ from body $p$ and vice versa for $\mathbf{b}_{p'}(\alpha_{pp'})$. The result agrees with the expected approximation (5.159). We also easily see that the general reciprocity relationship (5.172) is satisfied by approximation (5.233).

## 5.6 The Froude–Krylov Force and Small-Body Approximation

We may, in accordance with Eq. (5.221), decompose the excitation force vector $\mathbf{F}_{e,p}$ for body $p$ into two parts:

$$\hat{\mathbf{F}}_{e,p} = \hat{\mathbf{F}}_{FK,p} + \hat{\mathbf{F}}_{d,p}, \tag{5.234}$$

where

$$\hat{\mathbf{F}}_{FK,p} = i\omega\rho \iint_{S_p} \hat{\phi}_0 \mathbf{n}_p dS \tag{5.235}$$

is the Froude–Krylov force vector, with $\mathbf{n}_p = (\vec{n}_p, \vec{s}_p \times \vec{n}_p)$ [see Eqs. (5.5)–(5.6)], and

$$\hat{\mathbf{F}}_{d,p} = i\omega\rho \iint_{S_p} \hat{\phi}_d \mathbf{n}_p dS \tag{5.236}$$

is the diffraction force vector. Alternatively, we have from the Haskind relation [cf. Eq. (5.223)]

$$\hat{\mathbf{F}}_{e,p} = i\omega\rho \mathbf{I}(\hat{\phi}_0, \boldsymbol{\varphi}_p) = i\omega\rho \iint_S \left( \hat{\phi}_0 \frac{\partial \boldsymbol{\varphi}_p}{\partial n} - \boldsymbol{\varphi}_p \frac{\partial \hat{\phi}_0}{\partial n} \right) dS. \tag{5.237}$$

The first term represents the Froude–Krylov force, since $\partial \boldsymbol{\varphi}_p / \partial n = 0$ on $S_{p'}$ when $p' \neq p$, and $\partial \boldsymbol{\varphi}_p / \partial n = \mathbf{n}_p$ on $S_p$; see Eq. (5.166). Hence, we may compute the diffraction force vector

$$\hat{\mathbf{F}}_{d,p} = -i\omega\rho \iint_S \boldsymbol{\varphi}_p \frac{\partial \hat{\phi}_0}{\partial n} dS \tag{5.238}$$

by solving the radiation problem instead of solving the scattering (or diffraction) problem.

With a given incident wave and a given geometry for body $p$, the Froude–Krylov force vector $\hat{\mathbf{F}}_{FK,p}$ is obtained by direct integration. To compute the diffraction force, it is necessary to solve a boundary-value problem corresponding to either scattering (diffraction) or radiation. The latter problem is frequently, but not always, the easier one to solve.

### 5.6.1 The Froude–Krylov Force and Moment

Next, we shall consider the translational components of the Froude–Krylov force. For

$$i = 6(p - 1) + q \qquad (q = 1, 2, 3), \tag{5.239}$$

we write the three components of the Froude–Krylov force as

$$\hat{F}_{FK,i} = \hat{F}_{FK,pq} = i\omega\rho \iint_{S_p} \hat{\phi}_0 n_{pq} dS = i\omega\rho \iint_{S_p} \hat{\phi}_0 \vec{e}_q \cdot \vec{n}_p dS. \tag{5.240}$$

Here $\vec{e}_q$ $(q = 1, 2, 3)$ is the unit vector in the direction $x$, $y$ or $z$, for surge, sway or heave, respectively. The unit normal components $n_{pq}$ are as given by Eq. (5.5), but for body $p$. By making the union of the wet surface $S_p$ and the water plane area $S_{wp}$ of body $p$ (see Figure 5.14), we may decompose this integral into one integral over the closed surface $S_p + S_{wp}$ and one integral over $S_{wp}$:

$$\hat{F}_{FK,pq} = i\omega\rho \oiint_{S_p+S_{wp}} \hat{\phi}_0 \vec{e}_q \cdot \vec{n}_p dS - i\omega\rho \iint_{S_{wp}} \hat{\phi}_0 \delta_{q3} dS. \tag{5.241}$$

Note that the last term vanishes except for the heave mode ($q = 3$), because $n_q = \delta_{q3}$ on $S_{wp}$. We apply Gauss's theorem to the closed-surface integral and convert it to an integral over the volume $V_p$ of the displaced water:

$$i\omega\rho \oiint_{S_p+S_{wp}} \hat{\phi}_0 \vec{e}_q \cdot \vec{n}_p dS = i\omega\rho \iiint_{V_p} (\nabla\hat{\phi}_0) \cdot \vec{e}_q dV = i\omega\rho \iiint_{V_p} \frac{\partial\hat{\phi}_0}{\partial x_q} dV \tag{5.242}$$

($x_1 = x, x_2 = y, x_3 = z$). This may be obtained directly from Gauss's divergence theorem applied to the divergence $\nabla \cdot (\hat{\phi}_0 \vec{e}_q)$ where $\vec{e}_q$ is a constant (unit) vector. Thus, we have

$$\hat{F}_{FK,pq} = i\omega\rho \iiint_{V_p} \frac{\partial\hat{\phi}_0}{\partial x_q} dV + \delta_{q3}\rho g \iint_{S_{wp}} \hat{\eta}_0 dS, \tag{5.243}$$

where Eq. (4.39) also has been used. Note that $\hat{\phi}_0$ and $\hat{\eta}_0$ are the velocity potential and the wave elevation of the undisturbed incident wave—that is, what the wave would have been in the absence of all the $N$ bodies. Observing

Figure 5.14: Floating body with water plane area $S_{wp}$, submerged volume $V_p$ (volume of displaced water) and wet surface $S_p$ with unit normal $\vec{n}_p$. The vertical unit vector is indicated by $\vec{e}_3$.

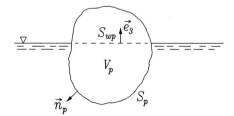

that [see Eq. (4.36)] $i\omega\partial\hat{\phi}_0/\partial x_q = i\omega\hat{v}_{0q} = \hat{a}_{0q}$, we may alternatively write Eq. (5.243) as

$$\hat{F}_{\text{FK},pq} = \rho \iiint_{V_p} \hat{a}_{0q}dV + \delta_{q3}\rho g \iint_{S_{wp}} \hat{\eta}_0 dS, \qquad (5.244)$$

where $\hat{a}_{0q}$ represents the $q$ component of the fluid acceleration due to the incident wave. In the particular case of a purely propagating incident plane wave, $\hat{\phi}_0$ is given by Eqs. (4.88) and (4.98). Then we have the vertical and horizontal components of the undisturbed fluid particle acceleration as [cf. Eqs. (4.90), (4.91) and (4.267)]

$$\hat{a}_{0z} = \hat{a}_{03} = i\omega\hat{v}_{03} = -gkAe'(kz)\exp\left[-ikr(\beta)\right]$$
$$= -\omega^2 A \frac{\sinh(kz+kh)}{\sinh(kh)}\exp\left[-ikr(\beta)\right], \qquad (5.245)$$

$$(\hat{a}_{0x}, \hat{a}_{0y}) = (\hat{a}_{01}, \hat{a}_{02}) = i\omega(\hat{v}_{01}, \hat{v}_{02}) = a_{0H}(\cos\beta, \sin\beta), \qquad (5.246)$$

where

$$a_{0H} = ikgAe(kz)\exp\left[-ikr(\beta)\right]. \qquad (5.247)$$

Finally, we consider the rotational modes of the Froude–Krylov 'force' (i.e., Froude–Krylov moments of roll, pitch and yaw) for body $p$. For

$$i = 6(p-1) + 3 + q \qquad (q = 1,2,3), \qquad (5.248)$$

set $F_i = M_{pq}$. Thus, according to Eqs. (5.6) and (5.19), the Froude–Krylov moment is given by

$$\hat{M}_{\text{FK},pq} = \hat{F}_{\text{FK},i} = i\omega\rho \iint_{S_p} \hat{\phi}_0 n_{pq} dS = i\omega\rho \iint_{S_p} \hat{\phi}_0 \vec{e}_q \cdot (\vec{s}_p \times \vec{n}_p) dS, \qquad (5.249)$$

where $\vec{s}_p$ is the vector from the body's reference point $(x_p, y_p, z_p)$ to the surface element $dS$ of the wet surface $S_p$ (cf. Figures 5.3 and 5.13). On the water plane area $S_{wp}$ (see Figure 5.14), we have $\vec{n}_p = \vec{e}_3$ and $\vec{e}_q \cdot (\vec{s}_p \times \vec{e}_3) = (y - y_p)\delta_{q1} - (x - x_p)\delta_{q2}$. Thus,

$$\hat{M}_{\text{FK},pq} = i\omega\rho \oiint_{S_p+S_{wp}} \hat{\phi}_0 \vec{e}_q \times \vec{s}_p \cdot \vec{n}_p dS$$
$$-i\omega\rho \iint_{S_{wp}} \hat{\phi}_0[(y - y_p)\delta_{q1} - (x - x_p)\delta_{q2}]dS. \qquad (5.250)$$

Note that, for the yaw mode ($j = 6$ or $q = 3$), there is no contribution from the last term. Applying Gauss's theorem on the closed-surface integral, we have

$$\hat{M}_{FK,pq} = i\omega\rho \iiint_{V_p} \nabla \cdot (\hat{\phi}_0 \vec{e}_q \times \vec{s}_p) dV$$

$$-i\omega\rho \iint_{S_{wp}} \hat{\phi}_0 [(y - y_p)\delta_{q1} - (x - x_p)\delta_{q2}] dS. \tag{5.251}$$

In the volume integral here, the integrand may be written in various ways:

$$\nabla \cdot (\hat{\phi}_0 \vec{e}_q \times \vec{s}_p) = \vec{s}_p \cdot [\nabla \times (\hat{\phi}_0 \vec{e}_q)] = -\hat{\phi}_0 \vec{e}_q \cdot (\nabla \times \vec{s}_p)$$

$$= \vec{s}_p \cdot (\nabla \hat{\phi}_0 \times \vec{e}_q) = \vec{e}_q \cdot (\vec{s}_p \times \nabla \hat{\phi}_0) = (\vec{s}_p \times \nabla \hat{\phi}_0)_q \tag{5.252}$$

$$= (\vec{s}_p \times \hat{\vec{v}}_0)_q = (\vec{s}_p \times \hat{\vec{a}}_0)_q \frac{1}{i\omega}.$$

Here, $\nabla \hat{\phi}_0 = \hat{\vec{v}}$ represents the fluid velocity $\vec{v}_0$ corresponding to the undisturbed velocity potential $\phi_0$, and $\vec{a}_0$ is the corresponding fluid acceleration.

### 5.6.2 The Diffraction Force

In the next subsection, we shall consider the Froude–Krylov force, the diffraction force and the excitation force for a small body. Then we shall make use of Eqs. (5.244), (5.251) and (5.252). We shall then also approximate the following equation for the diffraction force, which is still exact (within potential theory). Based on Eq. (5.238), we write the diffraction force as

$$\hat{\mathbf{F}}_{d,p} = -i\omega\rho \sum_{p'=1}^{N} \iint_{S_{p'}} \varphi_p \frac{\partial \hat{\phi}_0}{\partial n} dS. \tag{5.253}$$

On $S_{p'}$, we have

$$\frac{\partial \hat{\phi}_0}{\partial n} = \vec{n}_{p'} \cdot \nabla \hat{\phi}_0 = \vec{n}_{p'} \cdot \hat{\vec{v}}_0 = \frac{1}{i\omega} \vec{n}_{p'} \cdot \hat{\vec{a}}_0 = \frac{1}{i\omega} \sum_{q=1}^{3} n_{p'q} \hat{a}_{0q}, \tag{5.254}$$

and hence,

$$\hat{\mathbf{F}}_{d,p} = -\rho \sum_{q=1}^{3} \sum_{p'=1}^{N} \iint_{S_{p'}} \varphi_p n_{p'q} \hat{a}_{0q} dS. \tag{5.255}$$

Here the sum over $p'$ shows that the diffracted waves, not only from body $p$ but also from all the other bodies ($p \neq p'$), contribute to the diffraction force on body $p$.

### 5.6.3 Small-Body Approximation for a Group of Bodies

If the horizontal and vertical extensions of body $p$ are very much shorter than one wavelength, we may in Eqs. (5.244) and (5.251), as an approximation, take incident-wave quantities—such as $\hat{\phi}_0$, $\hat{\vec{v}}_0$ and $\hat{\vec{a}}_0$—outside the integral. Then (for $q = 1, 2, 3$) the Froude–Krylov force and moment components are given by

$$\hat{F}_{\mathrm{FK},pq} \approx \rho \hat{a}_{0pq} V_p + \delta_{q3} \rho g \hat{\eta}_{0p} S_{wp}, \tag{5.256}$$

$$\hat{M}_{\mathrm{FK},pq} \approx \rho(\vec{\Gamma}_p \times \hat{\vec{a}}_{0p})_q + \rho g \hat{\eta}_{0p}(S_{yp}\delta_{1q} - S_{xp}\delta_{2q}), \tag{5.257}$$

where $\vec{a}_{0p}$ is the fluid acceleration due to the undisturbed incident wave at the chosen reference point $(x, y, z) = (x_p, y_p, z_p)$—for instance, the centre of mass or the centre of displaced volume—for body $p$. Furthermore, $\hat{\eta}_{0p}$ is the same wave's elevation at $(x, y) = (x_p, y_p)$. Also,

$$\vec{\Gamma}_p = \iiint_{V_p} \vec{s}_p \, dV_p, \tag{5.258}$$

$$S_{xp} = \iint_{S_{wp}} (x - x_p) \, dS, \qquad S_{yp} = \iint_{S_{wp}} (y - y_p) \, dS \tag{5.259}$$

are moments of the displaced volume $V_p$ and of the water plane area $S_{wp}$ of body $p$. Equations (4.39) and (5.252) have been used to obtain Eq. (5.257).

The small-body approximation corresponds to replacing the velocity potential $\hat{\phi}_0$ and its derivatives by the first (zero-order) term of their Taylor expansions about the reference point $(x_p, y_p, z_p)$. Thus, assuming that $\hat{\phi}_0$ is as given by Eq. (4.83) or by Eqs. (4.88) and (4.98), we find it evident that the error of the approximation is small, of order $\mathcal{O}\{ka\}$ as $ka \to 0$, where

$$a = \max(|\vec{s}_p|). \tag{5.260}$$

Hence, we expect the approximation to be reasonably good if $ka \ll 1$—that is, if the linear extension of the body is very small compared to the wavelength. For this reason, we call it the small-body approximation or, alternatively, the long-wavelength approximation. The choice of the reference point is of little importance if $k \to 0$. However, in order for the approximation to be reasonably good also for cases when $k$ is not close to zero, particular choices may be better than others. To assist in a good selection of the reference point, comparisons with experiments or with numerical computations may be useful.

Before approximating the diffraction force, we make the further assumption that not only body $p$ but also each of the $N$ bodies have horizontal and vertical extensions very small compared with one wavelength. For each body $p'$, we may expand $\hat{a}_{0q}$ in the integrand of Eq. (5.255) as a Taylor series around the reference point $(x_{p'}, y_{p'}, z_{p'})$, and if we then neglect terms of order $\mathcal{O}\{ka\}$ as

$ka \to 0$, the $j$ component ($j = 1, 2, \ldots, 6$) of the diffraction force on body $p$ may be approximated to

$$\hat{F}_{d,i} \approx \frac{1}{i\omega} \sum_{p'=1}^{N} \sum_{q=1}^{3} Z_{ii'} \hat{a}_{0p'q} = \sum_{p'=1}^{N} \sum_{q=1}^{3} Z_{ii'} \hat{v}_{0p'q}, \tag{5.261}$$

where $i = 6(p-1) + j$, $i' = 6(p'-1) + q$ and

$$Z_{ii'} = -i\omega\rho \iint_{S_{p'}} \varphi_i n_{p'q} dS \tag{5.262}$$

is an element of the radiation impedance matrix. See Eqs. (5.37), (5.154) and (5.167).

The excitation force is, in accordance with Eq. (5.234), given as the sum of the Froude–Krylov force and the diffraction force. Thus, the $j$ component of the excitation force on body $p$ is given by

$$\hat{F}_{e,i} = \hat{F}_{FK,i} + \hat{F}_{d,i}, \tag{5.263}$$

where $\hat{F}_{d,i}$ is given by Eq. (5.261) and $\hat{F}_{FK,i}$ by Eq. (5.256) for a translational mode ($j = q = 1, 2, 3$) or by Eq. (5.257) for a rotational mode ($j = q + 3 = 4, 5, 6$).

### 5.6.4 Small-Body Approximation for a Single Body

In the remaining part of this section, let us consider the case of only one body, and thus, we have $N = 1$, $p = p' = 1$, $i = j$ and $i' = q$. Hence, we may omit the subscript $p$ on $\hat{a}_{0pq}$, $\hat{v}_{0pq}$ and $\hat{\eta}_{0p}$. Then the sum over $p'$ in Eq. (5.261) has only one term, and, hence, the diffraction force is given by

$$\hat{F}_{d,j} \approx \frac{1}{i\omega} \sum_{q=1}^{3} Z_{jq} \hat{a}_{0q} = \sum_{q=1}^{3} Z_{jq} \hat{v}_{0q}, \tag{5.264}$$

which, in view of Eq. (5.42), may be written as

$$\hat{F}_{d,j} \approx \sum_{q=1}^{3} (m_{jq} \hat{a}_{0q} + R_{jq} \hat{v}_{0q}). \tag{5.265}$$

As we observed in Section 5.2.4, the radiation impedance for a small body is usually dominated by the radiation reactance; that is,

$$|\omega m_{jq}| = |X_{jq}| \gg R_{jq}. \tag{5.266}$$

On deep water, it can be shown that (see the following paragraphs)

$$|R_{33}/\omega m_{33}| = \mathcal{O}\{ka\} \tag{5.267}$$

(and even smaller for other $jq$ combinations) as $ka \to 0$. Then the diffraction force is

$$\hat{F}_{d,j} \approx \sum_{q=1}^{3} m_{jq} \hat{a}_{0q}. \tag{5.268}$$

This is a consistent approximation when we neglect terms of order $\mathcal{O}\{ka\}$ relative to terms of $\mathcal{O}\{1\}$. Kyllingstad [64] has given a more consistent higher-order approximation than Eq. (5.265).

The free-surface boundary condition is, in the case of deep water,

$$0 = \left( -\frac{\omega^2}{g} + \frac{\partial}{\partial z} \right) \varphi_j = \left( -k + \frac{\partial}{\partial z} \right) \varphi_j. \tag{5.269}$$

When $k \to 0$, this condition becomes

$$\frac{\partial \varphi_j}{\partial z} = 0 \quad \text{on } z = 0. \tag{5.270}$$

That is, it corresponds to a stiff plate (or ice) on the water. In this case, no surface wave can propagate on the surface as a result of the body's (slow!) oscillation. However, there is still a finite added mass in the limit $k \to 0$, because there is some finite amount of kinetic energy in the fluid; see Eq. (5.190). Small-body approximations (5.256)–(5.257) and reciprocity relation (5.144) show explicitly for a body floating on deep water (with $D = 1$) that when $ka \to 0$, $R_{jq}/\omega$ is of $\mathcal{O}\{k\}$ and of $\mathcal{O}\{a^4\}$ for $(j,q) = (3,3)$ and even smaller for other possible $jq$ combinations. Moreover, in agreement with the text which follows Eq. (5.85) [also see Eq. (5.84)], it is reasonable to assume that $m_{jq}$ is of $\mathcal{O}\{k^0\}$ and of $\mathcal{O}\{a^3\}$. Hence, $|R_{jq}/\omega m_{jq}|$ is of $\mathcal{O}\{ka\}$ or smaller.

Neglecting terms of order $\mathcal{O}\{ka\}$, we have, for the translational modes ($j = 1, 2, 3$), the following approximation for the excitation force on a single small body:

$$\hat{F}_{e,j} \approx \rho g \hat{\eta}_0 S_w \delta_{j3} + \rho V \hat{a}_{0j} + \sum_{q=1}^{3} m_{jq} \hat{a}_{0q}. \tag{5.271}$$

For the rotational modes ($j = 4, 5, 6$), an approximation to the excitation moment of the small body is obtained by the addition of the two approximations (5.257) and (5.268).

Let us consider the excitation force on a body having two mutually orthogonal vertical planes of symmetry (e.g., the planes $x = 0$ and $y = 0$). A special case of this is an axisymmetric body (the $z$-axis being the axis of symmetry). Because in this case $\varphi_3$ is symmetric, whereas $n_1$ and $n_2$ are antisymmetric, it follows from definition (5.38) of the radiation impedance that $Z_{13} = 0$ and $Z_{23} = 0$, and hence, $m_{13} = m_{23} = 0$. Based on a similar argument, $m_{12} = 0$. For $j = 1, 2, 3$, the preceding approximation for $F_{e,j}$ simplifies to

$$\hat{F}_{e,j} \approx \rho g \hat{\eta}_0 S_w \delta_{j3} + \rho V (1 + \mu_{jj}) \hat{a}_{0j}, \tag{5.272}$$

where

$$\mu_{jj} = m_{jj}/\rho V \tag{5.273}$$

is the non-dimensionalised added-mass coefficient. If the undisturbed velocity potential represents a purely propagating wave, as given by Eqs. (4.88) and (4.98), and if the reference axis $p$–$p$ of the single body (see Figure 5.13) is the vertical line $(x, y) = (0, 0)$, then $\hat{\eta}_0 = A$. Inserting this into Eq. (5.272), we obtain, for the heave excitation force,

$$F_{e,3} = F_{\text{FK},3} + F_{d,3} \approx \left[ \rho g S_w + (\rho V + m_{33}) \frac{\hat{a}_{03}}{A} \right] A. \tag{5.274}$$

Using Eq. (5.245) gives

$$\hat{F}_{e,3} \approx \left[ \rho g S_w - \omega^2 (\rho V + m_{33}) \frac{\sinh(kz_1 + kh)}{\sinh(kh)} \right] A, \tag{5.275}$$

where $z_1$ is the $z$-coordinate of the reference point for the body. On deep water, we now have

$$\hat{F}_{e,3} \approx [\rho g S_w - \omega^2 \rho V (1 + \mu_{33}) \exp(kz_1)] A. \tag{5.276}$$

For a small floating body with $z_1 \approx 0$, this gives

$$\hat{F}_{e,3} \approx [\rho g S_w - \omega^2 \rho V (1 + \mu_{33})] A. \tag{5.277}$$

Note that as $\omega \to 0$, $\hat{F}_{e,3}/A \to \rho g S_w$, which is the hydrostatic (buoyancy) stiffness of the floating body (see Section 5.9.1).

For small values of $\omega$, the first term $\rho g S_w$ dominates over the second term $-\omega^2 \rho V (1 + \mu_{33})$. On deep water, where $\omega^2 = gk$, the second term is approximately linear in $k$ (forgetting the usually small frequency dependence of $\mu_{33}$). This corresponds to the tangent at $ka = 0$ of the curve indicated in Figure 5.15.

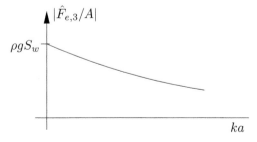

Figure 5.15: Typical variation of the amplitude of the heave force on a floating body for relatively small values of the angular repetency.

For most body shapes, the heave excitation force $\hat{F}_{e,3}$ is closely in phase with the undisturbed wave amplitude $A$ for those small values of $ka$, where the small-body approximation is applicable. An exception is a body shape with a relatively small water plane area, as indicated in Figure 5.16. Then the second term of the formula may become the dominating term even for a frequency interval within the range of validity of the small-body approximation. Then there is a phase shift of $\pi$ between $\hat{F}_{e,3}$ and $A$ within a narrow frequency interval near the angular frequency $\{gS_w/V(1 + \mu_{33})\}^{1/2}$.

For a submerged body ($S_w = 0$), we have

$$\hat{F}_{e,3} \approx -\omega^2 \rho V(1 + \mu_{33}) \exp(kz_1)A. \tag{5.278}$$

Now $\hat{F}_{e,3}$ and $A$ are in antiphase everywhere within the region of validity of the approximation. This antiphase may be simply explained by the fact that the hydrodynamic pressure, which is in phase with the wave elevation [cf. Eq. (4.89)], is smaller below the submerged body than above it. It should be observed that even if the heave excitation force in the small-body approximation is in antiphase with the wave elevation, the body's vertical motion is essentially in phase with the wave elevation, provided no other external forces are applied to the submerged body. This follows from the fact that the body's mechanical impedance is dominated by inertia. Then the acceleration and excursion are essentially in phase and antiphase, respectively, with the excitation force. (See also Sections 2.2 and 5.9.) For small values of $ka = \omega^2 a/g$, the preceding

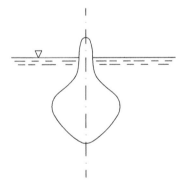

Figure 5.16: Floating body with a relatively small water plane area.

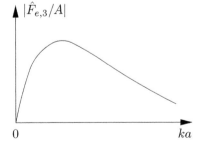

Figure 5.17: Typical variation of the amplitude of the heave force on a submerged body versus the angular repetency.

approximate formula (5.278) for $\hat{F}_{e,3}$ corresponds to the first increasing part of the force amplitude curve shown in Figure 5.17. The decreasing part of the curve for larger values of $ka$ is due to the exponential factor in formula (5.278).

For a horizontal cylinder whose centre is submerged at least one diameter below the free water surface, the added mass coefficients $\mu_{33} = \mu_{11}$ are close to 1, which is their exact value in an infinite fluid (without any free water surface) [23]. A similar approximation applies for the added-mass coefficients $\mu_{11}$, $\mu_{22}$ and $\mu_{33}$ for a sufficiently submerged sphere where the added-mass coefficient is $\frac{1}{2}$ in an infinite fluid (cf. [28], p. 144).

## 5.7 Axisymmetric Oscillating System

A group of concentric axisymmetric bodies is an axisymmetric system. A case with three bodies is shown in Figure 5.18. Concerning the sea bed, we have to assume either deep water or a sea-bed structure which is axisymmetric and concentric with the bodies. (The latter is true if the water depth is constant and equal to $h$.) Because the reference axis ($p$–$p$ in Figure 5.13) is the same for all bodies, we may choose it as the $z$-axis. Correspondingly, the global coordinates $(x, y, z) = (r \cos \theta, r \sin \theta, z)$ coincide with the local coordinates $(x_p, y_p, z) = (r_p \cos \theta_p, r_p \sin \theta_p, z)$ associated with each body $p$ (for all $p$).

For an axisymmetric body $p$, as shown in Figure 5.19, a wet-surface element $dS$ in position has a unit normal

$$\vec{n}_p = (n_{px}, n_{py}, n_{pz}) = (n_{pr} \cos \theta, n_{pr} \sin \theta, n_{pz}) = (n_{p1}, n_{p2}, n_{p3}). \qquad (5.279)$$

Further,

$$\vec{s}_p \times \vec{n}_p = (n_{p4}, n_{p5}, n_{p6}), \qquad (5.280)$$

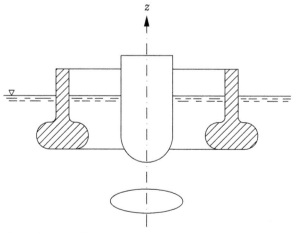

Figure 5.18: Axisymmetric system consisting of three concentric axisymmetric bodies: two floating and one submerged.

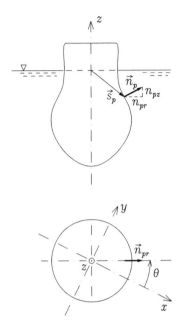

Figure 5.19: Side view and top view of a body symmetric with respect to the $z$-axis, with unit normal $\vec{n}_p$ on element $dS$ of the wet surface in position $\vec{s}_p$ from the reference point $(0, 0, z_p) = (0, 0, 0)$.

where

$$n_{p4} = yn_{pz} - zn_{py} = (rn_{pz} - zn_{pr})\sin\theta \equiv -n_{pM}\sin\theta$$
$$n_{p5} = zn_{px} - xn_{pz} = (zn_{pr} - rn_{pz})\cos\theta = n_{pM}\cos\theta \qquad (5.281)$$
$$n_{p6} = xn_{py} - yn_{px} = rn_{pr}(\cos\theta\sin\theta - \sin\theta\cos\theta) = 0.$$

To summarise, for $j = 1, \ldots, 6$, the components of the unit normal may be collected into the column vector

$$\mathbf{n}_p = (n_{pr}\cos\theta, n_{pr}\sin\theta, n_{pz}, -n_{pM}\sin\theta, n_{pM}\cos\theta, 0)^{\mathrm{T}}, \qquad (5.282)$$

where

$$n_{pM} = zn_{pr} - rn_{pz}, \qquad n_{pr} = \sqrt{n_{px}^2 + n_{py}^2}. \qquad (5.283)$$

Notice that for $j = 6$, boundary condition (5.166) is satisfied if $\varphi_{p6} = 0$, which was to be expected because an axisymmetric body yawing in an ideal fluid can generate no wave. The complex functions $\varphi_{pj}(j = 1, \ldots, 5)$ must satisfy the Laplace equation, and the usual homogeneous boundary conditions on the free surface $z = 0$ and on fixed surfaces such as the sea bed $z = -h$. Moreover, $\varphi_{pj}$ must satisfy the radiation condition.

For an axisymmetric body, it is easy to see that a particular solution of the type

$$\varphi_{pj} = \varphi_{pj}(r, \theta, z) = \varphi_{pj0}(r, z)\Theta_j(\theta) \qquad (5.284)$$

satisfies Laplace's equation, provided $\Theta_j(\theta)$ is a function of the type given by Eq. (4.200). If we choose

$$\Theta_1(\theta) = \Theta_5(\theta) = \cos\theta$$
$$\Theta_2(\theta) = \Theta_4(\theta) = \sin\theta \qquad (5.285)$$
$$\Theta_3(\theta) = \Theta_6(\theta) = 1,$$

then also the inhomogeneous boundary condition (5.166) on $S_p$ can be satisfied in accordance with Eq. (5.282). (Note, however, that $\varphi_{p60} \equiv 0$.)

In the far-field region, we have, asymptotically, according to Eqs. (4.211) and (4.214),

$$\varphi_{pj} \sim (\text{constant}) \times \Theta_j(\theta)e(kz)H_n^{(2)}(kr), \qquad (5.286)$$

with $n = 0$ for $j = 3$ and $n = 1$ for $j = 1, 2, 4$ and 5.

Since the solution for $\varphi_{pj}(r, \theta, z)$ is of the type of Eq. (5.284) also in the far-field region, we may write the Kochin functions and the far-field coefficients as follows:

$$h_{pj}(\theta) = h_{pj0}\Theta_j(\theta)$$
$$a_{pj}(\theta) = a_{pj0}\Theta_j(\theta). \qquad (5.287)$$

By comparing Eqs. (5.284) and (4.236), we see that in the far-field region, all the functions $\varphi_{pj0}(r, z)$ vary in the same way with $r$ and $z$, since asymptotically we have

$$\varphi_{pj0}(r, z) \sim a_{pj0}e(kz)(kr)^{-1/2}e^{-ikr} \qquad (5.288)$$

for $kr \to \infty$. Note that $a_{pj0}$ is independent of $r$, $\theta$ and $z$.

Similarly, we may, according to Haskind relation (5.227), write the excitation force coefficients as

$$f_{pj}(\beta) = f_{pj0}\Theta_j(\beta \pm \pi), \qquad (5.289)$$

where

$$f_{pj0} = \frac{\rho g D}{k}h_{pj0} = \frac{\rho g D}{k}\sqrt{2\pi}\,a_{pj0}\,e^{i\pi/4}. \qquad (5.290)$$

Note from Eq. (5.285) that $\Theta_j(\beta \pm \pi) = -\Theta_j(\beta)$ for $j = 1, 2, 4, 5$.

### 5.7.1 The Radiation Impedance

The radiation impedance matrix $\mathbf{Z}$ is, according to Eq. (5.158), partitioned into $6 \times 6$ matrices $\mathbf{Z}_{pp'}$ of which each element is of the type [see Eqs. (5.166) and (5.169)]

$$Z_{pj,p'j'} = -i\omega\rho \iint_{S_p} \varphi_{p'j'}\frac{\partial\varphi_{pj}^*}{\partial n}\,dS. \qquad (5.291)$$

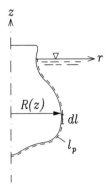

Figure 5.20: Wet surface $S_p$ of an axisymmetric body is a surface of revolution generated by curve $l_p$.

Let us consider the wet body surface $S_p$ as a surface of revolution generated by curve $l_p$, as shown in Figure 5.20. Then

$$Z_{pj,p'j'} = -i\omega\rho \int_0^{2\pi} d\theta \int_{l_p} dl\, R(z)\varphi_{p'j'}\frac{\partial\varphi_{pj}^*}{\partial n}$$

$$= -i\omega\rho \int_0^{2\pi} \Theta_{j'}(\theta)\Theta_j(\theta)d\theta \int_{l_p} \varphi_{p'j'0}\frac{\partial\varphi_{pj0}^*}{\partial n}R(z)dl. \tag{5.292}$$

Note that $\varphi_{pj0} = \varphi_{pj0}(r,z)$. Now, since

$$\int_0^{2\pi} (\cos\theta, \sin\theta, \sin\theta\cos\theta)d\theta = (0,0,0), \tag{5.293}$$

$$\int_0^{2\pi} (\cos^2\theta, \sin^2\theta, 1^2)d\theta = (\pi, \pi, 2\pi), \tag{5.294}$$

the radiation impedance matrix $\mathbf{Z}_{pp'}$ is of the form

$$\mathbf{Z}_{pp'} = \begin{bmatrix} Z_{p1p'1} & 0 & 0 & 0 & Z_{p1p'5} & 0 \\ 0 & Z_{p1p'1} & 0 & Z_{p1p'5} & 0 & 0 \\ 0 & 0 & Z_{p3p'3} & 0 & 0 & 0 \\ 0 & Z_{p1p'5} & 0 & Z_{p5p'5} & 0 & 0 \\ Z_{p1p'5} & 0 & 0 & 0 & Z_{p5p'5} & 0 \\ 0 & 0 & 0 & 0 & 0 & 0 \end{bmatrix}, \tag{5.295}$$

where we also utilised the fact that for the axisymmetric system, we have $Z_{p2p'2} = Z_{p1p'1}$, $Z_{p4p'4} = Z_{p5p'5}$, $Z_{p2p'4} = Z_{p1p'5}$ and $\varphi_{p6} = \varphi_{p60} \equiv 0$. Note that matrix $\mathbf{Z}_{pp'}$ is singular. Its rank is at most 5. [An $n \times n$ matrix is of rank $k$ ($k \le n$) if there in the matrix exists a non-vanishing minor of dimension $k \times k$, but no larger.]

The *non-vanishing* elements of $\mathbf{Z}$ may be expressed as

$$Z_{pj,p'j'} = -i\omega\rho(1 + \delta_{3j}\delta_{3j'})\pi \int_{l_p} \varphi_{p'j'0}\frac{\partial\varphi_{pj0}^*}{\partial n}R(z)dl, \tag{5.296}$$

where $\delta_{3j} = 1$ if $j = 3$ and $\delta_{3j} = 0$ otherwise. Thus, we have for the radiation impedance matrix,

$$Z_{pj,p'j'} = -i\omega\rho\sigma_{jj'}\pi \int_{l_p} \varphi_{p'j'0} \frac{\partial \varphi^*_{pj0}}{\partial n} R(z)dl, \tag{5.297}$$

where 27 of the 36 values of the numbers $\sigma_{jj'}$ vanish. There are just nine non-vanishing values, namely

$$\sigma_{33} = 2, \tag{5.298}$$

$$\sigma_{11} = \sigma_{22} = \sigma_{44} = \sigma_{55} = \sigma_{15} = \sigma_{51} = \sigma_{24} = \sigma_{42} = 1. \tag{5.299}$$

### 5.7.2 Radiation Resistance and Excitation Force

The radiation resistance matrix $\mathbf{R}_{pp'}$ is obtained by taking the real part of Eq. (5.295), which shows that at least 27 of the 36 elements vanish. To calculate the non-vanishing elements $R_{pj,p'j'}$, we take the integral along a generatrix $r =$ constant ($r \to \infty$) of the control cylinder $S_\infty$ in the far field (Figure 5.13) instead of along the curve $l_p$—that is, $r = R(z)$—as in Eq. (5.297) [cf. Eqs. (5.171) and (5.177)].

Thus, the radiation resistance matrix is

$$R_{pj,p'j'} = -i\omega\rho\sigma_{jj'}\pi \lim_{r\to\infty} \int_{-h}^{0} \varphi_{p'j'0} \frac{\partial \varphi^*_{pj0}}{\partial r} rdz. \tag{5.300}$$

From expression (5.288)—also see Eqs. (4.272), (5.284) and (5.287)—we have the asymptotic expression

$$\varphi_{pj0}(r,z) \sim h_{pj0}e(kz)(2\pi kr)^{-1/2}e^{-i(kr+\pi/4)}. \tag{5.301}$$

Thus, we have

$$\begin{aligned} R_{pj,p'j'} &= -i\omega\rho\sigma_{jj'}\pi h_{p'j'0}h^*_{pj0}\frac{ik}{2\pi k} \int_{-h}^{0} e^2(kz)dz \\ &= \frac{\omega\rho D(kh)}{4k}\sigma_{jj'}h_{p'j'0}h^*_{pj0} = \frac{\omega k}{4\rho g^2 D(kh)}\sigma_{jj'}f_{p'j'0}f^*_{pj0}, \end{aligned} \tag{5.302}$$

where we have used Eqs. (4.108) and (5.290). We note from Eq. (5.302) that the diagonal elements $R_{pj,pj}$ of the radiation resistance matrix are necessarily nonnegative. This is a result which was to be expected, because otherwise the principle of conservation of energy would be violated. If the only oscillation is mode $j$ of body $p$, the radiated power in accordance with Eq. (5.179) is $\frac{1}{2}R_{pj,pj}|\hat{u}_j|^2$, which, of course, cannot be negative.

Since the radiation resistance matrix is real, it follows from Eq. (5.302) that for those combinations of $j$ and $j'$ for which $\sigma_{jj'}$ does not vanish, we have that

$$f_{pj0}f^*_{p'j'0} = f^*_{pj0}f_{p'j'0} \tag{5.303}$$

is real. This is also in agreement with reciprocity relation (5.174), from which it follows that

$$R_{p'j',pj} = R_{pj,p'j'}.$$ (5.304)

Hence,

$$\frac{f_{p'j'0}}{f^*_{p'j'0}} = \frac{f_{pj0}}{f^*_{pj0}} \quad \text{or} \quad \frac{f^2_{p'j'0}}{|f_{p'j'0}|^2} = \frac{f^2_{pj0}}{|f_{pj0}|^2}.$$ (5.305)

Taking the square root gives

$$\frac{f_{p'j'0}}{|f_{p'j'0}|} = \pm\frac{f_{pj0}}{|f_{pj0}|},$$ (5.306)

which shows that the excitation force coefficients $f_{p'j'0}$ and $f_{pj0}$ have equal or opposite phases. Note that this applies only if $\sigma_{jj'} \neq 0$. Moreover, note from Eq. (5.302) that

$$R_{pj,p'j'} = \pm\sqrt{R_{pj,pj}R_{p'j',p'j'}}$$ (5.307)

is negative if and only if $f_{pj0}$ and $f_{p'j'0}$ have opposite phases.

Considering just body $p$, we find that the excitation surge force and pitch moment are in equal or opposite phases, and the same can be said about the excitation sway force and roll moment. The only non-vanishing off-diagonal elements of $\mathbf{R}_{pp}$ are

$$R_{p1,p5} = R_{p5,p1} = \pm\sqrt{R_{p1,p1}R_{p5,p5}},$$ (5.308)

$$R_{p2,p4} = R_{p4,p2} = \pm\sqrt{R_{p2,p2}R_{p4,p4}},$$ (5.309)

which are equal due to the axial symmetry. From this, it follows that $\mathbf{R}_{pp}$ is a more singular matrix than $\mathbf{Z}_{pp}$. Their ranks are at most 3 and 5, respectively (see Problem 5.13).

Let us now consider a system of two bodies. Then, since $\sigma_{33} = 2 \neq 0$, Eq. (5.306) shows that the heave excitation forces for the two bodies have either equal or opposite phases. If both bodies are floating (as in the upper part of Figure 5.18), we know that they have the same phase if the frequency is sufficiently low for the small-body approximation (5.272) to be applicable. Then $R_{p3,p'3}$ is positive. However, if one body is floating while the other is submerged (as if the largest, annular body were removed from Figure 5.18), then, since the latter has a vanishing water plane area, the heave excitation forces are in opposite phases for low frequencies where the small-body approximation is applicable. In this case, $R_{p3,p'3}$ is negative. Moreover, for this system, the excitation surge forces and pitch moments for the two bodies have equal or opposite phases. The same may be said about the excitation sway forces and roll moments for the two bodies.

For the two-body system, the radiation impedance matrix $\mathbf{Z}$, a $12 \times 12$ matrix, has the structure

$$\mathbf{Z} = \begin{bmatrix} \mathbf{Z}_{pp} & \mathbf{Z}_{pp'} \\ \mathbf{Z}_{p'p} & \mathbf{Z}_{p'p'} \end{bmatrix}. \tag{5.310}$$

We see from Eq. (5.295) that at least $4 \times 27 = 108$ of the $4 \times 36 = 144$ matrix elements vanish. Among the $4 \times 9 = 36$ remaining elements, no more than 23 different values are apparently possible because of the symmetry relation $\mathbf{Z}^T = \mathbf{Z}$; see Eq. (5.173). However, there are in fact only 13 different nonzero values if we do not forget that even though the matrix (5.295) has only 9 non-vanishing elements, these 9 elements share only 4 or 5 different values (for $p' = p$ or $p' \neq p$, respectively). The matrix $\mathbf{Z}$ is singular, and its rank is at most 10. However, as a result of relation (5.302), the rank of the radiation resistance matrix $\mathbf{R} = \mathrm{Re}\{\mathbf{Z}\}$ cannot be more than 3. A physical explanation of this fact will be given in the next chapter (see Section 6.5.2). Let us here just mention that the maximum rank of 3 is related to the fact that there is only three different linearly independent functions $\Theta_j(\theta)$, as given in Eq. (5.285).

If a third body is included in our axisymmetric system, the rank of the $18 \times 18$ matrix $\mathbf{Z}$ cannot be larger than 15, but the rank of the $18 \times 18$ matrix $\mathbf{R}$ is still no more than 3.

If the radiation resistance matrix is singular, it is possible to force the bodies to oscillate in still water, without radiating a wave in the far-field region. Such a cancellation is possible in the far-field region [see asymptotic approximation (5.288)], but not in general in the near-field region, where fluid motion is associated with the added-mass matrix [see Eqs. (5.189)–(5.190)]. As an example, consider surge and pitch oscillations of a single body. Since $\Theta_1(\theta) = \Theta_5(\theta) = \cos \theta$ according to Eq. (5.285), it is possible to choose a combination of $\hat{u}_{11}$ and $\hat{u}_{15}$ such that the far-field waves radiated by the surge mode and the pitch mode cancel each other. From Eqs. (5.163), (5.284), (5.288) and (5.290), this is achieved if

$$\frac{\hat{u}_{11}}{\hat{u}_{15}} = -\frac{a_{150}}{a_{110}} = -\frac{h_{150}}{h_{110}} = -\frac{f_{150}}{f_{110}}. \tag{5.311}$$

A similar interrelationship holds for a combined sway and roll oscillation that does not produce any wave in the far field. Hence, a single immersed axisymmetric body which has a complex velocity amplitude represented, for instance, by the six-dimensional column vector

$$\hat{\mathbf{u}} = \left[ -\frac{a_{150}}{a_{110}} \hat{u}_{15} \quad -\frac{a_{140}}{a_{120}} \hat{u}_{14} \quad 0 \quad \hat{u}_{14} \quad \hat{u}_{15} \quad \hat{u}_{16} \right]^T \tag{5.312}$$

does not produce any far-field radiated wave. Observe that the number of mutually independent velocities is three.

Note that for a single axisymmetric body, it is not possible for the far-field radiated wave to vanish unless $\hat{u}_{13} = 0$. However, it is possible for an axisymmetric system composed of two or more bodies to have nonzero heave velocities and a vanishing far-field radiated wave. For example, with a system of two bodies, the isotropic part of the far-field radiated wave is cancelled, provided the heave motions of the two bodies are related through

$$\frac{\hat{u}_{23}}{\hat{u}_{13}} = -\frac{a_{130}}{a_{230}}. \tag{5.313}$$

The $\cos\theta$ part is cancelled, provided the pitch and surge motions of the two bodies are related through

$$\hat{u}_{25} = -(a_{110}u_{11} + a_{150}u_{15} + a_{210}u_{21})/a_{250}. \tag{5.314}$$

Similarly, the $\sin\theta$ part is cancelled, provided the roll and sway motions of the two bodies are related through

$$\hat{u}_{24} = -(a_{120}u_{12} + a_{140}u_{14} + a_{220}u_{22})/a_{240}. \tag{5.315}$$

Moreover, any yaw motions of the two bodies do not radiate any waves. Thus, we have nine mutually independent velocities. Observe that the number of mutually independent velocities equals $6N - r$, the difference between dimensionality $6N$ and rank $r$ of the radiation resistance matrix **R**.

### 5.7.3 Example: Two-Body System

Consider two bodies with a common vertical axis of symmetry. One body (number 1) is floating, whereas the other (number 2) is submerged, and we take only the heave mode into account. Then, according to the numbering scheme of Eq. (5.154), the force-excitation coefficients of interest are $f_3(\omega)$ and $f_9(\omega)$, both of which are independent of the angle $\beta$ of wave incidence. Correspondingly, the radiation impedance matrix $\mathbf{Z}(\omega) = \mathbf{R}(\omega) + i\omega\mathbf{m}(\omega)$ has only the following elements of interest: $Z_{33}$, $Z_{99}$ and $Z_{39} = Z_{93}$. It follows from Eqs. (4.130), (5.298) and (5.302) that

$$\frac{f_i f_{i'}^*}{8R_{ii'}} = \frac{J}{k|A|^2} = \frac{J\lambda}{2\pi|A|^2} = \frac{\rho g v_g \lambda}{4\pi}, \tag{5.316}$$

where the subscripts $i$ and $i'$ are 3 or 9. Furthermore, $\lambda = 2\pi/k$ is the wavelength, and $v_g$ is the group velocity. (See also the last remark in Section 5.4.3.)

As an example, let us present some numerical results for the following geometry. The upper body is a floating cylinder with a hemisphere at the lower end. The radius is $a$, and the draft is $2.21a$ at equilibrium. The body displaces a water volume of $\pi a^3(1.21 + 2/3) \approx 5.89a^3$. The lower body is completely submerged, and it is shaped as an ellipsoid, with a radius of $2.27a$ and vertical extension $0.182a$. Its volume is $\pi a^3(2.27^2 \cdot 0.182)2/3 = 1.96a^3$. The centre of the ellipsoid is at a submergence $6.06a$, and the water depth is $h = 15a$.

For this geometry, hydrodynamic parameters have been computed using the computer code ©AQUADYN-2.1. The wet surfaces of the upper and lower bodies are approximated by 660 and 540 plane panels, respectively. The numerically computed hydrodynamic parameters are expected to have an accuracy of 2%–3% [65]. Excitation force coefficients $f_3(\omega)$ and $f_9(\omega)$ are shown in Figure 5.21. Observe that there is a frequency range, corresponding to $0.2 < ka < 0.4$, where the heave excitation forces on the two bodies are approximately equal but in opposite directions. We may refer to this as 'force compensation' [66, 67], and this has been utilised in some drilling platforms (semisubmersible floating vessels). Radiation resistances $R_{33}(\omega)$ and $R_{99}(\omega)$ are given by the curves in Figure 5.22. From Eq. (5.316), it follows that $R_{39}^2 = R_{93}^2 = R_{33}R_{99}$, and observing that $f_3/f_9 = -|f_3/f_9|$, we find that $R_{39} = R_{93}$ is negative:

$$R_{39} = R_{93} = -\sqrt{R_{33}R_{99}}. \tag{5.317}$$

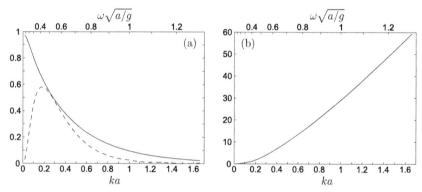

Figure 5.21: (a) Modulus of excitation force coefficients $|f_i(\omega)|/(\rho g \pi a^2)$ with $i = 3$ for the floating upper body (solid curve) and with $i = 9$ for the submerged lower body (dashed curve). (b) Phase angle $\angle f_3(\omega)$ (in degrees) for the heave excitation force coefficient of the floating upper body. It differs by $\pi$ (180 degrees) from the corresponding phase angle $\angle f_9(\omega)$ for the submerged lower body.

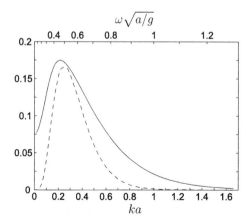

Figure 5.22: Normalised radiation resistances $R_{i,i'}(\omega)/(\omega \rho a^3 2\pi/3)$ for $(i, i') = (3, 3)$ (solid curve) and for $(i, i') = (9, 9)$ (dashed curve), versus normalised angular repetency $ka$ (lower scale) and normalised frequency $\omega \sqrt{a/g}$ (upper scale).

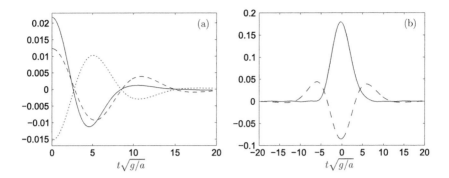

Figure 5.23: (a) Normalised radiation-problem impulse-response functions $k_{ii'}/(\pi\rho g a^2)$ versus normalised time $t\sqrt{g/a}$ for $(i,i') = (3,3)$ (solid curve), for $(i,i') = (9,9)$ (dashed curve) and for $(i,i') = (3,9)$ (dotted curve). (b) Normalised excitation-problem impulse-response function $f_{t,i}(t)(a/g)^{1/2}/(\pi\rho g a^2)$ versus normalised time $t(g/a)^{1/2}$ for $i = 3$ (solid curve) and for $i = 9$ (dashed curve).

The numerically computed values [65] of the added mass at infinite frequency are, in normalised quantity $m_{ii'}(\infty)/(\rho a^3 2\pi/3)$, given by 0.58, 15.0 and $-0.156$ for $(i,i')$ equal (3,3), (9,9) and (3,9), respectively. These numerical values are needed if the dynamics of the two-body system is to be investigated in the time domain [cf. Section 5.3.1 and also Eq. (5.365)]. For such an investigation, the causal impulse-response functions $k_{33}(t)$, $k_{99}(t)$ and $k_{39}(t) = k_{93}(t)$ are also needed. For $t > 0$, they are, according to Eq. (5.110), equal to twice the inverse Fourier transform of $R_{33}(\omega)$, $R_{99}(\omega)$ and $R_{39}(\omega) = R_{93}(\omega)$, respectively. The curves in Figure 5.23(a) show computed numerical results for these impulse-response functions. Observe that they, together with the numerical values of $m_{ii'}(\infty)$ stated earlier, constitute, in principle, the same quantitative information as the radiation impedances $Z_{33}(\omega)$, $Z_{99}(\omega)$ and $Z_{39}(\omega)$.

Impulse-response functions $f_{t,3}(t)$ and $f_{t,9}(t)$ are inverse Fourier transforms of the excitation-force coefficients $f_3(\omega)$ and $f_9(\omega)$, respectively. The curves in Figure 5.23(b) show computed numerical values for the axisymmetric two-body example. Observe that these impulse-response functions do not vanish for $t < 0$, and hence, they are noncausal, as explained in Section 5.3.2.

## 5.8 Two-Dimensional System

For a two-dimensional system, there is no variation in the $y$-direction, and wave propagation is in the $x$-direction. Each body has three degrees of freedom (or modes of motion)—namely, surge, heave and pitch—corresponding to $j = 1$, $j = 3$ and $j = 5$, respectively.

Let us use a prime symbol on variables to denote physical quantities per unit width. Thus, $F'$, $m'$, $Z'$ and $R'$ denote force, mass, impedance and resistance, respectively, per unit width of the two-dimensional system. The total wave force on body $p$ is per unit width given by

$$\hat{\mathbf{F}}'_{pt} = \hat{\mathbf{F}}'_{pe} + \hat{\mathbf{F}}'_{pr} = \hat{\mathbf{F}}'_{pe} - \sum_{p'} \mathbf{Z}'_{pp'} \hat{\mathbf{u}}_{p'}, \tag{5.318}$$

which is a three-dimensional column vector. The two terms result from the incident wave and from the bodies' radiation, respectively. The $3 \times 3$ coefficient matrix $\mathbf{Z}'_{pp'}$ represents the radiative interaction from body $p'$ on body $p$.

In analogy with Eq. (5.220), we may define $3N$-dimensional force vectors $\mathbf{F}'_t$ and $\mathbf{F}'_e$ and correspondingly $3N$-dimensional velocity vector $\mathbf{u}$, which are related by

$$\hat{\mathbf{F}}'_t = \hat{\mathbf{F}}'_e - \mathbf{Z}'\hat{\mathbf{u}}, \tag{5.319}$$

where the $3N \times 3N$ matrix $\mathbf{Z}'$ represents the radiation impedance per unit width. The radiated wave has a velocity potential given by

$$\hat{\phi}_r = \sum_{p=1}^{N} \boldsymbol{\varphi}_p^{\mathrm{T}} \hat{\mathbf{u}}_p = \boldsymbol{\varphi}^{\mathrm{T}}\hat{\mathbf{u}}, \tag{5.320}$$

where $\boldsymbol{\varphi}_p$ and $\boldsymbol{\varphi}$ are coefficient column vectors of dimension 3 and $3N$, respectively. In accordance with Eq. (4.246), the far-field (asymptotic) approximation may be written as

$$\boldsymbol{\varphi} \sim -\frac{g}{i\omega}\mathbf{a}^{\pm}e(kz)e^{\mp ikx} \quad \text{as } x \to \pm\infty. \tag{5.321}$$

Moreover, taking the complex conjugate of Eq. (4.248), we find

$$\mathbf{I}'(\boldsymbol{\varphi}^*, \boldsymbol{\varphi}^{\mathrm{T}}) = -i\left(\frac{g}{\omega}\right)^2 D(kh)(\mathbf{a}^{+*}\mathbf{a}^{+T} + \mathbf{a}^{-*}\mathbf{a}^{-T}), \tag{5.322}$$

and Eq. (4.249) gives

$$\mathbf{I}'(\boldsymbol{\varphi}, \boldsymbol{\varphi}^{\mathrm{T}}) = 0. \tag{5.323}$$

The two-dimensional Kochin function for radiation is given by

$$\mathbf{h}'(0) = -\frac{k}{D(kh)}\mathbf{I}'[e(kz)e^{ikx}, \boldsymbol{\varphi}] = -\frac{gk}{\omega}\mathbf{a}^{+}, \tag{5.324}$$

$$\mathbf{h}'(\pi) = -\frac{k}{D(kh)}\mathbf{I}'[e(kz)e^{-ikx}, \boldsymbol{\varphi}] = -\frac{gk}{\omega}\mathbf{a}^{-}, \tag{5.325}$$

in agreement with Eqs. (4.268), (4.285) and (4.286). The Kochin function $H'_d(\beta)$ ($\beta = 0, \pi$) for the diffracted wave is given by analogous expressions—with $\boldsymbol{\varphi}$, $\mathbf{a}^{\pm}$ and $\mathbf{h}'$ replaced by $\hat{\phi}_d$, $A_d^{\pm}$ and $H'_d$, respectively—in Eqs. (5.321)–(5.325). For the diffraction problem (that is, when the bodies are not moving), we may define the reflection coefficient

$$\Gamma_d = A_d^{-}/A \tag{5.326}$$

[see Eq. (4.84)] and the transmission coefficient

$$T_d = 1 + A_d^+/A. \tag{5.327}$$

These coefficients quantitatively describe how the incident wave of amplitude $|A|$ is divided into a reflected wave and a transmitted wave, which is propagating beyond the bodies. From energy conservation arguments, it is expected that

$$|\Gamma_d|^2 + |T_d|^2 = 1. \tag{5.328}$$

This can indeed be proved mathematically (see Problem 5.14).

Now let us consider some reciprocity relations involving the radiation resistance and the excitation force for a two-dimensional system. The radiation resistance per unit width is given by the $3N \times 3N$ matrix

$$\begin{aligned}
\mathbf{R}' &= \frac{1}{2}i\omega\rho\mathbf{I}'(\boldsymbol{\varphi}^*, \boldsymbol{\varphi}^T) = \frac{\rho g^2 D(kh)}{2\omega}(\mathbf{a}^{+*}\mathbf{a}^{+T} + \mathbf{a}^{-*}\mathbf{a}^{-T}) \\
&= \frac{\omega\rho D(kh)}{2k^2}[\mathbf{h}'^*(0)\mathbf{h}'^T(0) + \mathbf{h}'^*(\pi)\mathbf{h}'^T(\pi)] \tag{5.329} \\
&= \frac{\omega\rho D(kh)}{2k^2}[\mathbf{h}'(0)\mathbf{h}'^\dagger(0) + \mathbf{h}'(\pi)\mathbf{h}'^\dagger(\pi)].
\end{aligned}$$

Compare Eqs. (5.176), (5.322), (5.324) and (5.325). Observe that because $\mathbf{R}'$ is real, we may, if we wish, take the complex conjugate of Eq. (5.329) without changing the result. The excitation force per unit width is

$$\mathbf{F}'_e = \mathbf{f}'(\beta)A = i\omega\rho\mathbf{I}'(\hat{\phi}_0, \boldsymbol{\varphi}), \tag{5.330}$$

where, for $\beta = 0$ or $\beta = \pi$,

$$\mathbf{f}'(\beta) = \frac{\rho g D(kh)}{k}\mathbf{h}'(\beta \pm \pi). \tag{5.331}$$

That is,

$$\mathbf{f}'(0) = \frac{\rho g D(kh)}{k}\mathbf{h}'(\pi) = -\frac{\rho g^2 D(kh)}{\omega}\mathbf{a}^-, \tag{5.332}$$

$$\mathbf{f}'(\pi) = \frac{\rho g D(kh)}{k}\mathbf{h}'(0) = -\frac{\rho g^2 D(kh)}{\omega}\mathbf{a}^+. \tag{5.333}$$

Compare Eqs. (5.222), (5.324) and (5.325). Also, the two-dimensional version of Eq. (5.224) has been used. Note that the excitation force on a surging piston in a wave channel, derived previously [see Eq. (5.80)], agrees with the general equation (5.330) combined with Eq. (5.332).

For a single symmetric body oscillating in heave, the wave radiation is symmetric, and hence,

$$\begin{bmatrix} a_3^- \\ h_3'(\pi) \\ f_3'(0) \end{bmatrix} = \begin{bmatrix} a_3^+ \\ h_3'(0) \\ f_3'(\pi) \end{bmatrix}. \tag{5.334}$$

If it oscillates in surge and/or pitch, the radiation is antisymmetric, and hence,

$$\begin{bmatrix} a_j^- \\ h_j'(\pi) \\ f_j'(0) \end{bmatrix} = - \begin{bmatrix} a_j^+ \\ h_j'(0) \\ f_j'(\pi) \end{bmatrix} \quad \text{for } j = 1 \text{ and } j = 5. \tag{5.335}$$

Thus, according to Eq. (5.329), the radiation resistance matrix for the two-dimensional symmetric body is

$$\mathbf{R}' = \frac{\rho g^2 D(kh)}{\omega} \begin{bmatrix} |a_1^+|^2 & 0 & a_5^+ a_1^{+*} \\ 0 & |a_3^+|^2 & 0 \\ a_1^+ a_5^{+*} & 0 & |a_5^+|^2 \end{bmatrix}. \tag{5.336}$$

In this matrix, it is possible to find a non-vanishing minor of dimension $2 \times 2$, but not of dimension $3 \times 3$. Hence, $\mathbf{R}'$ is a singular matrix of rank no more than 2.

Since $\mathbf{R}'$ is a symmetrical and real matrix, we may use Eqs. (5.332) and (5.333) and then argue as we did in connection with Eqs. (5.303)–(5.308). We then conclude that the off-diagonal element of $\mathbf{R}'$ is positive or negative when the excitation force coefficients for surge and pitch for the symmetric two-dimensional body are in the same phase or in opposite phases, respectively.

It can be shown, in general, for a two-dimensional system of oscillating bodies, that the system's radiation resistance matrix has a rank which is at most 2. Let $\mathbf{x}$ and $\mathbf{y}$ be two arbitrary column vectors of order $3N$. Then the matrix $\mathbf{xy}^T$ is of rank no more than 1 [16, p. 239]. The matrix $\mathbf{R}'$ in Eq. (5.329) is a sum of two such matrices. Hence, $\mathbf{R}'$ is a matrix of rank 2 or smaller. (The rank is smaller than 2 if the column vectors $\mathbf{a}^+$ and $\mathbf{a}^-$ are linearly dependent.) A physical explanation of why the rank is bounded is given in Section 8.3.

## 5.9 Motion Response

When an immersed body is not moving, it is subjected to an excitation force $\mathbf{F}_e$ if an incident wave exists. If the body is oscillating, it is subjected to two other forces—namely the radiation force $\mathbf{F}_r$, which we have already considered [see Eq. (5.36) or (5.104)], and a hydrostatic buoyancy force $\mathbf{F}_b$. The latter force originates from the static-pressure term $-\rho g z$ in Eq. (4.12), because the body's wet surface experiences varying hydrostatic pressure as a result of its oscillation. However, oscillation in surge, sway or yaw modes of motion does not produce any hydrostatic force. In the following paragraphs, let us limit ourselves to the study of heave motion in more detail. For a more general treatment of hydrostatic buoyancy forces and moments, refer to textbooks by Newman [28, chapter 6] or Mei et al. [1, chapter 8]. In linear theory, we shall assume that the hydrostatic buoyancy force $\mathbf{F}_b$ is proportional to the excursion $\mathbf{s}$ of the body from its equilibrium position. We write

$$\mathbf{F}_b = -\mathbf{S}_b \mathbf{s}, \tag{5.337}$$

where the elements of the 'buoyancy stiffness' matrix $\mathbf{S}_b$ are coefficients of proportionality. Many of the 36 elements—for instance, all elements associated with surge or sway—vanish. Moreover, we note that the velocity is

$$\mathbf{u} = \frac{d}{dt}\mathbf{s}. \tag{5.338}$$

There may be additional forces acting on the oscillating body, such as a viscous force $\mathbf{F}_v$ and a control or load force $\mathbf{F}_u$ from a control mechanism or from a mechanism which delivers or absorbs energy. There may also be a friction force $\mathbf{F}_f$ due to this mechanism or other means for transmission of forces. Moreover, there may be mooring forces, which we shall not consider explicitly. We may assume that mooring forces are included in $\mathbf{F}_u$ (or in $\mathbf{F}_b$, if the mooring is flexible and linear theory applicable).

From Newton's law, the dynamic equation for the oscillating body may be written as

$$\mathbf{m}_m\frac{d^2}{dt^2}\mathbf{s} = \mathbf{F}_e + \mathbf{F}_r + \mathbf{F}_b + \mathbf{F}_v + \mathbf{F}_f + \mathbf{F}_u. \tag{5.339}$$

Matrix $\mathbf{m}_m$ represents the inertia of the oscillating body. The three first diagonal elements are $m_{m11} = m_{m22} = m_{m33} = m_m$, where $m_m$ is the mass of the body. Some of the off-diagonal elements of $\mathbf{m}_m$ involving rotary modes of motion may be different from zero (see [28, chapter 6] or [1, chapter 8]). For $j$ and $j'$ equal to 4, 5 or 6, $m_{mjj'}$ is of dimension inertia moment.

The term $\mathbf{F}_u$ is introduced to serve some purpose. It may be a control force intended to reduce the oscillation of a ship or floating platform in rough seas, for instance, or it may represent a load force necessary for conversion of ocean-wave energy. The terms $\mathbf{F}_v$ and $\mathbf{F}_f$ represent unavoidable viscous and friction effects. Thus, although we in Section 4.1 assumed an ideal fluid, in this time-domain formulation, we may make a practical correction by introducing nonzero loss forces in Eq. (5.339). In general, such forces may depend explicitly on time as well as on $\mathbf{s}$ and $\mathbf{u}$. For mathematical convenience, however, we shall assume that $\mathbf{F}_f$ may be written as

$$\mathbf{F}_f = -\mathbf{R}_f\mathbf{u} = -\mathbf{R}_f\frac{d\mathbf{s}}{dt}, \tag{5.340}$$

where $\mathbf{R}_f$ is a constant friction resistance matrix. A viscosity resistance matrix $\mathbf{R}_v$ may be defined in an analogous manner.

### 5.9.1 Dynamics of a Floating Body in Heave

In this subsection, we assume that the body is constrained to oscillate in the heave mode only. Then we may consider the matrices in Eqs. (5.337), (5.339) and (5.340) as scalars $S_b$, $m_m$ and $R_f$.

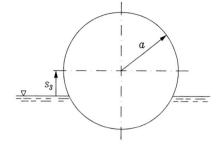

Figure 5.24: Floating sphere displaced a distance $s_3$ above its equilibrium semisubmerged position.

Let us first determine the buoyancy stiffness $S_b$. As an example, consider a sphere of radius $a$ and mass $m_m = \rho 2\pi a^3/3$. Thus, in equilibrium, it is semisubmerged. When the sphere is displaced upwards a distance $s_3$, as shown in Figure 5.24, the volume of displaced water is $\pi(2a^3 - 3a^2 s_3 + s_3^3)/3$, and hence, there is a restoring force $F_b = -\rho g \pi (a^2 s_3 - s_3^3/3)$, from which we see that the buoyancy stiffness is

$$S_b = \pi \rho g a^2 (1 - s_3^2/3a^2) \tag{5.341}$$

for $|s_3| < a$. For small excursions, such that $|s_3| \ll a$, we have

$$S_b \approx \pi \rho g a^2, \tag{5.342}$$

which is independent of $s_3$. We note that it is $\rho g$ multiplied by the water plane area.

For a floating body, in general, the buoyancy stiffness (for small heave excursion $s_3$) is

$$S_b = \rho g S_w, \tag{5.343}$$

where $S_w$ is the (equilibrium) water plane area of the body. For a floating vertical cylinder of radius $a$, we have $S_w = \pi a^2$. (For such a cylinder, the water plane area is independent of the heave position $s_3$. Then the hydrostatic stiffness $S_b$ is constant even if $s_3$ is not very small.) For a freely floating body, the weight equals $\rho g V$, where $V$ is the volume of displaced water at equilibrium.

For a body constrained to oscillate in the heave mode only, the dynamic equation (5.339) becomes

$$m_m \ddot{s}_3 = F_{e3} + F_{r3} + F_{b3} + F_{v3} + F_{f3} + F_{u3}. \tag{5.344}$$

In terms of complex amplitudes, this may be written as

$$\{-\omega^2[m_m + m_{33}(\omega)] + i\omega[R_v + R_f + R_{33}(\omega)] + S_b\}\hat{s}_3 = \hat{F}_{e3} + \hat{F}_{u3}, \tag{5.345}$$

when use has also been made of Eqs. (5.37), (5.42), (5.337), (5.338) and (5.340). Note that Eq. (5.345) corresponds to the Fourier transform of the time-domain equation (5.344) [cf. Eqs. (2.158) and (2.166)]. Introducing the 'intrinsic' transfer function

$$G_i(\omega) = S_b + i\omega R_l - \omega^2 m_m + i\omega Z_{33}(\omega), \tag{5.346}$$

where $R_l = R_v + R_f$ is the loss resistance, we write Eq. (5.345) simply as

$$G_i(\omega)\hat{s}_3 = \hat{F}_{e3} + \hat{F}_{u3} \equiv \hat{F}_{\text{ext}}. \tag{5.347}$$

We have adopted the adjective 'intrinsic' because all terms in $G_i$ relate to the properties of the oscillating system, and Eq. (5.347) describes quantitatively how the system responds to an external force $F_{\text{ext}} = F_{e3} + F_{u3}$, of which the first term $F_{e3}$ is given by an incident wave and the second term $F_{u3}$ is determined by an operator or by a control device assisting the operation of the oscillating body. In particular cases, $F_{e3}$ or $F_{u3}$ may vanish. In the radiation problem, for instance, there is no incident wave, and $F_{e3} = 0$ ($F_{\text{ext}} = F_{u3}$).

We also introduce the 'intrinsic mechanical impedance'

$$Z_i(\omega) = Z_{33}(\omega) + R_l + i\omega m_m + S_b[1/i\omega + \pi\delta(\omega)] \tag{5.348}$$

such that

$$G_i(\omega) = i\omega Z_i(\omega) \tag{5.349}$$

and

$$Z_i(\omega)\hat{u}_3 = \hat{F}_{e3} + \hat{F}_{u3} = \hat{F}_{\text{ext}}. \tag{5.350}$$

The last term with $\delta(\omega)$ in Eq. (5.348) secures that the inverse Fourier transform of $Z_i(\omega)$ is causal (see Problem 2.15). This term, which is of no significance if $\omega \neq 0$, may be of importance if the input and output functions in Eq. (5.350) were Fourier transforms rather than complex amplitudes. Note that this term does not explicitly show up in the expression (5.346) for $G_i(\omega)$ since $i\omega\delta(\omega) = 0$.

With given external force $\hat{F}_{e3} + \hat{F}_{u3}$, the solution of Eq. (5.350), for $\omega \neq 0$, is

$$\hat{u}_3 = \frac{\hat{F}_{\text{ext}}}{Z_i(\omega)} = \frac{\hat{F}_{\text{ext}}}{R_{33}(\omega) + R_l + i[\omega m_{33}(\omega) + \omega m_m - S_b/\omega]}, \tag{5.351}$$

where we have used Eq. (5.97) in Eq. (5.348). When the intrinsic reactance vanishes, we say that the system is in resonance. This happens for a frequency $\omega = \omega_0$ for which

$$\text{Im}\{Z_i(\omega_0)\} = 0 \quad \text{or} \quad \omega_0 = \sqrt{S_b/[m_m + m_{33}(\omega_0)]}. \tag{5.352}$$

Then the heave velocity is in phase with the external force. If $R_{33}(\omega)$ has a relatively slow variation with $\omega$, we also have maximum velocity amplitude response $|\hat{u}_3/\hat{F}_{\text{ext}}|$ at a frequency close to $\omega_0$.

Let us now for a moment consider a floating sphere of radius $a$, having a semisubmerged equilibrium position. When this sphere is heaving on deep water, the radiation impedance depends on the frequency, in accordance with Figure 5.6. The hydrostatic stiffness is $S_b = \rho g \pi a^2$, and the mass is $m_m = \rho(2\pi/3)a^3$. Furthermore, the added mass is $m_{33}(\omega) = m_m \mu_{33}$, where the dimensionless parameter $\mu_{33} = \mu_{33}(ka)$ is in the region $0.38 < \mu_{33} < 0.9$, as appears from Figure 5.6. Note that $\mu_{33} \rightarrow 0.5$ as $ka \rightarrow \infty$. Insertion into Eq. (5.352) gives the following condition for resonance of the heaving semisubmerged sphere:

$$\omega_0 = \sqrt{\frac{3g}{2a[1 + \mu_{33}(\omega_0)]}}. \tag{5.353}$$

We may solve for $\omega_0$ by using iteration. If, as a first approximation, we assume a value $\mu_{33} \approx 0.5$, we have $\omega_0 \approx \sqrt{g/a}$. Next, we may improve the approximation. For $\omega_0 = \sqrt{g/a}$—that is, $k_0 a = \omega_0^2 a/g = 1$—we have $\mu_{33} = 0.43$. Then we have the following corrected value for the angular eigenfrequency:

$$\omega_0 \approx \sqrt{3g/2a(1 + 0.43)} = 1.025\sqrt{g/a}, \tag{5.354}$$

and, correspondingly, $k_0 a = \omega_0^2 a/g \approx 1.05$. For a sphere of 20 m diameter ($a = 10$ m), the resonance frequency is given by $\omega_0 \approx 1.01$ rad/s, $f_0 = \omega_0/2\pi \approx 0.161$ Hz. The resonance period is $T_0 = 1/f_0 \approx 6.2$ s.

As another example, let us consider a floating slender cylinder of radius $a$ and depth of submergence $l$, as shown in Figure 5.25. We assume that $a \ll l$. The hydrostatic stiffness is $S_b = \rho g \pi a^2$, and the mass is $m_m = \rho \pi a^2 l$. Furthermore, the added mass is $m_{33} = m \mu_{33}$, where the dimensionless added-mass coefficient $\mu_{33}$ depends on $ka$, $l/a$ and $kh$. From [68, p. 49], we take the value

$$m_{33} = 0.167\rho(2a)^3 = 0.64(2\pi/3)a^3\rho, \tag{5.355}$$

which is the high-frequency limit ($ka \gg 1$) on deep water ($kh \gg 1$) when the cylinder is relatively tall ($l/a \gg 1$). For the floating, truncated, vertical cylinder discussed in Section 5.2.4 (cf. Figure 5.7), condition $l/a \gg 1$ is not satisfied, and the added mass appears to exceed the value given by Eq. (5.355) by a

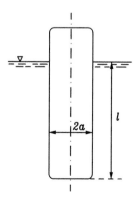

Figure 5.25: Floating vertical cylinder. The diameter is $2a$ and the depth of submergence is $l$.

factor between 1.4 and 1.7. If $l \gg a$, we have $m_m \gg m_{33}$, and the angular eigenfrequency (natural frequency) is

$$\omega_0 = \sqrt{S_b/(m_m + m_{33})} \approx \sqrt{S_b/m_m} = \sqrt{g/l}. \tag{5.356}$$

Correspondingly, the resonance period is $T_0 \approx 2\pi\sqrt{l/g}$, which amounts to $T_0 = 10$ s for a cylinder of draft $l = 25$ m.

Looking at Eqs. (5.347) and (5.350), we interpret $G_i(\omega)$ and $Z_i(\omega)$ as transfer functions of linear systems where the input functions are the heave excursion and heave velocity, respectively. For both systems, the output is the force, which we call external force. These linear systems 'produce' radiation force, friction force, restoring force, and inertial force, which are balanced by the external force.

However, if we consider the heave oscillation to be caused by the external force, it would seem more natural to consider linear systems in which input and output roles have been interchanged. Mathematically, these systems may be represented by

$$s_3(\omega) = H_i(\omega)F_{\text{ext}}(\omega), \tag{5.357}$$
$$u_3(\omega) = Y_i(\omega)F_{\text{ext}}(\omega), \tag{5.358}$$

where we have written Fourier transforms rather than complex amplitudes for the input and output quantities [cf. Eqs. (2.158) and (2.166)]. The transfer functions for these linear systems are

$$H_i(\omega) = 1/G_i(\omega), \tag{5.359}$$
$$Y_i(\omega) = i\omega H_i(\omega). \tag{5.360}$$

[Note that we could not define $Y_i$ as the inverse of $Z_i$, since the inverse of $\delta(\omega)$ in Eq. (5.348) does not exist; see also Problem 2.15.]

By taking inverse Fourier transforms, we have the following time-domain representation of the four considered linear systems:

$$g_i(t) * s_3(t) = z_i(t) * u_3(t) = F_{\text{ext}}(t), \tag{5.361}$$
$$h_i(t) * F_{\text{ext}}(t) = s_3(t), \qquad y_i(t) * F_{\text{ext}}(t) = u_3(t). \tag{5.362}$$

The impulse response functions $g_i(t)$, $z_i(t)$, $h_i(t)$ and $y_i(t)$ entering into these convolutions are inverse Fourier transforms of $G_i(\omega)$, $Z_i(\omega)$, $H_i(\omega)$ and $Y_i(\omega)$, respectively. Using Table 2.2 and Eqs. (5.103) and (5.107) with Eqs. (5.348) and (5.349), we find

$$z_i(t) = k_{33}(t) + R_l\delta(t) + [m_{33}(\infty) + m_m]\dot{\delta}(t) + S_b U(t), \tag{5.363}$$
$$g_i(t) = \dot{z}_i(t) = \dot{k}_{33}(t) + R_l\dot{\delta}(t) + [m_{33}(\infty) + m_m]\ddot{\delta}(t) + S_b\delta(t). \tag{5.364}$$

As explained in Section 5.3, we know that the impulse response function $k_{33}(t)$ for radiation is causal. Then it is obvious that $z_i(t)$ and $g_i(t)$ are also causal.

Note that without the term with $\delta(\omega)$ in Eq. (5.348), we would have had $\frac{1}{2}\mathrm{sgn}(t)$ instead of the causal unit step function $U(t)$ in Eq. (5.363), but Eq. (5.364) would have remained unchanged. Combining Eqs. (5.361) and (5.363), we may write the time-domain dynamic equation as

$$(m_{33}(\infty) + m_m)\ddot{s}_3(t) + R_l\dot{s}_3(t) + k_{33}(t) * \dot{s}_3(t) + S_b s_3(t) = F_{\mathrm{ext}}(t) \qquad (5.365)$$

when we observe that $\dot{k}_{33}(t) * s_3(t) = k_{33}(t) * \dot{s}_3(t)$, since $k_{33}(t) \to 0$ as $t \to \infty$. [This may easily be shown by partial integration. See also Eq. (2.146).] Similarly, $\dot{\delta}(t) * s_3(t) = \delta(t) * \dot{s}_3(t) = \dot{s}_3(t)$ and $\ddot{\delta}(t) * s_3(t) = \dot{\delta}(t) * \dot{s}_3(t) = \ddot{s}_3(t)$.

It is reasonable to assume that impulse response functions $h_i(t)$ and $y_i(t)$ are also causal; this view may obtain support by studying Problem 2.15. A rigorous proof has, however, been delivered by Wehausen [69].

## Problems

### Problem 5.1: Surge and Pitch of a Hinged Plate

A rectangular plate is placed in a vertical position with its upper edge above a free water surface, while its lower edge is hinged below the water surface, at depth $c$. We choose a coordinate system such that the plate is in the vertical plane $x = 0$, and the hinge at the horizontal line $x = 0$, $z = -c$. Further, for this immersed plate, we choose a reference point $(x, y, z) = (0, 0, z_0)$; see Figure 5.3. The plate performs oscillatory sinusoidal rotary motion with respect to the hinge. The angular amplitude is $\beta_5$ (assumed to be small), and the angular frequency is $\omega$. Determine the six-dimensional vectors $\mathbf{u}$ and $\mathbf{n}$, defined in Section 5.1.1, for the three cases:

(a) $z_0 = -c$ (reference point at the hinge)
(b) $z_0 = 0$ (reference point at the free water surface)
(c) $z_0 = -c/2$ (reference point midway between hinge and $z = 0$)

### Problem 5.2: Excitation Force on a Surging Piston in a Wave Channel

A wave channel is assumed to extend from $x = 0$ to $x \to -\infty$. The water depth is $h$, and the channel width is $d$. Choose a coordinate system such that the wet surface of the end wall is $x = 0$, $0 < y < d$, $0 > z > -h$.

In the vertical end wall, there is a rectangular piston of width $d_1$ and height $a_2 - a_1$, corresponding to the surface

$$x = 0, \quad b < y < c, \quad -a_2 < z < -a_1,$$

where $c - b = d_1$ $(0 < b < c < d)$. In a situation in which the piston is not oscillating, the velocity potential is as given by Eq. (5.30). Derive an expression for the surge excitation force on the piston in terms of $A, \rho, g, k, h, a_1, a_2$ and $d_1$.

## Problem 5.3: Radiation Resistance for a Vertical Plate

A wave channel of width $d$ and water depth $h$ has in one end ($x = 0$) a wave generator in the form of a stiff vertical plate (rectangular piston) which oscillates harmonically with velocity amplitude $\hat{u}_1$ (pure surge motion) and with angular frequency $\omega$. In the opposite end of the wave channel, there is an ideal absorber. That is, we could consider the channel to be of infinite length ($0 < x < \infty$).

(a) Find an exact expression for the velocity potential of the generated wave in terms of an infinite series, where some parameters are defined implicitly through a transcendental equation.
(b) By means of elementary functions, express the radiation resistance $R_{11}$ in terms of the angular repetency $k$, the fluid density $\rho$ and the lengths $h$ and $d$. Draw a curve for $(R_{11}\omega^3/\rho g^2 d)$ versus $kh$ in a diagram.
(c) Also solve the same problem for the case when the plate is hinged at its lower edge (at $z = -h$), and let $\hat{u}_1$ now represent the horizontal component of the velocity amplitude at the mean water level.

## Problem 5.4: Flap as a Wave Generator

An incident wave $\hat{\eta} = Ae^{-ikx}$ hits a vertical plate hinged with its lower end at water depth $z = -c$ at a horizontal position $x = 0$, where $0 < c < h$. At $x = 0$, $-h < z < -c$ (below the hinge), there is a fixed vertical wall. Consider the system as a wave channel of width $\Delta y = d$.

(a) Determine the excitation pitch moment $\hat{M}_y = \hat{F}_5$ expressed by $\rho, g, k, A, h$ and $d$. Give simpler expressions for the two special cases $h \to \infty$ and $c = h$.
(b) Solve the radiation problem by obtaining an expression for $\varphi_5 = \hat{\phi}_r/\hat{u}_5$. Further derive a formula for the radiation impedance $Z_{55} = R_{55} + i\omega m_{55}$.
(c) Check that $R_{55} = |\hat{F}_5/A|^2\omega/(2\rho g^2 Dd)$. Check also that the excitation pitch moment $\hat{F}_5$ is recovered through application of the Haskind relation.

## Problem 5.5: Pivoting Vertical Plate as a Wave Generator

This is the same as Problem 5.4, but with the plate extending down to $z = -h$, while the hinge is still at $z = -c$. Show for this case that, with proper choice of $c/h$ ($0 < c < h/2$), it is possible to find a frequency where the radiated wave vanishes—that is, $R_{55} = 0$. [Hint: to find the repetency (wave number) corresponding to this frequency, a transcendental equation involving hyperbolic functions has to be solved. The solution may be graphically represented as the intersection of a straight line with the graph for $\tanh (kh/2)$.]

## Problem 5.6: Circular Wave Generator

A vertical cylinder of radius $a$ and height larger than the water depth stands on the sea bed, $z = -h$. The cylinder wall is pulsating in the radial direction with a complex velocity amplitude $\hat{u}_r(z) = \hat{u}_0 c(z)$, where the constant $\hat{u}_0 = \hat{u}_r(0)$ represents the pulsating velocity at the mean water level $z = 0$. Note that $c(0) = 1$. Assume that the system is axisymmetric; that is, there is no variation with the coordinate angle $\theta$.

(a) Express the velocity potential $\hat{\phi} = \varphi \hat{u}_0$ for the generated wave as an infinite series in terms of the normalised vertical eigenfunctions $\{Z_n(z)\}$ and the corresponding eigenvalues $\lambda_0 = k^2$, $\lambda_1 = -m_1^2$, $\lambda_2 = -m_2^2$, ....

(b) Express the radiation impedance $Z_{00}$ as an infinite series, and find a simple expression for the radiation resistance $R_{00}$.

[Hint: for Bessel functions, we have the relations $(d/dx)J_0(x) = -J_1(x)$ and correspondingly for $Y_0(x)$, $H_0^{(2)}(x)$ and $K_0(x)$, whereas $(d/dx)I_0(x) = I_1(x)$. Here $I_n$ and $K_n$ are modified Bessel functions of order $n$ and of the first and second kinds, respectively. Asymptotic expressions are $I_0(x) \to e^x/\sqrt{2\pi x}$ and $K_0(x) \to \pi e^{-x}/\sqrt{2\pi x}$ as $x \to +\infty$. Moreover, we have $J_1(x)Y_0(x) - J_0(x)Y_1(x) = 2/\pi x$.]

## Problem 5.7: Radiation Resistance in Terms of Far-Field Coefficients

The time-average mechanical power radiated from an oscillating body may be expressed as

$$P_r = \sum_{j=1}^{6}\sum_{j'=1}^{6} \frac{1}{2} R_{jj'}\hat{u}_j\hat{u}_{j'}^*.$$

In an ideal fluid, this time-average power may be recovered in the radiated wave-power transport—for instance, in the far-field region where the formula for wave-power transport with a plane wave is applicable. In the far-field region, the radiated wave is

$$\hat{\phi}_r = \hat{\phi}_l(r,\theta,z) + A_r(\theta)e(kz)(kr)^{-1/2}e^{-ikr},$$

where the near-field part $\hat{\phi}_l$ is negligible, because $\hat{\phi}_l = \mathcal{O}\{r^{-1}\}$ as $r \to \infty$. Use the decomposition

$$A_r(\theta) = \sum_{j=1}^{6} a_j(\theta)\hat{u}_j$$

in deriving an expression for the radiation resistance matrix $R_{jj'}$ in terms of the far-field coefficient vector $a_j(\theta)$.

## Problem 5.8: Excitation-Force Impulse Response

Consider the surge excitation force per unit width of a horizontal strip of a vertical wall normal to the wave incidence. The strip of height $\Delta z$ covers the interval $(z - \Delta z, z)$ where $z < 0$. Show that for the excitation-force impulse-response function, we have

$$\Delta f'_{t,1}(t) = \rho g (g/\pi |z|)^{1/2} e^{gt^2/4z} \Delta z \to 2\rho g \delta(t) \Delta z$$

as $z \to 0$. Compare Sections 4.9.2 and 5.3.2.

## Problem 5.9: Froude–Krylov Force for an Axisymmetric Body

The radius of the wet surface of an axisymmetric body with vertical axis $x = y = 0$ is given as the function $a = a(z)$. Show that the heave component of the Froude–Krylov force due to a plane incident wave $\hat{\eta} = A e^{-ikx}$ is given by

$$f_{FK,3} = F_{FK,3}/A = \pi \rho g \int_{-b}^{-c} J_0(ka) \frac{da^2(z)}{dz} e(kz)\,dz,$$

where $J_0$ is the zero-order Bessel function of the first kind. Furthermore, $b$ and $c$ are the depth of submergence of the wet surface's bottom and top, respectively. If the body is floating, then $c = 0$. The lower end of the body is at $z = -b$.

For a floating vertical cylinder with hemispherical bottom, we have

$$a^2(z) = \begin{cases} a_0^2 & \text{for } -l < z < 0 \\ a_0^2 - (z+l)^2 & \text{for } -l - a_0 < z < -l, \end{cases}$$

where $l = b - a_0$ is the length of the cylindrical part of the wet surface. We may expand the above exact formula for $f_{FK,3}$ as a power series in $ka_0$ and $kl$. Assume deep water ($kh \gg 1$), and find the first few terms of this expansion. Compare with the Froude–Krylov force as obtained from the small-body approximation. [Hint: assume as known the following:

$$\frac{1}{2\pi} \int_0^{2\pi} e^{-ix\cos\theta}\,d\theta = J_0(x) = 1 - \frac{(x/2)^2}{(1!)^2} + \frac{(x/2)^4}{(2!)^2} - + \cdots .]$$

## Problem 5.10: Radiation-Resistance Limit for Zero Frequency

For a floating axisymmetric body, we write the heave-mode radiation resistance as

$$R_{33} = \epsilon_{33} \omega \rho a^3 2\pi/3,$$

where $a$ is the radius of the water plane area. Show that if the water depth is infinite, then

$$\epsilon_{33} \to ka3\pi/4 \quad \text{as } ka \to 0,$$

while for finite water depth $h$,

$$\epsilon_{33} \to 3\pi a/(8h) \quad \text{as } ka \to 0.$$

[Hint: make use of Eqs. (5.277) and (5.302).]

## Problem 5.11: Heave Excitation Force for a Semisubmerged Sphere

For a semisubmerged floating sphere of radius $a$, we write the heave excitation force as $F_3 = \kappa SA = \kappa \rho g \pi a^2 A$, where $S$ is the buoyancy stiffness and $A$ the complex amplitude of the undisturbed incident wave at the centre of the sphere. Further, $\kappa$ is the non-dimensionalised excitation force. The radiation impedance is

$$Z_{33} = R_{33} + i\omega m_{33} = \omega \rho \tfrac{2}{3} \pi a^3 (\epsilon + i\mu).$$

For deep water ($kh \gg 1$), values for $\epsilon = \epsilon(ka)$ and $\mu = \mu(ka)$ computed by Hulme (1982) are given by the following table:

| $ka$ | 0 | 0.05 | 0.1 | 0.2 | 0.3 | 0.4 |
|------|-----|--------|--------|--------|--------|--------|
| $\mu$ | 0.8310 | 0.8764 | 0.8627 | 0.7938 | 0.7157 | 0.6452 |
| $\epsilon$ | 0 | 0.1036 | 0.1816 | 0.2793 | 0.3254 | 0.3410 |

| $ka$ | 0.5 | 0.6 | 0.7 | 0.8 | 0.9 | 1 |
|------|--------|--------|--------|--------|--------|--------|
| $\mu$ | 0.5861 | 0.5381 | 0.4999 | 0.4698 | 0.4464 | 0.4284 |
| $\epsilon$ | 0.3391 | 0.3271 | 0.3098 | 0.2899 | 0.2691 | 0.2484 |

Compute numerical values of $|\kappa|$ from the values of $\epsilon$ by using the exact reciprocity relation between the radiation resistance and the excitation force. Also compute approximate values of $\kappa$ from the values of $\mu$ by using the small-body approximation for the excitation force. Compare the values by drawing a curve for the exact $|\kappa|$ and the approximate $\kappa$ in the same diagram, versus $ka$ ($0 < ka < 1$).

## Problem 5.12: Radiation Resistance for a Cylindrical Body

A vertical cylinder of diameter $2a$ and of height $2l$ is (in order to reduce viscous losses) extended with a hemisphere in its lower end. When the cylinder is floating in equilibrium position, it has a draught of $l + a$ (and the cylinder's top surface is a distance of $l$ above the free water surface). We shall assume deep water in the present problem.

If we assume, for simplicity, that the heave excitation force is

$$\hat{F}_3 \equiv F(l) \approx e^{-kl} F(0), \tag{1}$$

where $F(0)$ is the heave excitation force for a semisubmerged sphere of radius $a$, then it follows from the reciprocity relationship between the radiation resistance and the excitation force [cf. Eq. (4) in the following] that the radiation resistance for heave is

$$R_{33} \equiv R_0 \approx R_H e^{-2kl}, \tag{2}$$

where $R_H$ is the radiation resistance for the semisubmerged sphere:

$$R_H = \omega \rho (2\pi/3) a^3 \epsilon. \tag{3}$$

Here, $\epsilon = \epsilon(ka)$ is Havelock's dimensionless damping coefficient as computed by Hulme [47]. Compare the following table or the table given in Problem 5.11. [It may be observed from the solution of Problem 5.9 that a relationship like (1) is valid exactly for the Froude–Krylov force. In the small-body case, approximation (1) is valid for the diffraction force if the variation of the added mass with $l$ is neglected. Note, however, that this assumption is not exactly true, because a modification of the radiation resistance according to approximation (2) is associated with some modification of the added mass, according to the Kramers–Kronig relations. See Section 5.3.1.]

For an axisymmetric body on deep water, we have the reciprocity relation

$$R_0 = R_{33} = (\omega k/2\rho g^2)|f_3|^2, \tag{4}$$

where $f_3 = \hat{F}_3/A$ is the heave excitation force coefficient and $A$ is the incident wave amplitude (complex amplitude at the origin). Express the dimensionless parameters $|f_3|/S$ and $\omega R_{33}/S$ in terms of $\epsilon$, $ka$ and $kl$, where $S = \rho g \pi a^2$ is the hydrostatic stiffness of the body. Compute numerical values for both these parameters with $l/a = 0, 4/3$ and $8/3$ for each of the $ka$ values from the following table:

| $ka$ | 0.01 | 0.05 | 0.1 | 0.15 | 0.2 | 0.3 |
|------|------|------|-----|------|-----|-----|
| $\epsilon$ | 0.023 | 0.1036 | 0.1816 | 0.24 | 0.2793 | 0.3254 |

## Problem 5.13: Eigenvalues of a Radiation Resistance Matrix

Given that the diagonal matrix elements are nonnegative, show that the $6 \times 6$ radiation resistance matrix for an axisymmetric body has at least three vanishing eigenvalues (or, expressed differently, one triple eigenvalue equal to zero), and hence, its rank is at most $6 - 3 = 3$. Also show that the three remaining eigenvalues are positive (or zero in exceptional cases) and that two of them have to be equal (or, expressed differently, there is a double eigenvalue which is not necessarily zero). [Hint: consider a real matrix of the form

$$\mathbf{r} = \begin{bmatrix} r_{11} & 0 & 0 & 0 & r_{15} & 0 \\ 0 & r_{22} & 0 & r_{24} & 0 & 0 \\ 0 & 0 & r_{33} & 0 & 0 & 0 \\ 0 & r_{24} & 0 & r_{44} & 0 & 0 \\ r_{15} & 0 & 0 & 0 & r_{55} & 0 \\ 0 & 0 & 0 & 0 & 0 & 0 \end{bmatrix},$$

where $r_{33} \geq 0$, $r_{22} = r_{11} \geq 0$, $r_{44} = r_{55} \geq 0$, $r_{24} = r_{15}$ and $r_{15}^2 = r_{11}r_{55}$; cf. Eqs. (5.295), (5.302) and (5.309). The eigenvalues $\lambda$ are the values which make the determinant of $\mathbf{r} - \lambda\mathbf{I}$ equal to zero, where $\mathbf{I}$ is the identity matrix. Factorising the determinant, which is a sixth-degree polynomial in $\lambda$, gives the eigenvalues in a straightforward manner.]

## Problem 5.14: Reciprocity Relations between Diffracted Waves

In agreement with Eq. (5.168), let $S$ be the totality of wet surfaces of oscillating bodies, as indicated in Figure 5.13. It follows from boundary condition (5.165) and definition (4.230) that

$$I(\hat{\phi}_i, \hat{\phi}_j) = 0, \qquad I(\hat{\phi}_i^*, \hat{\phi}_j^*) = 0$$

if $\hat{\phi}_i$ and $\hat{\phi}_j$ are sums of an incident wave and a diffracted wave, namely

$$\hat{\phi}_j = (-g/i\omega)A_j e(kz) \exp[-ikr(\beta_j)] + \hat{\phi}_{d,j},$$

and similarly for $\hat{\phi}_i$ with subscript $j$ replaced by $i$. (It is assumed that the bodies indicated in Figure 5.13 are held fixed.) Let us consider two-dimensional cases with two waves either from opposite directions ($\beta_i = \beta_1 = 0$ and $\beta_j = \beta_2 = \pi$, that is, $r(\beta_i) = x$ and $r(\beta_j) = -x$, respectively) or from the same direction ($\beta_i = \beta_j$).

(a) With Kochin functions $H'(\beta) = H'_d(\beta)$ for the diffracted waves, show that

$$A_i H_j'^*(\beta_i) + A_j^* H_i'(\beta_j) = \frac{\omega}{gk}[H_i'(0)H_j'^*(0) + H_i'(\pi)H_j'^*(\pi)].$$

   [Hint: use Eq. (4.291) and the two-dimensional version of Eq. (4.295).]
(b) By considering the diffracted plane waves in the far-field region, let us define transmission coefficients $T$ and reflection coefficients $\Gamma$ as follows:

$$T_1 = 1 + A_1^+/A_1, \qquad\qquad \Gamma_1 = A_1^-/A_1,$$
$$T_2 = 1 + A_2^-/A_2, \qquad\qquad \Gamma_2 = A_2^+/A_2,$$

where the relationship between the far-field coefficients $A_j^\pm$ and the Kochin functions $H_j'$ are given by Eqs. (4.285) and (4.286). Derive the following three relations:

$$\Gamma_j \Gamma_j^* + T_j T_j^* = 1 \quad \text{for } j = 1, 2, \qquad T_1 \Gamma_2^* + T_2^* \Gamma_1 = 0.$$

   (The former of these relations represents energy conservation. For a geometrically symmetrical case, the latter relation reduces to a result derived in Problem 4.13.) [Hint: consider the three cases when $(i, j)$ equals $(1,1)$, $(2,2)$ and $(1,2)$.]
(c) Use the two-dimensional version of Eq. (4.271) to show that $T_2 = T_1$, as derived by Newman [34].

## Problem 5.15: Diffracted and Radiated Two-Dimensional Waves

According to the boundary condition (5.166), $\partial\varphi/\partial n$ is real on $S$, the totality of wet surfaces of wave-generating bodies. Hence, we may replace $\varphi$ by $\varphi^*$ in

Eq. (5.222) as well as in Eqs. (5.136)–(5.138). By combining Eqs. (5.222) and (5.224), this gives

$$\mathbf{I}'(\hat{\phi}_0 + \hat{\phi}_d, \varphi^*) = \frac{gD(kh)}{i\omega k}\mathbf{h}'(\beta \pm \pi)A$$

for the two-dimensional case, where $\hat{\phi}_0 + \hat{\phi}_d = \hat{\phi}$ is as given in Problem 5.14 (when the subscript $j$ is omitted).

(a) Use Eq. (4.291) and the two-dimensional version of Eq. (4.295) to show that (for $\beta = 0$)

$$\Gamma_d \mathbf{h}'^*(\pi) + T_d \mathbf{h}'^*(0) = \mathbf{h}'(\pi),$$

where $\Gamma_d = \Gamma_1$ and $T_d = T_1$. (Here the reflection coefficient $\Gamma_1$ and the transmission coefficient $T_1$ are defined in Problem 5.14.) [Hint: in Eqs. (4.291) and (4.295), set $H_i' = H_d'$, $H_j' = \mathbf{h}'$, $A_i = A$ and $A_j = 0$; cf. Newman [34].]

(b) Consider now a single (two-dimensional) body which has a symmetry plane $x = 0$ and which oscillates in the heave mode only. Show that in the far-field region beyond the body ($x \to \infty$), there is no propagating wave, provided the heave velocity is given as

$$\hat{u}_3 = T_d(gk/\omega h_3')A,$$

since the radiated wave then cancels the transmitted wave for the diffraction problem.

(c) Show that for maximum absorbed wave power, the heave velocity is given by

$$\hat{u}_3 = \hat{u}_{3,\mathrm{OPT}} = (gk/2\omega h_3'^*)A.$$

[Hint: use Eqs. (6.11) and (5.329)–(5.332).] What is the absorbed power if $\hat{u}_3 = 2\hat{u}_{3,\mathrm{OPT}}$?

(d) Find a possible value of $\hat{u}_3$ for which the resulting far-field wave propagating in the negative $x$-direction ($x \to -\infty$) is as large as the incident wave. Note that this value of $\hat{u}_3$ is not unique, since the phase of the wave propagating in the negative $x$-direction has not been specified.

(e) Make a comparative discussion of the various results for $\hat{u}_3$ obtained earlier, when we assume that the body is so small that $\Gamma_d \to 0$. Can you say something about the aforementioned unspecified phase for the two 'opposite' cases $\Gamma_d \to 0$ and $\Gamma_d \to 1$? (In the latter case, the body has to be large.)

## Problem 5.16: Rectangular Piston in the End Wall of Wave Channel

A wave channel is assumed to extend from $x = 0$ to $x \to -\infty$. The water depth is $h$, and the channel width is $d$. Choose a coordinate system such that the wet surface of the end wall is $0 > z > -h$, $0 < y < d$.

In the vertical end wall, there is a rectangular piston of width $d_1$ and height $a_2 - a_1$, corresponding to the surface $x = 0$, $b < y < c$, $-a_2 < z < -a_1$, where $c - b = d_1$. The piston has a surging velocity of complex amplitude $\hat{u}_1$.

The velocity potential has a complex amplitude $\hat{\phi} = \varphi_1 \hat{u}_1$, where

$$\varphi_1 = \sum_{n=0}^{\infty} \sum_{q=0}^{\infty} b_{nq} \exp(\gamma_{nq} x) \cos\left(\frac{q\pi y}{d}\right) Z_n(z).$$

Moreover,

$$\gamma_{nq} = [m_n^2 + (q\pi/d)^2]^{1/2},$$
$$Z_n(z) = N_n^{-1/2} \cos[m_n(z+h)],$$
$$N_n = \tfrac{1}{2} + \sin(2m_n h)/(4m_n h),$$

and $m_n$ is the $(n+1)$th solution of

$$\omega^2 = -gm\tan(mh)$$

such that, with $m_0 = ik$,

$$k^2 = -m_0^2 > -m_1^2 > -m_2^2 > \ldots > -m_n^2 > \ldots.$$

Show that $\hat{\phi}$ satisfies Laplace's equation in the fluid, the radiation condition at $x = -\infty$, and the homogeneous boundary conditions on the free surface $z = 0$, on the bottom $z = -h$ and on the walls $y = 0$ and $y = d$ of the wave channel. Further, show that the boundary conditions on the wall $x = 0$ are satisfied if the unknown coefficients are given by

$$b_{nq} = \sigma_q (d_1/d) t_q N_n^{-1/2} s_n/\gamma_{nq},$$

where $\sigma_0 = 1$ and $\sigma_q = 2$ for $q \geq 1$, and

$$t_q = (d/q\pi d_1)[\sin(q\pi c/d) - \sin(q\pi b/d)],$$
$$s_n = \{\sin[m_n(h - a_1)] - \sin[m_n(h - a_2)]\}/m_n h.$$

Moreover, show that the radiation resistance is

$$R_{11} = \frac{\omega \rho s_0^2 d_1^2 h}{k N_0 d}\left\{1 + \sum_{q=1}^{q_0} \frac{2k t_q^2}{[k^2 - (q\pi/d)^2]^{1/2}}\right\},$$

where $q_0$ is the integer part of $kd/\pi$. Finally, determine the added mass $m_{11}$ expressed as a double sum over $n$ and $q$.

# CHAPTER SIX

# Oscillating-Body Wave-Energy Converters

This chapter starts with a consideration of wave absorption as a destructive interference between an incident wave and a wave radiated by an immersed oscillating-body wave-energy converter (WEC). Then, various WECs are classified according to size and orientation with respect to wavelength and propagation direction, respectively, of the incident wave. Next, for a given incident sinusoidal wave, mathematical expressions are derived for the wave power which is absorbed by a single-mode oscillating WEC body, as a function of its velocity amplitude and phase. A particular subject of interest is the optimum motion for maximising the power absorbed from the sea wave, as is also control methods to obtain optimum WEC oscillation.

Concerning the WEC's size, only its maximum horizontal extension has been considered thus far. However, for a particular WEC discussed in Section 6.4, not only the WEC body's volume but also its geometrical shape are considered.

## 6.1 Wave Absorption as Wave Interference

Absorbing wave energy means that energy has to be removed from the waves. Hence, there must be a cancellation or reduction of waves which are passing a WEC or are being reflected from it. Such a cancellation or reduction of waves can be realised by the oscillating WEC, provided it generates waves which oppose (are in counterphase with) the passing and/or reflected waves. In other words, the generated wave has to interfere destructively with the other waves.

Within classical theories for microphones or receiving antennae, it is a well-known point of view that to absorb an acoustic or electromagnetic wave means to radiate a wave that interferes destructively with the incident wave. Concerning absorption of an incoming ocean wave by means of an immersed oscillating body, Budal may have been the first to explain such a view, which he stated as follows: 'a secondary, ring-shaped, outgoing wave is generated, which interferes with the incoming wave in such a way that the resulting transmitted wave carries

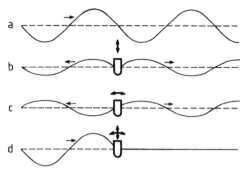

Figure 6.1: To absorb waves means to generate waves. Curve *a* represents an undisturbed incident wave moving from left to right on water in a narrow wave channel. Curve *b* illustrates symmetric wave generation (on otherwise calm water) by a floating symmetric body oscillating in the heave mode (up and down). Curve *c* illustrates antisymmetric wave generation by the same body oscillating in the surge and/or pitch mode. Curve *d*, which represents the superposition (sum) of the three waves, illustrates complete absorption of the incident wave energy.

with it less energy than the incoming wave does' [70]. To radiate an outgoing wave, the WEC needs to displace water, in an oscillatory manner.

A simple example to illustrate this fact is shown in Figure 6.1. Here, a floating symmetric body is immersed in a narrow flume (water between two parallel glass walls). Complete wave-energy absorption is possible, theoretically, provided the body is simultaneously performing optimal oscillation in the vertical direction as well as in the horizontal direction. Because power is proportional to amplitude squared, the waves shown by curves *a*, *b* and *c* have amplitudes in ratios 2:1:1, but powers in ratios 4:1:1, respectively. Thus, if the body oscillates only in the vertical direction or only in the horizontal direction, then one-half of the incident wave energy is received by the oscillating body. The remaining half is separated into equally large parts propagating in opposite directions, away from the oscillating body. This situation is related to the symmetry of the oscillating body in Figure 6.1.

Note that not only the amplitude but also the phase of the oscillation need to be optimum. If, say, the *b* and *c* curves had had opposite phases, then the right-hand half of the shown curve *d* would have had twice the amplitude of curve *a*. Then the immersed oscillating system would have functioned as an energy-delivering wave generator instead of an energy-receiving wave absorber.

It may perhaps seem like a paradox that a WEC need to produce an outgoing wave in order to receive incoming energy from the incident wave, but, as we have seen from the preceding illustration, this *is* a necessity if the WEC is to absorb any wave energy.

If, in contrast to the floating body indicated in Figure 6.1, the oscillating body is a submerged horizontal cylinder, its optimum oscillation requires the horizontal and vertical motions to have equal amplitudes, with phases differing by one quarter of a wave period. This is the principle utilised by the famous 'rotating Bristol cylinder' WEC. The cylinder axis performs a circular motion [71, 72].

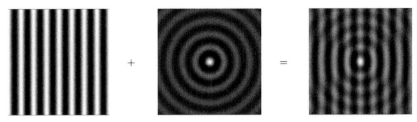

Figure 6.2: Snapshots of wave pattern of two interfering waves as seen from above. A plane wave, incident from the left-hand side, is interfering with an outgoing circular wave generated by an isotropically radiating system.

The sense of rotation of the cylinder axis determines whether the system acts as a device receiving energy from the incident wave or as a device supplying additional wave energy to the sea.

Contrary to the aforementioned symmetric WECs, if an immersed large body is sufficiently nonsymmetric, it may, theoretically, absorb up to 100% and performs with optimum oscillation in only one mode of motion. An example is the famous Salter Duck WEC, oscillating in the pitch mode [73].

The aforementioned examples correspond to the so-called two-dimensional WECs, which are discussed in more detail in Chapter 8.

The simplest three-dimensional WEC example that we may think of is an isotropically radiating WEC which is interacting with an incident plane wave, as illustrated by Figure 6.2. As with the two-dimensional examples discussed earlier, whether the wave interaction corresponds to energy removal or energy delivery depends on the relative differences of the amplitudes and phases of the two (plane and circular) waves. It has been shown [34, 70, 74, 75] that, with an incident plane wave, for which the wave-power level is $J$, as given by Eq. (4.130), the maximum converted power is

$$P_{\text{MAX}} = J/k = d_{a,\text{MAX}}J, \tag{6.1}$$

under conditions of optimum amplitude and phase relationships between the plane wave and the circular wave. Here,

$$d_{a,\text{MAX}} = 1/k = \lambda/2\pi \tag{6.2}$$

is the maximum *absorption width* for the isotropically radiating WEC. The quantity absorption width is defined as the ratio between $P$ and $J$ and is more commonly called *capture width*. A derivation of this result is given later in the present chapter; see Eqs. (6.13)–(6.15) and (6.103)–(6.104). Compare also Problem 4.16, where the interaction between a plane wave and a circular wave is discussed.

### 6.1.1  Classification of WEC Bodies

The isotropically radiating system in Figure 6.2 can for example be a heaving axisymmetric body. Such a WEC body was, in 1975, by Budal & Falnes [70] called a *point absorber*, provided 'its horizontal extension is much smaller than one wavelength'. To justify their assumption of negligible wave diffraction on the point absorber, we need to adopt Brian Count's more precise definition of point absorbers as 'structures that are small in comparison to the incident wavelength (say less than 1/20th of a wavelength)' [76].

In contrast to point-absorber WECs, a *line-absorber* WEC has one of its two horizontal extensions at least as large as one wavelength, while the second horizontal extension is very much smaller than one wavelength. Traditionally, a line absorber is called a *terminator* if it is aligned perpendicular to the predominant propagation direction of the incident wave, and it is called an *attenuator* if it is parallel. Early examples of terminators are the Salter Duck and the Bristol Cylinder, while early examples of attenuators are the ship-like Kaimei device and the flexible-bag device [76]. The two last-mentioned devices belong to the type of WECs discussed in Chapter 7.

So far, we have classified WECs having horizontal extension less than 1/20 of a wavelength and one wavelength or more as point absorbers and line absorbers, respectively. To bridge the gap between them, WECs which have a horizontal extension between 5% to 100% of a wavelength may be classified as *quasi-point absorbers (QPAs)* [77]. In contrast to point absorbers, wave diffraction is not negligible for QPAs.

If a semisubmerged body is oscillating in the heave mode, it displaces a water volume $V_{water}$ when it moves downwards. When the body returns upwards, an equally large amount of water is sucked back. As a result of this local water motion, an outgoing wave is generated. This heaving body is a *monopole* or a 'source' type of wave radiator. If the body oscillates in the surge or pitch mode, it is a *dipole* type of wave radiator, corresponding to two monopoles of opposite signs, displaced horizontally from each other in the $x$-direction (see Figure 5.2). Also, sway and roll modes are dipole-type of wave radiators corresponding to two opposite monopoles displaced horizontally from each other in the $y$-direction. If the body oscillates in the yaw mode, it is a *quadrupole* type of wave radiator. This corresponds to two, horizontally displaced, dipole-type of wave radiators of opposite polarities. A fully submerged heaving body is a monopole type of wave radiator. Its net monopole strength, however, results from two partial monopole contributions of opposite signs, but displaced vertically from each other. The upper contribution to the net monopole strength is larger than the opposite lower contribution. The lower contribution is smaller.

Observe that point absorbers may preferably be monopole radiators, for example, heaving semisubmerged bodies or volume-varying submerged bodies. Depending on its horizontal extension, a QPA—in contrast to a point absorber—may possibly be a dipole radiator, operating in, for example, surge and/or pitch modes.

Our aim is to study the very important primary wave-energy conversion, the process of ocean-wave energy being converted to some unspecified mechanical energy to be stored by a WEC. It is outside the scope of this book to discuss the secondary conversion of the absorbed energy to be used for chemical, mechanical or electrical applications, or the need to smooth out the large minute-to-minute variation of the incident wave energy, for which case an energy store is essential, for example, when delivering electric energy to a weak grid [5]. For recent reviews of practical wave-energy conversion efforts, see, for example, [9–11].

## 6.2 WEC Body Oscillating in One Mode

For simplicity, let us first consider a single body which has only one degree of motion, only heave ($j = 3$) or only pitch ($j = 5$), for example. An incident wave produces an excitation force $F_{e,j}$. The body responds to this force by oscillating with a velocity $u_j$. The total force is given by Eq. (5.155), which in this case simplifies to

$$\hat{F}_{t,j} = \hat{F}_{e,j} - Z_{jj}\hat{u}_j = f_j A - Z_{jj}\hat{u}_j, \tag{6.3}$$

where $Z_{jj}$ is the body's radiation impedance, $f_j$ is the excitation-force coefficient and $A$ is the complex amplitude of the incident wave's elevation at the origin $(x, y) = (0, 0)$. If we multiply through this equation by $\frac{1}{2}\hat{u}_j^*$ and take the real part, we obtain, according to Eq. (2.76), the time-averaged *absorbed power*

$$P = \tfrac{1}{2}\mathrm{Re}\{\hat{F}_{t,j}\hat{u}_j^*\} = P_e - P_r, \tag{6.4}$$

where

$$P_e = \tfrac{1}{2}\mathrm{Re}\{\hat{F}_{e,j}\hat{u}_j^*\} = \tfrac{1}{4}(f_j A \hat{u}_j^* + f_j^* A^* \hat{u}_j) \tag{6.5}$$

is the *excitation power* from the incident wave, and

$$P_r = \tfrac{1}{2}\mathrm{Re}\{Z_{jj}\hat{u}_j\hat{u}_j^*\} = \tfrac{1}{2}R_{jj}|\hat{u}_j|^2 \tag{6.6}$$

is the *radiated power* power due to the WEC body's oscillation. Here $R_{jj} = \mathrm{Re}\{Z_{jj}\}$ is the radiation resistance.

Note that the excitation power

$$P_e = \overline{F_{e,j}(t)u_j(t)} = \tfrac{1}{2}\mathrm{Re}\{\hat{F}_{e,j}\hat{u}_j^*\} = \tfrac{1}{2}|\hat{F}_{e,j}||\hat{u}_j|\cos\gamma_j, \tag{6.7}$$

where $\gamma_j = \varphi_u - \varphi_F$ is the phase difference between $\hat{u}_j$ and $\hat{F}_{e,j}$, is linear in $u_j$, whereas the radiated power

$$P_r = -\overline{F_{r,j}(t)u_j(t)} = \tfrac{1}{2}R_{jj}|\hat{u}_j|^2 \tag{6.8}$$

is quadratic in $u_j$, since $F_{r,j}$ is linear in $u_j$. Thus, a graph of $P$ versus $|\hat{u}_j|$ is a parabola, as indicated in Figure 6.3. We observe that in order to absorb wave power (that is, $P > 0$, or, $P_e > P_r$), it is necessary that $|\gamma_j| < \pi/2$ and also that $0 < |\hat{u}_j| < (|\hat{F}_{e,j}|/R_{jj})\cos\gamma_j$. The optimum phase angle is $\gamma_{j,\mathrm{opt}} = 0$.

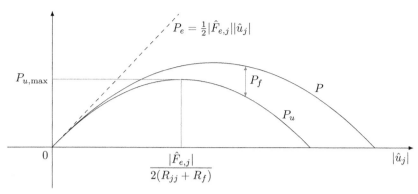

Figure 6.3: Curves showing absorbed wave power $P$, useful power $P_u$ and excitation power $P_e$ versus velocity amplitude $|\hat{u}_j|$ at optimum phase angle $\gamma_{j,\mathrm{opt}} = 0$.

Hence, a certain fraction of the excitation power $P_e$ arriving at the oscillating body is necessarily returned to the sea as radiated power $P_r$. Wave interference is a clue to the explanation of this fact. The radiated wave interferes destructively with the incident wave. The resulting wave, which propagates beyond the wave absorber, transports less power than the incident wave. The radiated power $P_r$ should be considered as a necessity rather than as a loss.

### 6.2.1 Maximum Absorbed Wave Power

For a given body and a given incident wave, the excitation force $F_{e,j}$ is a given quantity. Let us discuss how the absorbed power $P$ depends on the oscillating velocity $u_j$.

Observing Eqs. (6.7)–(6.8), we note the following. The optimum velocity amplitude is

$$|\hat{u}_{j,\mathrm{opt}}| = \frac{|\hat{F}_{e,j}|}{2R_{jj}} \cos \gamma_j, \tag{6.9}$$

giving the maximum absorbed power

$$P_{\max} = \frac{|\hat{F}_{e,j}|^2}{8R_{jj}} \cos^2 \gamma_j = P_{r,\mathrm{opt}} = \frac{1}{2} P_{e,\mathrm{opt}}. \tag{6.10}$$

With this optimum condition, the radiated power is as large as the absorbed power. Note that, for the case of resonance ($\gamma_j = 0$), we write (as in the following paragraphs) $P_{\mathrm{MAX}}$ and $\hat{u}_{j,\mathrm{OPT}}$ instead of $P_{\max}$ and $\hat{u}_{j,\mathrm{opt}}$.

It can be shown (see Problem 6.5) that $P_{\max}$ as given by Eq. (6.10) is greater than $P_{a,\max}$, as given by Eq. (3.41), except that they are equally large when $\gamma_j = 0$—that is, at resonance. The reason why $P_{a,\max}$ may be smaller than $P_{\max}$ is that in the former case, the oscillation amplitude is restricted by the simple dynamic equation (3.30), whereas $\hat{u}_j$ in the present section is considered as a quantity to be freely selected. Thus, in order to realise the optimum $\hat{u}_j$, it may be necessary to include a control device in the oscillation system. On this assumption, we may consider $\hat{u}_j$ to be an independent variable. The *optimum* value $\hat{u}_{j,\mathrm{OPT}}$, however, depends on the excitation force $\hat{F}_{e,j} = f_j(\beta)A$, and, consequently, on the incident-wave parameters $A$ and $\beta$.

Assuming that, in addition to selecting the optimum amplitude according to Eq. (6.9), we also select the optimum phase $\gamma_j = 0$—that is, an optimum complex amplitude

$$\hat{u}_{j,\mathrm{OPT}} = \frac{\hat{F}_{e,j}(\beta)}{2R_{jj}} = \frac{f_j(\beta)A}{2R_{jj}}, \tag{6.11}$$

then the maximum absorbed power is

$$P_{\mathrm{MAX}} = \frac{|\hat{F}_{e,j}(\beta)|^2}{8R_{jj}} = P_{r,\mathrm{OPT}} = \frac{1}{2}P_{e,\mathrm{OPT}}. \tag{6.12}$$

Note that here $P_{\mathrm{MAX}}$ agrees with Eq. (3.45), which is applicable at resonance.

If the oscillation amplitude is twice that of the optimum value given by Eq. (6.11), then the absorbed wave power is zero, $P = 0$, and $P_r = P_e = 4P_{\mathrm{MAX}}$. This is the situation when a linear array of resonant (that is, $\gamma_j = 0$) heaving slender bodies—or one single body in a narrow wave channel (see Figure 6.1)—is used as a dynamic wave reflector [78]. Then the radiated wave cancels the transmitted wave on the downwave side of the oscillating body. On the upwave side, a standing wave occurs, resulting from superposition of the radiated and incident waves with approximately equal amplitudes but opposite propagation directions.

We may now use Eq. (5.145) to express the radiation resistance in terms of the excitation force. Then we have

$$P_{\mathrm{MAX}} = \frac{|\hat{F}_{e,j}(\beta)|^2}{8R_{jj}} = \frac{2\pi J|\hat{F}_{e,j}(\beta)|^2}{k\int_{-\pi}^{\pi}|\hat{F}_{e,j}(\beta')|^2\,d\beta'} = \frac{J}{k}G_j(\beta) = \frac{\lambda}{2\pi}JG_j(\beta), \tag{6.13}$$

where

$$G_j(\beta) = \frac{2\pi|\hat{F}_{e,j}(\beta)|^2}{\int_{-\pi}^{\pi}|\hat{F}_{e,j}(\beta')|^2\,d\beta'} = \frac{2\pi|f_j(\beta)|^2}{\int_{-\pi}^{\pi}|f_j(\beta')|^2\,d\beta'} = \frac{2\pi|h_j(\beta\pm\pi)|^2}{\int_{-\pi}^{\pi}|h_j(\beta'\pm\pi)|^2\,d\beta'}. \tag{6.14}$$

Here, in the last step, we have made use of Eq. (5.142). Observe that this *optimum power-gain function* $G_j(\beta)$ is applicable only when $P = P_{\mathrm{MAX}}$, and not, in general, when $P < P_{\mathrm{MAX}}$.

For an immersed body with a vertical axis of symmetry, it follows, from Eqs. (5.285)–(5.287) and (6.14), that the optimum power-gain functions are $G_1(\beta) = G_5(\beta) = 2\cos^2\beta$ for the surge and pitch modes, $G_2(\beta) = G_4(\beta) = 2\sin^2\beta$ for the sway and roll modes, and $G_3(\beta) \equiv 1$ for the heave mode. Except for the heave mode, which is the simplest case, these optimum power-gain functions were first derived by Newman [34].

Averaging the optimum power-gain function $G_j(\beta)$ as defined by Eq. (6.14) over all horizontal wave-incidence angles $\beta$, we obtain

$$G_{j,\text{average}} = 1, \tag{6.15}$$

except for singular cases where $G_{j,\text{average}} = 0$. We have such an exceptional case, for instance, with the yaw mode ($j = 6$) for an immersed body that has a vertical axis of symmetry. Moreover, if such a body has a shape with a relatively small water-plane area, as indicated on Figure 5.16, we may, for one particular frequency, have $G_{3,\text{average}} = 0$ because $h_3 = 0$ for this frequency.

For the case of deep water, the wavelength $\lambda$ and the wave-power level $J$ are given by Eqs. (4.100) and (4.131), respectively. Thus, for this case, we find from Eq. (6.13) the simple formula

$$P_{\text{MAX}} = c_j H^2 T^3, \tag{6.16}$$

where

$$c_j = c_j(\beta) = G_j(\beta)\rho g^3/128\pi^3 = (245 \,\text{W}\,\text{m}^{-2}\text{s}^{-3})\, G_j(\beta), \tag{6.17}$$

for the immersed WEC body oscillating in mode $j$.

## 6.2.2 Converted Useful Power

The wave power $P$ which is converted by a WEC body is divided between converted useful power $P_u$ and an unavoidable lost power $P_f$ due to viscous effects, friction and non-ideal energy-conversion equipment.

For simplicity, we consider a power takeoff system with linear characteristics, as indicated schematically in Figure 6.4, where resistance $R_f$ is associated with the lost power [see Eq. (5.340)]. In a similar way, we introduce a load resistance $R_u$, corresponding to the useful power, such that the load force $F_{u,j}$ [see Eqs. (5.339) and (5.344)] is

$$F_{u,j} = -R_u u_j. \tag{6.18}$$

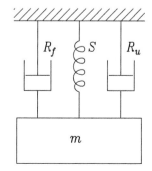

Figure 6.4: Schematic model of a linear power takeoff system of an oscillating system. Mechanical resistances $R_u$ and $R_f$ account for converted useful power and power lost by friction and viscous damping, respectively.

Mass $m$ includes the mass of the oscillating body, and stiffness $S$ includes the hydrostatic buoyancy effect. Radiation resistance $R_{jj}$ and added mass $m_{jj}$ are not shown in the schematic diagram of Figure 6.4.

The lost power is

$$P_f = \tfrac{1}{2}R_f|\hat{u}_j|^2, \tag{6.19}$$

and the *useful power* is

$$
\begin{aligned}
P_u &= \tfrac{1}{2}R_u|\hat{u}_j|^2 = P - P_f = P_e - (P_r + P_f) \\
&= \tfrac{1}{2}|\hat{F}_{e,j}||\hat{u}_j|\cos\gamma_j - \tfrac{1}{2}(R_{jj} + R_f)|\hat{u}_j|^2.
\end{aligned}
\tag{6.20}
$$

For the particular case of $\gamma_j = 0$, corresponding to optimum phase, a graphical representation of $P_u$, $P$, $P_e$ and $P_f$ versus $|u_j|$ are given in the diagram of Figure 6.3. If

$$\hat{u}_j = \hat{F}_{e,j}/2(R_{jj} + R_f), \tag{6.21}$$

that is, if $\gamma_j = 0$ and $|\hat{u}_j| = |\hat{F}_{e,j}|/2(R_{jj} + R_f)$, then we have the maximum useful power

$$P_{u,\text{MAX}} = |\hat{F}_{e,j}|^2/8(R_{jj} + R_f). \tag{6.22}$$

Note that a lower oscillation amplitude is required to obtain maximum useful power than to obtain the maximum absorbed wave power. The phase condition $\gamma_j = 0$ for optimum is the same in both cases. The velocity has to be in phase with the excitation force.

In the preceding discussion, we have assumed that $\hat{u}_j$ is a variable at our disposal; that is, we may choose $|\hat{u}_j|$ and $\gamma_j$ as we like. To achieve the desired value(s), it may be necessary to provide the dynamic system with a certain device for power takeoff and oscillation control. This device supplies a 'load' force $F_{u,j}$ in addition to the excitation force $F_{e,j}$. In analogy with Eq. (5.350), the oscillation velocity has to obey the equation

$$Z_i(\omega)\hat{u}_j = \hat{F}_{e,j} + \hat{F}_{u,j}, \tag{6.23}$$

where $Z_i(\omega)$ is the intrinsic mechanical impedance for oscillation mode $j$. It is the mechanical impedance of the oscillating system, and it includes the radiation impedance but not the effects of the mentioned device for control and power takeoff. These effects are represented by the last term $\hat{F}_{u,j}$ in Eq. (6.23). As an extension of Eq. (6.18), let us now assume that the load force may be written as

$$\hat{F}_{u,j} = -Z_u(\omega)\hat{u}_j, \tag{6.24}$$

where $Z_u(\omega)$ is a load impedance. Introducing this into Eq. (6.23) gives

$$[Z_i(\omega) + Z_u(\omega)]\hat{u}_j = \hat{F}_{e,j}. \tag{6.25}$$

The converted useful power is

$$
\begin{aligned}
P_u &= \tfrac{1}{2}\mathrm{Re}\{-\hat{F}_u\hat{u}_j^*\} = \tfrac{1}{2}\mathrm{Re}\{Z_u(\omega)\}|\hat{u}_j|^2 \\
&= \frac{1}{2}\mathrm{Re}\{Z_u(\omega)\}\frac{|\hat{F}_{e,j}|^2}{|Z_i(\omega) + Z_u(\omega)|^2} \\
&= \frac{R_u(\omega)|\hat{F}_{e,j}|^2/2}{(R_i(\omega) + R_u(\omega))^2 + (X_i(\omega) + X_u(\omega))^2},
\end{aligned}
\tag{6.26}
$$

where we, in the last step, have split the impedances $Z(\omega)$ into real parts $R(\omega)$ and imaginary parts $X(\omega)$. Note that the last expression here agrees with Eq. (3.39). Utilising the results of Eqs. (3.42) and (3.44), we find that the optimum values for $R_u(\omega)$ and $X_u(\omega)$ are $R_i(\omega)$ and $-X_i(\omega)$, respectively. Hence, for

$$
Z_u(\omega) = Z_i^*(\omega) \equiv Z_{u,\mathrm{OPT}}(\omega),
\tag{6.27}
$$

we have the maximum useful power [in agreement with Eq. (6.22)]

$$
P_u = \frac{|\hat{F}_{e,j}|^2}{8R_i} \equiv P_{u,\mathrm{MAX}}.
\tag{6.28}
$$

Note that the reactive part of the total impedance is cancelled under the optimum condition. This is automatically fulfilled in the case of resonance. Moreover, under the optimum condition, we have resistance matching, that is, $R_u(\omega) = R_i(\omega)$.

### 6.2.3 The Wave-Power 'Island'

We consider an incident plane wave propagating on deep water or on water of constant depth. Let its complex wave-elevation amplitude be

$$
\hat{\eta} = Ae^{-ik(x\cos\beta + y\sin\beta)} = Ae^{-ikr\cos(\beta-\theta)},
\tag{6.29}
$$

as expressed in Cartesian coordinates $(x, y, z)$ and, alternatively, in cylindrical coordinates $(r, \theta, z)$; thus, $x = r\cos\theta$ and $y = r\sin\theta$. The plane wave's propagation direction is at an angle $\beta$ with respect to the $x$-axis, and its complex elevation amplitude at the origin is $A$.

Let us now assume that an immersed WEC body is oscillating in one mode, mode $j$, for instance. We define two quantities $U$ and $E(\beta)$, as follows:

$$
|U|^2 = UU^* = R_{jj}\hat{u}_j\hat{u}_j^*/2 = P_r = \frac{\omega\rho v_p v_g}{4\pi g}\int_0^{2\pi} h_j(\theta)h_j^*(\theta)d\theta\,\hat{u}_j\hat{u}_j^*
\tag{6.30}
$$

$$
E(\beta) = f_{e,j}(\beta)\hat{u}_j^*/4 = (\rho v_p v_g/2)\,h_j(\beta + \pi)\,\hat{u}_j^*,
\tag{6.31}
$$

where we in the last step have made use of Eqs. (5.135), (5.142) and (4.107).

In Eq. (6.30), the nonnegative quantity $P_r$ represents the radiated wave power (caused by any forced oscillation of the immersed body). Note that the introduced complex quantity $U = \sqrt{P_r}e^{i\delta}$ may have any arbitrary phase angle $\delta$ in the interval $-\pi < \delta \leq \pi$ of the complex plane. We shall find it convenient, however, to choose $U$ to have the same phase angle as $A^*E^*(\beta)$. Then $A^*E^*(\beta)/U$ is a real positive quantity, which, notably, is independent of the complex velocity amplitude $\hat{u}_j$.

We may now simplify Eqs. (6.4)–(6.6) for the time-average absorbed wave power to

$$P = P_e - P_r = AE(\beta) + A^*E^*(\beta) - |U|^2. \tag{6.32}$$

By simple algebraic manipulation of Eq. (6.32), we may show that

$$P = |AE(\beta)/U^*|^2 - |U - AE(\beta)/U^*|^2, \tag{6.33}$$

from which simply by inspection we see that the first term equals the maximum possible absorbed power, provided the last term vanishes—that is, if the quantities $U$ and $E(\beta)$ have optimum values $U_0$ and $E_0(\beta)$ satisfying the optimum condition

$$U_0 - AE_0(\beta)/U_0^* = 0, \quad \text{that is,} \quad AE_0(\beta) = |U_0|^2 = A^*E_0^*(\beta). \tag{6.34}$$

Thus, evidently, we have several different alternative expressions for the maximum absorbed power—for example,

$$P_{\text{MAX}} = P_{r,\text{OPT}} = |U_0|^2 = P_{e,\text{OPT}}/2 = AE_0(\beta) = A^*E_0^*(\beta) = |AE_0(\beta)|; \tag{6.35}$$

see also Eq. (6.12).

As we have chosen $A^*E^*(\beta)/U$ to be a real positive quantity which is independent of $u_i$ and, therefore, also of $U$, we have

$$\frac{A^*E^*(\beta)}{U} = \frac{A^*E_0^*(\beta)}{U_0} = U_0^* = U_0^*(\beta) = |U_0(\beta)| = U_0(\beta), \tag{6.36}$$

where we have made use of the optimum condition (6.34). In general, we shall consider $U$ to be an independent complex oscillation-state variable, while the optimum value $U_0(\beta)$ is real and positive, because we have chosen $U$ to have the same phase as $A^*E^*(\beta)$. According to Eq. (6.36), we have $A^*E^*(\beta) = U_0^*U$ and $AE(\beta) = U_0U^*$. If we insert this into Eq. (6.32) and also use Eq. (6.35), we obtain the simple equation

$$P_{\text{MAX}} - P = U_0U_0^* - U_0U^* - U_0^*U + UU^* = |U_0 - U|^2 = |U_0(\beta) - U|^2, \tag{6.37}$$

which, for a fixed value $P < P_{\text{MAX}}$, corresponds to the equation of a circle of radius $\sqrt{P_{\text{MAX}} - P}$ centred at $U_0(\beta)$ in the complex $U$ plane. Equation (6.37) may be illustrated as an axisymmetric paraboloid, the wave-power 'island', in a diagram where a vertical real $P$ axis is erected on a horizontal complex $U$ plane, as shown in Figure 6.5. The upper parabola shown in Figure 6.3 corresponds

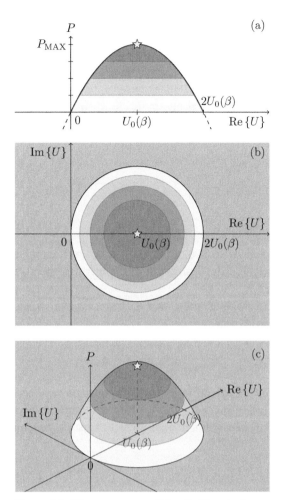

Figure 6.5: The wave-power 'island'. Absorbed wave power $P$ as a function of the complex collective oscillation amplitude $U = \mathrm{Re}\{U\} + i\,\mathrm{Re}\{U\} = |U|e^{i\delta}$, where the phase $\delta = \arg\{U\}$ is chosen to satisfy Eq. (6.36), and where $|U|$ is given by Eq. (6.30) for the one-mode oscillating-body case, and by Eq. (8.74) for the case of a general WEC array. The largest possible absorbed wave power $P_{\mathrm{MAX}}$ is indicated by a star on the top of the axisymmetric paraboloid, and $U_0$ is the optimum collective oscillation amplitude. Colour changes indicate levels where $P/P_{\mathrm{MAX}}$ equals 1/4, 1/2 and 3/4. (a) Side view. (b) Top view. (c) Inclined view. For a colour version of the figure, see figure 1 of [63], accessible at https://doi.org/10.1098/rsos.140305.

to an intersection between the paraboloid and a vertical plane through the real vertical $P$ axis and the real axis of the horizontal complex $U$ plane. If this vertical plane makes an angle $\gamma_j$ with the real $U$ axis, then the intersection is a lower parabola with a maximum corresponding to Eq. (6.10).

As will be shown later in this book, by generalising the definition of the complex quantities $U$ and $E(\beta)$, the simple formula (6.37), as well as the wave-power 'island', is applicable to oscillating water columns (OWCs) and even to arrays of WEC bodies and OWCs.

### 6.3 Optimum Control of a WEC Body

The purpose of this section is to study wave-energy conversion when the waves are not sinusoidal and to discuss conditions for maximising the converted energy. For simplicity, we are discussing oscillation in only one mode—for instance, heave ($j = 3$) or pitch ($j = 5$). Thus, we shall omit the subscript $j$ in the present section. Replacing the complex amplitudes $\hat{u}$, $\hat{F}_e$, and $\hat{F}_u$ by the corresponding Fourier transforms $u(\omega)$, $F_e(\omega)$ and $F_u(\omega)$, we may write Eq. (6.23) as

$$Z_i(\omega)u(\omega) = F_e(\omega) + F_u(\omega) = F_{\text{ext}}(\omega). \tag{6.38}$$

We rewrite the corresponding time-domain equation as

$$z_i(t) * u_t(t) = F_{e,t}(t) + F_{u,t}(t), \tag{6.39}$$

which for the heave mode specialises to Eq. (5.361). The causal impulse-response function $z_i(t)$ is the inverse Fourier transform of the intrinsic mechanical impedance $Z_i(\omega)$; see Section 5.9.

In analogy with Eq. (6.8), the average useful power is $P_u = -\overline{F_{u,t}(t)u_t(t)}$, and thus, the converted useful energy is

$$W_u = -\int_{-\infty}^{\infty} F_{u,t}(t)u_t(t)\,dt. \tag{6.40}$$

(This integral exists if the integrand tends sufficiently fast to zero as $t \to \pm\infty$.) By applying the frequency convolution theorem (2.148) for $\omega = 0$ and utilising the fact that $F_{u,t}(t)$ and $u_t(t)$ are real, we obtain, in analogy with Eqs. (2.201) and (2.205), that

$$W_u = \frac{1}{2\pi}\int_0^{\infty} \{-F_u(\omega)u^*(\omega) - F_u^*(\omega)u(\omega)\}\,d\omega. \tag{6.41}$$

By algebraic manipulation of the integrand, we may rewrite this as

$$\begin{aligned}
W_u &= \frac{1}{2\pi}\int_0^{\infty}\left\{\frac{|F_e(\omega)|^2}{2R_i(\omega)} - \left[\frac{F_e(\omega)F_e^*(\omega)}{2R_i(\omega)} + F_u(\omega)u^*(\omega) + F_u^*(\omega)u(\omega)\right]\right\}d\omega \\
&= \frac{2}{\pi}\int_0^{\infty}\left\{\frac{|F_e(\omega)|^2}{8R_i(\omega)} - \frac{\alpha(\omega)}{8R_i(\omega)}\right\}d\omega,
\end{aligned} \tag{6.42}$$

where

$$\alpha(\omega) = F_e(\omega)F_e^*(\omega) + 2R_i(\omega)[F_u(\omega)u^*(\omega) + F_u^*(\omega)u(\omega)]. \tag{6.43}$$

Here $F_e(\omega)$ is the Fourier transform of the excitation force $F_{e,t}(t)$, and

$$R_i(\omega) = \frac{Z_i(\omega) + Z_i^*(\omega)}{2} \tag{6.44}$$

is the intrinsic mechanical resistance—that is, the real part of the intrinsic mechanical impedance. In the following, we shall prove that $\alpha(\omega) \geq 0$, and we shall examine under which optimum conditions we have $\alpha(\omega) \equiv 0$. Then, since $R_i(\omega) > 0$, Eq. (6.42) means that the maximum converted useful energy is

$$W_{u,\text{MAX}} = \frac{2}{\pi} \int_0^\infty \frac{|F_e(\omega)|^2}{8R_i(\omega)} \, d\omega \tag{6.45}$$

under the condition that $\alpha(\omega) = 0$ for all $\omega$. Otherwise, we have, in general,

$$W_u = W_{u,\text{MAX}} - W_{u,P}, \tag{6.46}$$

where

$$W_{u,P} = \frac{2}{\pi} \int_0^\infty \frac{\alpha(\omega)}{8R_i(\omega)} \, d\omega, \tag{6.47}$$

which is a lost-energy penalty for not operating at optimum. Observe that $W_{u,P} \geq 0$. We wish to minimise $W_{u,P}$ and, if possible, reduce it to zero. It can be shown that Eq. (6.28), which applies for the case of a sinusoidal wave, is in agreement with Eq. (6.45); see Problem 2.14.

In order to show that $\alpha \geq 0$, we shall make use of the dynamic equation (6.38) to eliminate either $F_e(\omega)$ or $F_u(\omega)$ from Eq. (6.43).

Omitting for a while the argument $\omega$, we find that Eq. (6.38) gives $F_e = Z_i u - F_u$, and thus,

$$F_e F_e^* = F_u F_u^* + Z_i Z_i^* u u^* - F_u Z_i^* u^* - F_u^* Z_i u. \tag{6.48}$$

Further, using Eq. (6.44), we have

$$\begin{aligned} 2R_i(F_u u^* + F_u^* u) &= (Z_i + Z_i^*)(F_u u^* + F_u^* u) \\ &= F_u Z_i u^* + F_u^* Z_i^* u + F_u Z_i^* u^* + F_u^* Z_i u. \end{aligned} \tag{6.49}$$

Inserting these two equations into Eq. (6.43) gives

$$\alpha = F_u F_u^* + Z_i Z_i^* u u^* + F_u Z_i u^* + F_u^* Z_i^* u = (F_u + Z_i^* u)(F_u^* + Z_i u^*). \tag{6.50}$$

Hence, we have

$$\alpha(\omega) = |F_u(\omega) + Z_i^*(\omega)u(\omega)|^2 \geq 0, \tag{6.51}$$

and from this, we see that the optimum condition is

$$F_u(\omega) = -Z_i^*(\omega)u(\omega), \tag{6.52}$$

which for sinusoidal waves and oscillations agrees with Eqs. (6.24) and (6.27).

Eliminating differently, we have $F_u = Z_i u - F_e$ from the dynamic equation (6.38), which by insertion into Eq. (6.43) gives

$$\begin{aligned} \alpha &= F_e F_e^* + 2R_i(Z_i u u^* - F_e u^* + Z_i^* u u^* - F_e^* u) \\ &= F_e F_e^* + (2R_i)^2 u u^* - F_e 2R_i u^* - F_e^* 2R_i u \\ &= (F_e - 2R_i u)(F_e^* - 2R_i u^*) \end{aligned} \tag{6.53}$$

when Eq. (6.44) has also been observed. Hence, we have

$$\alpha(\omega) = |F_e(\omega) - 2R_i(\omega)u(\omega)|^2 \geq 0. \tag{6.54}$$

From this second proof of the relation $\alpha \geq 0$, we obtain the optimum condition written in the alternative way:

$$u(\omega) = \frac{F_e(\omega)}{2R_i(\omega)}, \tag{6.55}$$

which for sinusoidal waves and oscillations agrees with Eq. (6.11).

### 6.3.1 Methods for Optimum Control

The two alternative ways to express the optimum condition, as Eq. (6.52) or as Eq. (6.55), have been used for quite some time. To achieve the optimum condition (6.52) has been called *reactive control* [79] or *complex-conjugate control* [80], and it was already in the mid-1970s tested experimentally in sinusoidal waves [81]. 'Reactive control' refers to the fact that the optimum reactance $X_u$ (the imaginary part of $Z_u = -F_u/u$) cancels the reactance $X_i$ (the imaginary part of $Z_i$). 'Complex-conjugate control' refers to the fact that the optimum load impedance $Z_u$ equals the complex conjugate of the intrinsic impedance $Z_i$. On the other hand, to achieve the optimum condition (6.55) has been called *phase control* and *amplitude control* [74] because condition (6.55) means, firstly, that the oscillation velocity $u$ must be in phase with the excitation force $F_e$ and, secondly, that the velocity amplitude $|u|$ must equal $|F_e|/2R_i$. A similar method was proposed and investigated before 1970 [82]. Then an attempt was made to achieve optimum oscillation on the basis of wave measurement at an 'upwave' position in a wave channel. In this particular case, the object was not to utilise the wave energy but to achieve complete wave absorption in a wave channel.

We should note that in order to obtain the optimum condition $\alpha(\omega) \equiv 0$, unless $Z_i$ is real, it is necessary that reactive power is involved to achieve optimum. This means that the load-and-control machinery which supplies the load force $F_u$ not only receives energy but also must return some energy during part of the oscillation cycle. (The final part of Section 2.3.1 helps us understand this fact.) Obviously, it is desirable that this machinery has a very high energy-conversion efficiency (preferably close to 1). If, for some reason, we do not want or are unable to return the necessary amount of energy, then we have a suboptimal control, for which $\alpha(\omega) \neq 0$ and, hence, $W_{u,P} > 0$ [see Eq. (6.47)].

An example of such a suboptimal method is phase control by latching, a principle [78, 83] that is illustrated by Figure 6.6 for a heaving body which is so small that the excitation force is in phase with the incident wave elevation (sinusoidal in this case). By comparing curves $b$ and $c$ in Figure 6.6, it is obvious that if the resonance (curve b) represents optimum in agreement with condition (6.55), then the latching phase control (curve c) is necessarily suboptimal.

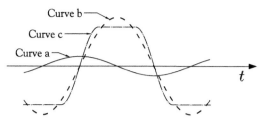

Figure 6.6: Resonance and phase control. The curves indicate incident wave elevation and vertical displacement of (different versions of) a heaving body as functions of time. Curve a: Elevation of the water surface due to the incident wave (at the position of the body). This would also represent the vertical position of a body with negligible mass. For a body of diameter very small compared with the wavelength, curve a also represents the wave's heave force on the body. Curve b: Vertical displacement of heaving body whose mass is so large that its natural period is equal to the wave period (resonance). Curve c: Vertical displacement of a body with smaller mass and, hence, shorter natural period. Phase control is then obtained by keeping the body in a fixed vertical position during certain time intervals.

With latching-phase-control experiments in [84, 85], means have not been provided to enable the power takeoff machinery to return any energy during parts of the oscillation cycle.

An overview of the problem of optimum control for maximising the useful energy output from a physical dynamic system represented by Eq. (6.38) is illustrated by the block diagram in Figure 6.7. The oscillation velocity $u$ is the system's response to the external force input $F_{ext} = F_e + F_u$. A switch in the block diagram indicates the possibility to select between two alternatives to determine the optimum load force $F_u$. One alternative is the reactive-control (or complex-conjugate-control) method corresponding to optimum condition (6.52). The other alternative is the amplitude-phase-control method corresponding to optimum condition (6.55), for which knowledge on the wave excitation force $F_e$ is used to determine the optimum oscillation velocity $u_{opt} = F_e/2R_i$, which is compared with the measured actual velocity $u$. The deviation between $u$ and $u_{opt}$ is used as input to an indicated (but here not specified) controller, which provides the optimum load force $F_u$. This alternative control method, contrary to the reactive-control method, requires knowledge on the excitation force $F_e$.

The other switch in the block diagram of Figure 6.7 indicates the possibility to select between two alternatives to determine the excitation force from measurement either of the total wave force $F_w = F_e + F_r = F_e - Z_r u$ acting on the dynamic system (Figure 6.8) or of the incident wave elevation $A$ at a reference position some distance away from the system (Figure 6.9). Even if this position is upstream (referred to the wave propagation direction), the excitation force coefficient $f_e(\omega)$ may correspond to a noncausal impulse-response function [37]. (See also Section 5.3.2.)

The two switches for choice of alternatives indicated in the block diagram of Figure 6.7 represent a redundancy if equipment is installed so that two methods are available both to determine $F_e$ and to determine $F_u$. Obtained alternative

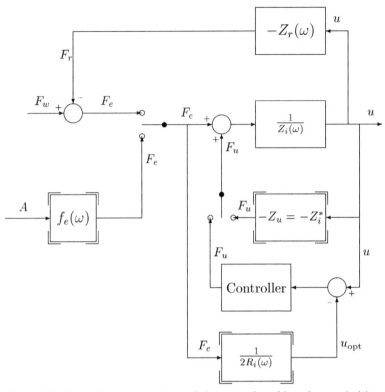

Figure 6.7: Block diagram overview of the control problem for maximising the converted energy by a system represented by the transfer function $1/Z_i(\omega)$, where the external force $F_{\text{ext}} = F_e + F_u$ is the input and the velocity $u$ the response. Two possibilities are indicated to determine both the excitation force $F_e = f_e A = F_w - F_r$ and the optimum load force. Three of the blocks, which are marked by a hook on each corner, represent noncausal systems.

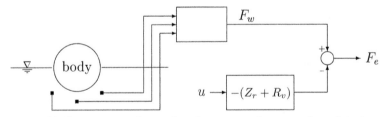

Figure 6.8: Pressure transducers placed on or near the wet surface of the immersed body may be used for measurement of the hydrodynamic pressure in order to estimate the total wave force $F_w$ on the body. Deducting from $F_w$ a wave force $-(Z_r + R_v)u$ due to the body oscillation, the wave excitation force $F_e$ is obtained.

results may then be compared. Such a redundancy may be useful if it is difficult to achieve a satisfactory optimum control in practice.

Figure 6.9 illustrates the situation if a measurement of wave elevation is chosen to be the method used for determining the excitation force $F_e$ and if the amplitude-and-phase-control method is chosen in order to maximise the useful power.

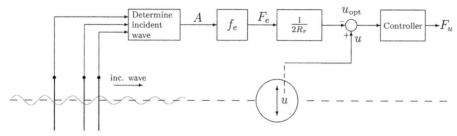

Figure 6.9: Measurement of incident wave elevation $A$, at some distance from the immersed body, for determination of the excitation force $F_e = f_e A$ and of the optimum oscillation velocity $u_{opt} = F_e/2R_i$, whose deviation from the measured actual velocity $u$ is utilised as a signal to a controller for supplying the correct load force $F_u$ to the immersed body.

A method for a more direct measurement of the total wave force $F_w$ is indicated in Figure 6.8. An array of pressure transducers is placed on the wet body surface or in fixed positions near this surface. From measurement of the pressure in the fluid, the force $F_w$ may be estimated using Eq. (5.21). If the body velocity $u$ is also measured, the wave excitation force $F_e$ may be estimated from

$$F_e = F_w - F_r - F_v + [-F_b] = F_w + Z_r u + R_v u + [S_b s], \tag{6.56}$$

where $Z_r$ is the radiation impedance (as in the uppermost block of Figure 6.7) and $R_v$ is the viscous resistance (as in Section 5.9). Moreover, $s(\omega) = u(\omega)/i\omega$ is the Fourier transform of the displacement. Observe that, compared with the uppermost block in Figure 6.7, we have in Figure 6.8 made the generalisation of replacing $Z_r$ with $Z_r + R_v$. The terms within the brackets in Eq. (6.56) should be included if the pressure transducers indicated in Figure 6.8 are placed on the body's wet surface and excluded if they are placed in fixed position in the water.

Let us next discuss the significance of the optimum conditions (6.52) and (6.55) in the time domain. Obviously $z_i(t)$, the inverse Fourier transform of $Z_i(\omega)$, is a causal function (cf. Subsection 5.9.1). Thus, in agreement with Eqs. (2.176)–(2.179), we have the following. Let

$$\mathcal{F}^{-1}\{Z_i(\omega)\} = z_i(t) = r_i(t) + x_i(t) = \begin{cases} 2r_i(t) & \text{for } t > 0 \\ r_i(0) & \text{for } t = 0, \\ 0 & \text{for } t < 0 \end{cases} \tag{6.57}$$

where the even and odd parts of $z_i(t)$—namely, $r_i(t)$ and $x_i(t)$—are the inverse Fourier transforms of $R_i(\omega) = \text{Re}\{Z_i(\omega)\}$ and of $iX_i(\omega) = i\text{Im}\{Z_i(\omega)\}$, respectively. Consequently, the inverse Fourier transform of $Z_i^*(\omega)$ is

$$\mathcal{F}^{-1}\{Z_i^*(\omega)\} = r_i(t) - x_i(t) = r_i(-t) + x_i(-t) = z_i(-t). \tag{6.58}$$

Thus, whereas $Z_i(\omega)$ corresponds to a causal impulse-response function $z_i(t)$, its conjugate $Z_i^*(\omega)$ corresponds to a noncausal (and, in particular, an 'anti-causal') impulse-response function $z_i(-t)$, which vanishes for $t > 0$. Thus, apart from exceptional cases (such as, e.g., when the incident wave is purely sinusoidal), the optimum condition (6.52) cannot be exactly satisfied in practice. Also, the optimum condition (6.55) cannot in general be realised exactly, because the transfer function $R_i(\omega)$ or $1/R_i(\omega)$ is even in $\omega$ and consequently corresponds to an even (and, hence, noncausal) impulse-response function of time.

Moreover, also the transfer function $f_e(\omega)$ corresponds to a noncausal impulse-response function $f_t(t)$, as explained in Subsection 5.3.2. Thus, three of the blocks shown in Figure 6.7 represent noncausal impulse-response functions, and consequently, they cannot be realised exactly. As an example, let us now discuss in some detail the relation $F_e(\omega) = f_e(\omega)A(\omega)$ in the time domain. Using Eq. (5.111), we have for the excitation force

$$F_{e,t}(t) = \int_{-\infty}^{\infty} f_t(t - \tau)a(\tau)\,d\tau, \tag{6.59}$$

where $f_t(t)$ and $a(t)$ are the inverse Fourier transforms of $f_e(\omega)$ and $A(\omega)$, respectively. Now, since $f_t$ is noncausal—that is, $f_t(t - \tau) \neq 0$ also for $\tau > t$—there is a finite contribution to the integral from the interval $t < \tau < \infty$. This means that future values of the incident wave elevation $a(t)$ are required to compute the present-time excitation force $F_{e,t}(t)$. Similarly, since $-Z_i^*(\omega)$ and $1/R_i(\omega)$ represent noncausal impulse-response functions, future values of the actual body velocity $u(t)$ and of the actual excitation force $F_{e,t}(t)$ are needed to compute the present-time values of the optimum load force $F_{u,t}(t)$ and the optimum body velocity $u_t(t)$, respectively.

Two different strategies may be attempted to solve these non-realisable (noncausal) optimum-control problems in an approximate manner. We may refer to these approximate strategies as suboptimal control methods. One strategy is to predict the relevant input quantities a certain time distance into the future, which is straightforward for purely sinusoidal waves and oscillations, and more difficult for a broad-band than for a narrow-band wave spectrum associated with a real-sea situation. The other strategy is to replace the transfer functions $f_e(\omega)$, $-Z_i^*(\omega)$ and $1/2R_i(\omega)$ by practically realisable transfer functions, which represent causal impulse-response functions, and which approximate the ideal but non-realisable transfer functions. The aim is to make the approximation good at the frequencies which are important—that is, the interval of frequencies which contains most of the wave energy. Outside this interval, the chosen realisable transfer function may deviate very much from the corresponding ideal but non-realisable transfer function. With this version of a suboptimal strategy, too, better performance may be expected for a narrowband than for a broadband wave spectrum.

Several investigations have been carried out on future predictions of the wave elevation or hydrodynamic pressure at a certain point [40, 41] and also on the utilisation of future predictions for optimum wave-energy conversion [84, 85]. The method of replacing a non-realisable ideal transfer function by a realisable but suboptimal transfer function have also been investigated to some extent [86–89]. The following paragraph, as an example, gives an outline of a causalising approximation, as proposed by Perdigão and Sarmento [86].

Let us discuss the approximation where the transfer function $1/2R_i(\omega)$ in Eq. (6.55) has been replaced by a rational function $N(\omega)/D(\omega)$, where $N(\omega)$ and $D(\omega)$ are polynomials of order $n$ and $m$, respectively. We want to ensure that the corresponding impulse-response function is causal and that the control is stable. For this reason, we impose the requirements that $n < m$ and that $D(\omega)$ has all its zeros in the upper half of the complex $\omega$ plane. In accordance with this choice, we replace Eq. (6.55) with

$$u(\omega) = F_e(\omega)N(\omega)/D(\omega), \tag{6.60}$$

and a corresponding replacement may be made to change the appropriate non-realisable block in Figure 6.7 with a realisable one. Inserting Eq. (6.60) into Eq. (6.54) gives

$$\alpha(\omega) = |F_e(\omega)|^2|1 - 2R_i(\omega)N(\omega)/D(\omega)|^2. \tag{6.61}$$

If this is inserted into the expression (6.47) for $W_{u,P}$, it becomes obvious that we, for a general $F_e(\omega)$, have $W_{u,P} > 0$. We now need to choose the unknown complex coefficients of the two polynomials in order to minimise $W_{u,P}$. For instance, if we had chosen $m = 3$ and $n = 2$, that is,

$$\frac{N(\omega)}{D(\omega)} = \frac{b_0 + b_1\omega + b_2\omega^2}{1 + a_1\omega + a_2\omega^2 + a_3\omega^3}, \tag{6.62}$$

then six complex coefficients $a_1$, $a_2$, $a_3$, $b_0$, $b_1$ and $b_2$ need to be determined. Thus, $W_{u,P}$ becomes a function of the unknown coefficients, which are determined such as to minimise $W_{u,P}$ (observing the constraint that all zeros of $D(\omega)$ have positive imaginary parts). We see that the wave spectrum, represented by the quantity $|F_e(\omega)|^2$ in Eq. (6.61), will function as a weight factor in the integral (6.47), which is to be minimised. Hence, we expect the approximation to be good in the frequency interval where $|F_e(\omega)|$ is large—in particular, if the wave spectrum is narrow. We observe that for a different wave spectrum—that is, for a different function $F_e(\omega)$—a new determination of the best polynomials $N(\omega)$ and $D(\omega)$ has to be made. Note, however, that we are not, in real time, able to determine the Fourier transform $F_e(\omega)$ exactly, which would require integration over the interval $-\infty < t < \infty$ [see Eq. (2.135)]. In practice, we have to determine $F_e(\omega)$ only approximately based on measurement during the last minutes or hours.

## 6.4 The Budal Upper Bound (BUB)

Note from Eq. (6.12) that, when the velocity amplitude $|\hat{u}_j| = |\hat{u}_j|_{\text{OPT}} = \omega|\hat{s}_j|_{\text{OPT}}$, which is proportional to the elevation amplitude $|A|$ of the incident wave, then half of the incident excitation power $P_e$ is absorbed while the remaining half is radiated, in order to produce the wave that is needed to interfere destructively with the incident wave. This may be the situation when the incident wave amplitude is rather low.

On the other hand, for larger wave amplitudes, the situation may be that $|\hat{u}_j|_{\mathrm{OPT}} \gg \omega|\hat{s}_j|_{\max}$, where $|\hat{s}_j|_{\max}$ is the design displacement amplitude of the WEC body. Then, since $|\hat{u}_j| \ll |\hat{u}_j|_{\mathrm{OPT}}$, we have $P_r \ll P \approx P_e$ (cf. Figure 6.3). That is, the excitation power is essentially absorbed by the WEC, and relatively little power is then radiated. However, in this case, only a small fraction of the wave power in the ocean is absorbed. Most of the wave power remains in the sea. Note that this situation, from an economic point of view, sometimes may be desirable for a WEC, particularly in situations with large wave heights. Because wave power in the ocean is free, whereas the realisation of a large body-velocity amplitude requires economical expenditure, it may be advantageous to aim at a design value of the velocity amplitude $|\hat{u}_j| = \omega|\hat{s}_j|_{\max}$ which is substantially smaller than $|\hat{u}_j|_{\mathrm{OPT}}$, except during wave conditions when the wave amplitude $|A|$ and, hence, also $|\hat{F}_{e,j}| = |f_{e,j}A|$ and $|\hat{u}_j|_{\mathrm{OPT}}$ are rather small.

What should be maximised in practice is the ratio between the energy delivered by the WEC and the total cost including investment, maintenance and operation. However, it is a rather complicated and large task to provide a reliable information on this problem. A much simpler problem, studied four decades ago by Budal, is to consider the ratio between the absorbed power $P$ and the volume $V$ of a heaving semisubmerged WEC body. In the following, let us discuss an upper bound [83, 90] to this ratio. Firstly, however, we consider a few other upper bounds for the absorbed wave power $P$.

The converted useful power is at most equal to the absorbed power $P$. For the absorbed power, we have from Eqs. (6.4), (6.6) and (6.7) that

$$P < \tfrac{1}{2}|\hat{F}_{e,j}| \cdot |\hat{u}_j| \cos\gamma_j \le \tfrac{1}{2}|\hat{F}_{e,j}| \cdot |\hat{u}_j|. \tag{6.63}$$

The two inequalities here approach equality if $|\hat{u}_j| \ll |\hat{u}_j|_{\mathrm{OPT}}$ and if the phase is approximately optimum, respectively.

For certain WEC bodies and modes, this design amplitude may be expressed as a simple function of the available hull volume $V$—for example, $|\hat{s}_3|_{\max} = V/2S_w$ for a heaving tall cylindrical body with water-plane area equaling $S_w$. In general, however, the design amplitude for an oscillating WEC body may be differently related to the size of the hull volume. The WEC body's useful oscillation amplitude may also be limited by, for example, the Keulegan–Carpenter number (see Section 6.4.4).

The design amplitude for an oscillating WEC body is somehow directly related to the volume of water displaced by the body. As explained earlier (for example, in Section 6.1), such an oscillating water-displacement volume is a necessity in order for the oscillating WEC body to absorb energy from an incident wave.

The excitation force amplitude is proportional to the wave amplitude $|A|$ and is bounded by $|f_{e,j}|_{max} |A|$. Thus, we have

$$|\hat{u}_j| \leq \omega |\hat{s}_j|_{max} \quad \text{and} \quad |\hat{F}_{e,j}| \leq |f_{e,j}|_{max} |A|. \tag{6.64}$$

Then, from Eq. (6.63), we have

$$P < P_K = c_K |s_j|_{max} H/T, \tag{6.65}$$

where

$$c_K = \pi |f_{e,j}|_{max}/2 \quad \text{and} \quad H = 2|A|. \tag{6.66}$$

In the present section, we are, in the mathematical sense, considering upper bounds rather than upper limits. Next, we shall discuss the Budal upper bound (BUB). As a preliminary, expressions (6.63) and (6.65) are considered to be examples of upper bounds—not upper limits (in the mathematical sense)—for the absorbed wave power. [To emphasise this, we remark that these two expressions are upper bounds for the absorbed wave power $P$ even if we should wish to multiply the right-hand sides by any arbitrary factor that is larger than 1!]

Let us now assume that a semisubmerged body of volume $V$ and water-plane area $S_w$ is oscillating in heave. Thus, with $j = 3$, it is reasonable (at least for a body of cylindrical shape) to assume that the design amplitude $|s_3|_{max}$ for heave excursion does not exceed $V/2S_w$. Then we have the further inequalities

$$|\hat{u}_3| < \omega V/2S_w \quad \text{and} \quad |\hat{F}_{e,3}| < \rho g S_w |A|, \tag{6.67}$$

where we have made use of the small-body approximation (5.277). The latter inequality approaches equality if the horizontal extension of the body is very small compared to the wavelength. Combining inequalities (6.67) with the last of inequalities (6.63), we obtain

$$P < \frac{\rho g \omega V |A|}{4} = \frac{\pi \rho g V |A|}{2T} = \frac{\pi \rho g V H}{4T}, \tag{6.68}$$

where $H = 2|A|$ is the wave height and $T = 2\pi/\omega$ is the period of the wave as well as of the heave oscillation. Here, on the right-hand side, $T$ and $H$ are wave parameters, and the hull volume $V$ is the only WEC-body parameter. Its size may perhaps serve as a rough indicator of parts of the investment cost of the WEC. We may expect that the simple inequality (6.68) has a larger right-hand side than inequalities (6.63) and (6.65). It might again be emphasised that these mathematical inequalities are upper bounds and not upper limits.

When deriving inequality (6.68), Budal [83, 90] considered a tall cylindrical body with a relatively small water-plane area. Then, with optimum phase ($\gamma_3 = 0$), the heave amplitude $|\hat{s}_3|$ may be significantly larger than the wave-elevation amplitude $|A|$. Inequalities (6.67) are based on this assumption. However, for a wave-interacting low cylindrical body with a relatively large water-plane area, the heave amplitude should not exceed the wave amplitude.

Moreover, the excitation force amplitude is bounded by the body's buoyancy force at equilibrium in still water. In this case, inequalities (6.67) are to be replaced with

$$|\hat{u}_3| < \omega|A| \quad \text{and} \quad |\hat{F}_{e,3}| < \rho g V/2, \tag{6.69}$$

as suggested by Rainey [91]. This alternative to inequalities (6.67) leads, however, to the same inequality (6.68).

Taking as typical values $T = 8$ s and $H = 2$ m, we find that inequality (6.68) gives an upper bound $P/V < 2.0$ kW/m$^3$. From the preceding discussion, it becomes evident that to approach this upper bound, it is necessary that the body volume $V$ tends to zero—which is, of course, not very practical for a WEC. However, $V$ has to be as small as practical if we are to maximize the utilisation of the WEC body volume, which is what upper bound (6.68) is about. In Problem 6.2, we discuss a case of a heaving axisymmetric body of diameter $2a = 6$ m, which is rather small compared to the 100 m wavelength for the considered 8 s wave. Although the body volume is as small as 283 m$^3$, we obtain in this case only 0.8 kW/m$^3$ for $P/V$. This is significantly less than the upper bound 2.0 kW/m$^3$. A performance closer to this upper bound is, however, obtained with larger wavelengths and, thus, longer wave periods, as discussed in Problem 6.2.

It should be emphasised that inequality (6.68) was derived based on the assumption of sinusoidal wave and oscillation. With latching control, however, a larger upper bound is applicable [84]. Let us assume that the immersed body's vertical motion between its two extreme positions takes place momentarily at the instants of extreme heave force (cf. Figure 6.6). Then the heave position, as a function of time, corresponds to a periodic square-wave function, for which the first harmonic amplitude is a factor $4/\pi \approx 1.27$ larger than its physical excursion from equilibrium.

Thus, assuming a sinusoidal wave, but only a periodic WEC-body motion, we may generalise the inequality (6.68) to

$$P < P_B = c_B V H/T, \tag{6.70}$$

where $P_B$ is the *Budal upper bound (BUB)*. Further,

$$c_B = \sigma \rho g = \sigma \times 10.0 \times 10^3 \text{W m}^{-4} \text{s}, \tag{6.71}$$

where $\rho = 1030$ kg/m$^3$ is the mass density of seawater, and $g = 9.81$ m/s$^2$ is the acceleration of gravity. Moreover,

$$\sigma = \begin{cases} \sigma_s = \pi/4 & \text{for sinusoidal oscillation,} \\ \sigma_p = 1 & \text{for periodic oscillation.} \end{cases} \tag{6.72}$$

In both cases, a sinusoidal wave of period $T = 2\pi/\omega$ and elevation amplitude $|A| = H/2$ is assumed. For a typical case with $T = 8$ s and $H = 2$ m, we find an upper bound of $P/V < 2.5$ kW/m$^3$ for the case of periodic oscillation.

The Budal upper bound, given by Eq. (6.70), was derived by considering a heaving semisubmerged WEC body. As explained earlier, in Section 6.1, for an oscillating WEC body to absorb energy from an incident wave, it is required that the WEC body radiates a wave which interferes destructively with the incident wave. When the heaving, semisubmerged body of volume $V$ moves downwards from its uppermost position to its lowest position, it pushes a volume of water $V_{water} = V$ downwards and sideways. When it moves upwards, it sucks back an equally large amount of water. It is by this back-and-forth water motion that the WEC body is radiating the wave which is needed to remove energy from the incident wave. Thus, if we write

$$P < P_B = c_B V_{water} H/T, \tag{6.73}$$

then BUB is applicable also to certain other kinds of WECs, such as an OWC. These WECs are of the monopole-type of wave radiators/absorbers. When we have replaced the heaving-body volume $V$ with the oscillating water volume $V_{water}$, the BUB in the form of inequality (6.73) becomes a universal upper bound.

We may understand this if we consider a dipole-type of wave radiator, such as an immersed thin horizontal cylinder performing surging oscillation in its axial direction. Let its vertical cross-sectional area be of a size equalling the horizontal cross-sectional area of the just considered heaving body. However, we let the surging, horizontal-axis cylinder be longer in the horizontal direction: we let the surging cylinder length be one-half of a wavelength. Thus, we may compare this surging cylinder to a set of two heaving, vertical-axis, cylindrical bodies, separated by half a wavelength in the direction of the incident-wave propagation. The size of displaced water amount $V_{water}$ may, however, be larger for the surging WEC body than for the two heaving WEC bodies. The universal relationship (6.73), where the WEC body volume $V$ has been replaced by the water volume amplitude $V_{water}$, is applicable in both cases. It is another manifestation of the principle that energy may be removed from an incoming wave by radiating a wave that interferes destructively with it. To radiate a wave, the WEC body needs to alternately push and suck a certain amount of water during each oscillation cycle.

### 6.4.1 Two Different Upper Bounds for Absorbed Power

The BUB inequality (6.70) represents an upper bound for the wave power which may be absorbed from a sinusoidal wave by a heaving semisubmerged body. Observe that the BUB is an *upper bound*, and not, in the mathematical sense, an upper limit.

In contrast, according to Eq. (6.13), another upper bound, $P \leq P_A = G(\beta)J/k$, which is valid for constant water depth $h$, may, in certain ideal cases, even correspond to an upper limit. For the mathematically simpler case of deep water, this upper-bound relationship may, according to Eq. (6.16), be simplified to

$$P \leq P_A = c_A H^2 T^3. \tag{6.74}$$

The coefficient $c_A = c_{A,j}$ is given by Eq. (6.17) for each of the WEC body's oscillation modes, $j = 1, \ldots, 6$. In the following, discussing an axisymmetric body in the heave mode, we shall, for convenience, omit the subscript $j = 3$. Then, since $G_3(\beta) \equiv 1$,

$$c_A = c_{A,3} = \rho(g/\pi)^3/128 = 245\,\mathrm{W\,m^{-2}s^{-3}}, \tag{6.75}$$

according to Eq. (6.17).

Note that this latter upper bound, inequality (6.74), depends on the WEC body's shape and oscillation mode, but not on its size. It corresponds to an ideal situation where the available sea-wave energy is exploited as much as possible. On the other hand, the BUB, inequality (6.70), as well as its generalisations (6.73) and (6.65), concerns an optimum utilisation of the deployed WEC hull size, or maximum stroke. We may call inequality (6.74) theoretical upper bound and inequality (6.70), (6.73) or (6.65) practical upper bound. For wave periods $T \leq T_c$, for example, upper bound (6.74) governs, whereas for $T \geq T_c$, upper bound (6.70), (6.73) or (6.65) governs. A WEC should exploit the available wave energy as much as possible for $T \leq T_c$, since at these periods the hull size or maximum stroke is not a restriction. On the other hand, a WEC should utilise its hull size or design stroke as much as possible for $T \geq T_c$, because at these periods, these limits govern the maximum energy that can practically be absorbed by the WEC.

Notice that the dependence on the wave period $T$ is very much different in the two upper bounds. The power of $T^{-1}$ in BUB reflects the fact that the WEC body performs more oscillation strokes per hour for shorter wave periods, while the power of $T^3$ in Eq. (6.74) reflects the fact that with increasing $T$, firstly, the wave motion extends more deeply below the sea surface and, secondly, the group velocity is larger.

Based on the two upper bounds and on wave data for a particular ocean site, we shall, in the following, make a reasonable decision upon sizes of PTO capacity and of WEC hull volume for a simple axisymmetric example. The two upper-bound relations (6.70) and (6.74) and also some additional upper-bound relations for WECs have during recent years been discussed in more detail in several papers [77, 92, 93].

### 6.4.2 Axisymmetrically Radiating WEC Body

For a WEC body radiating isotropically, the two different upper-bound curves, one monotonically ascending curve for $P_A$ and one monotonically descending curve for $P_B$, are illustrated in the $P$-versus-$T$ diagram of Figure 6.10—for example, where $V = 524\,\mathrm{m}^3$ and $H = 2.26$ m. The two curves intersect at a point $(T_c, P_c) = (T_{c,s}, P_{c,s}) = (9.3\,\mathrm{s}, 1.0\,\mathrm{MW})$, for which $P_A = P_B \equiv P_c$. From Eqs. (6.70) and (6.74), we now find the intersection-point relations

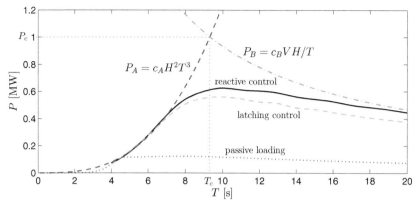

Figure 6.10: The two upper, monotonically increasing and decreasing curves show, according to Eqs. (6.74) and (6.70), two different upper bounds—$P_A$ and $P_B$, respectively—for the maximum possible power $P$ which may be absorbed from a sinusoidal wave of height $H$ and period $T$ by means of an immersed heaving axisymmetric WEC of volume stroke $V$. The shown scales for $T$ and $P$ are applicable when $H = 2.26$ m, $V = 524$ m$^3$ and $\sigma = \pi/4$. The two curves' intersection point is $(T_c, P_c) = (T_{c,s}, P_{c,s}) = (9.3$ s$, 1.0$ MW$)$, as indicated by two straight dotted lines. Below the two upper-bound curves for $P_A$ and $P_B$, three numerically computed theoretical curves are shown, where it has been assumed that the WEC is a semisubmerged sphere, which is heaving with optimum amplitude, but with three different phase-control methods, as indicated on the diagram. The sphere has a diameter of $2a = 10$ m, and its heave amplitude is load-constrained to not exceed 3 m. (The figure is a variant of figure 6 in [77]).

$$T_c^4 = \frac{c_B V}{c_A H} \quad \text{and} \quad P_c^4 = c_A c_B^3 V^5 H^3. \tag{6.76}$$

Thus, the 'WEC parameters' $P_c$ and $V$ can be expressed in terms of 'wave parameters' $H$ and $T_c$ as

$$P_c = c_A H^2 T_c^3 = (245 \text{ W m}^{-2} \text{ s}^{-3}) H^2 T_{c,s}^3, \tag{6.77}$$

$$V = (c_A/c_B)HT_c^4 = (0.0310 \text{ W m}^2 \text{ s}^{-4}) HT_{c,s}^4, \tag{6.78}$$

where $\sigma = \sigma_s = \pi/4$ is implied with the numerical value of the intersection-point wave period $T_{c,s}$; cf. Eqs. (6.72).

Below these two upper-bound curves, Figure 6.10 contains three curves, which represent computed values for the wave power absorbed by the heaving WEC body, assumed to be a semisubmerged sphere of diameter $2a = 10$ m and volume $V = 524$ m$^3$. These three curves—corresponding to reactive control, latching control and passive operation—show maxima significantly below the intersection-point power $P_c$. The passive-loading curve touches the $P_A$ curve at the heave-resonance period $T \approx 4.3$ s. The reactive-control curve follows the $P_A$ curve from the same resonance period 4.3 s to a period about 6.8 s, where $P_B > 2P_A$. For the wave height $H$ in question, the $P_A$ curve is unreachable when $P_B < 2P_A$.

For reactive control and for latching control, the oscillation is not sinusoidal. Thus, for these two cases, the BUB with $\sigma = \sigma_u = 1$ is applicable. This is not shown in Figure 6.10, but it would correspond to an intersection point $(T_c, P_c) = (T_{c,u}, P_{c,u}) = (9.9\,\mathrm{s}, 1.2\,\mathrm{MW})$. Looking at the right-hand portions of the two curves corresponding to reactive control and latching control in Figure 6.10, one might expect that they, if extrapolated to larger wave periods, may extend above the shown monotonically declining BUB curve, corresponding to $\sigma = \sigma_s = \pi/4$.

Based on a diagram shown in Figure 6.10 and several other similar diagrams [77, 92, 93], it may be concluded that the *maximum* absorbed wave power $P_{\max}$ relative to the intersection-point power $P_{c,s}$ is roughly only a fraction of about

$$
P_{\max}/P_{c,s} \approx \begin{cases} 0.6 & \text{for reactive control} \\ 0.5 & \text{for latching control} \\ 0.1 & \text{for passive system,} \end{cases} \tag{6.79}
$$

corresponding to power takeoff (PTO) capacity of 0.6 MW, 0.5 MW or 0.1 MW respectively.

For a wave period of $T = 8\,\mathrm{s}$, the chosen wave height $H = 2.26\,\mathrm{m}$ has a rather high probability of occurrence off the European Atlantic coast but a significantly less probability for wave periods above 10 s. Thus, it may be favorable to choose a much smaller—and hence less expensive—WEC body than the present one, for which the diameter is 10 m. This would also lower both intersection-point values $T_c$ and $P_c$. The WEC unit may then be better matched to our oceans' wave climate.

According to definitions in Section 6.1.1, this heaving 10 m diameter WEC body is a quasi-point absorber (QPA) for wavelengths $\lambda < 200$ m—that is, for wave periods $T < 11$ s and, thus, for the substantial part of our oceans' wind-generated wave energy. Next, we consider a significantly smaller WEC body, designed for converting energy from waves of periods in the range of 6 to 11 s.

### 6.4.3 A Point Absorber

North Atlantic wave data off the west coast of Scotland [94, p. 80] indicate that the wave-power level $J$ exceeds about 40 kW/m during about one-third of the year. Thus, it is reasonable to base the design of a WEC body on this level. Moreover, it appears that wave states occur rather rarely outside the wave-period region from 6 s to 11 s and most frequently for wave periods $T$ within the interval $7\,\mathrm{s} < T < 9\,\mathrm{s}$.

Observing the absorbed-power curves on Figure 6.10, we note that the response stays high in a much wider wave-period range above rather than below the intersection wave period $T_c$. For this reason, we shall choose an intersection-point period significantly below 8 s. Then the BUB curve for $P_B$ in the diagram of Figure 6.10 has to be lowered, corresponding to a smaller WEC-body volume $V$. In this $P$-versus-$T$ diagram, the wave height $H$ and the WEC-body volume $V$ are constant, given parameters. If the volume is reduced, the $P_B$ curve is lowered in the same proportion, but the $P_A$ curve remains unchanged. The intersection point $(T_c, P_c)$ will slide down along this $P_A$ curve. The immersed WEC body volume is better utilised, but less energy will be removed from the sea waves by each WEC unit. Thus, if many smaller WEC bodies are deployed in a WEC array, the distance between adjacent WEC-body units may need to be shorter.

As an example, let us consider a single WEC body which is a heaving semisubmerged vertical cylinder of radius $a$ and height $b$. As argued later, in Section 6.4.4, we shall choose $b \geq 2a$. Let us assume that its diameter is $2a = 5$ m. Then, the cylindrical WEC body has a volume of $V' = \pi a^2 b \geq 2\pi a^3 = 98$ m$^3$. It is a point absorber (PA) for wavelengths $\lambda > 20 \times 2a = 100$ m; thus, according to Eq. (4.100), for wave periods $T > 8.0$ s. Otherwise, for smaller wave periods, it will be a QPA.

Because of the smaller volume $V'$ (here assumed to be 98 m$^3$), the intersection point corresponds to smaller values $(T_c', P_c') = (T_{c,s}', P_{c,s}') = (6.1 \text{ s}, 0.29 \text{ MW})$, as obtained from the two equations (6.76). The coarse approximations (6.79) concerning a WEC's PTO capacity are presumably still valid. Correspondingly, if each WEC unit is equipped for optimum-control operation, then its PTO capacity should be about 0.15 MW.

The vertical cylindrical body should, at its lower end, be extended with a hemisphere, of the same diameter $2a = 5$ m. The main purpose is to reduce the risk of energy loss from the shedding of vortices. (See the following Section 6.4.4.) An additional purpose is to provide the possibility of lowering the WEC body's centre of gravity. The intended maximum heave amplitude is $b$, and the equilibrium draft of the axisymmetric WEC body is then $a + b$.

### 6.4.4 Importance of the Keulegan–Carpenter Number

Figure 6.1(b) may serve as an illustration of the 5 m diameter WEC body considered in Section 6.4.3 when downscaled and tested in a wave channel.

If such a reduced-size laboratory model has a diameter of $2a = 150$ mm, the length scale is $1 : 33.3$. Then the time scale is $1 : \sqrt{33.3} = 1 : 5.77$, because the acceleration of gravity is $g = 9.81$ m/s$^2$ in model scale as well as in full scale. The volume scale is $1 : 33.3^3 = 1 : 37.0 \times 10^3$, as is also the force scale if we neglect the difference between sea-water and fresh-water mass densities. Then the energy scale is $1 : 33.3^4 = 1 : 1.23 \times 10^6$, and the power scale is $1 : 33.3^{7/2} = 1 : 214 \times 10^3$.

Such a laboratory model has been arranged to realise resonant heave oscillation of wave period $T = 1.49$ s, corresponding to full-scale period $T = 8.6$ s. The width of the wave channel is 0.50 m, and the water depth is $h = 1.50$ m. Thus, according to Eq. (4.54), the wavelength is $\lambda = 3.4$ m, corresponding to a full-scale value of 115 m. As the diameter is $2a < \lambda/20$, the model may be classified as a point absorber in the performed model experiments [95].

An important part of these laboratory experiments is the linearity tests, with results as shown in Figure 6.11, where the horizontal scale shows wave amplitude, while the vertical scale shows heave amplitude, under conditions when the

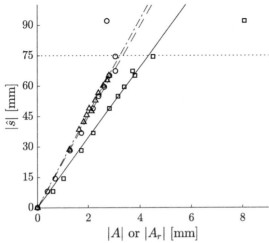

Figure 6.11: Heave oscillation versus wave elevation. Three sets of linearity tests on a heaving vertical, 150 mm diameter cylinder, which is extended with a hemisphere on its lower end. It is immersed to an equilibrium draft of $(100 + 75)$ mm $= 0.175$ m, in a 0.50 m wide wave channel, where the water depth is $h = 1.50$ m. Sinusoidal oscillations have a period of $T = 1.49$ s. Square plots show resonant heave amplitude $|\hat{s}|$ responding to incident-wave amplitude $|A| = H/2$, when the mechanical load is minimised to include only unavoidable friction and viscosity losses. Triangular plots show radiated wave amplitude responding to the body's heave amplitude when there is no incident wave. Circular plots show the radiated wave amplitude $|A_r|$ responding to the body's heave amplitude $|\hat{s}|$ caused by the incident-wave amplitude $|A|$. Note that then an almost negligible diffracted-wave contribution is included in this $|A_r|$. The dotted horizontal line indicates where the Keulegan-Carpenter number equals $\pi$. The inclined straight lines through the origin are matched to the experimental points, excluding points for which $|\hat{s}| > 70$ mm. These three lines have steepnesses 23, 22 and 17. (This figure is a variant of figure 4 in [95])

damping load of oscillation is minimised to include only unavoidable viscosity and friction losses. Three inclined straight lines of steepnesses 17, 22 and 23 are matched to the experimental points for which the heave amplitude is less than 70 mm. The lines with steepnesses 17 and 22 correspond to the right-hand zero crossing of the lower and upper parabola, respectively, as shown in Figure 6.3.

The experimental results shown by square plots in Figure 6.11 show the body's heave amplitude $|\hat{s}|$ resulting from a sinusoidal wave of amplitude $|A|$ propagating along the wave flume. Deviation from linearity indicates that the wave's ability to move the immersed body is seriously impoverished if the heave amplitude exceeds 75 mm, which here equals the minimum radius of curvature of the immersed WEC-body surface. This demonstrates that for oscillatory flow it may be very important to consider the Keulegan–Carpenter number $N_{KC}$.

We suggest that numerical oscillating-fluid results based on potential theory may be applicable if the *relative* oscillation amplitude $|\hat{s}_b|$ of the water, along the immersed WEC-body surface, nowhere exceeds the local radius of curvature $\rho_b$ on this surface. Then the Keulegan–Carpenter number [96]

$$N_{KC} \equiv \pi|\hat{v}_b|/(\omega\rho_b) = \pi|\hat{s}_b|/\rho_b < \pi \tag{6.80}$$

is small enough for avoidance of vortex shedding and corresponding energy loss. Here $|\hat{v}_b|$ is the tangential relative water velocity along the WEC-body's wet surface at its minimum radius of curvature, $\rho_b$. This number is related to possible loss of energy related to the occurrence of vortices being shed from the oscillating body. A lesson to learn is that sharp edges or corners on a WEC body's wet surface should be avoided.

While the square plots in Figure 6.11 show the primary heave response caused by the incident wave, a secondary response is a radiated wave—including an almost negligible diffracted-wave contribution—as shown by the circular plots. Note that, in this case, when the Keulegan–Carpenter number approaches, or exceeds $\pi$, the oscillating body reduces its ability both to radiate a wave and to convert wave energy. Remember Budal's formulation: to absorb a wave means to radiate a wave!

Concerning the triangular plots in Figure 6.11, there is no diffracted-wave contribution but only a radiated wave. The waves were generated by exciting the body to oscillate, while the wave channel's own wavemaker was not used. Thus, for these triangular plots, there is only a radiated wave without any diffracted-wave contribution.

## 6.5 Several WEC-Body Modes

As discussed in Section 5.1, an immersed body may oscillate in up to six different modes $j$ ($j = 1, 2, \ldots, 6$). Moreover, if there is an incident wave, an excitation force $F_{e,j}$ results, as discussed also in Sections 5.5.1 and 5.5.5. As a response to the excitation force, the immersed body may attain a velocity $u_j$ whereby an additional wave force, the radiation force, $F_{r,j}$, is set up. In accordance with Eq. (5.155), the total wave force in oscillating mode $j$ has a complex amplitude

$$\hat{F}_{t,j} = \hat{F}_{e,j} + \hat{F}_{r,j} = f_j(\beta)A - \sum_{j'=1}^{6} Z_{jj'}\hat{u}_{j'} \tag{6.81}$$

where $Z_{jj'}$ is an element of the radiation impedance matrix, which is discussed in Sections 5.2.1, 5.5.1 and 5.5.3. Further, $\beta$ is the angle of wave incidence, and $A$ is the complex wave-elevation amplitude of the undisturbed incident wave at the chosen sea-surface reference point for the immersed WEC body. In accordance with Eq. (2.76), the (time-average) power absorbed by oscillating mode $j$ is

$$P_j = \tfrac{1}{2}\mathrm{Re}\{\hat{F}_{t,j}\hat{u}_j^*\} = \tfrac{1}{4}(\hat{F}_{t,j}\hat{u}_j^* + \hat{F}_{t,j}^*\hat{u}_j), \tag{6.82}$$

which, in view of Eq. (6.81), may be written as

$$P_j = \frac{1}{4}\left[f_j(\beta)A\hat{u}_j^* + f_j^*(\beta)A^*\hat{u}_j\right] - \frac{1}{4}\sum_{j'=1}^{6}\left(Z_{jj'}\hat{u}_{j'}\hat{u}_j^* + Z_{jj'}^*\hat{u}_{j'}^*\hat{u}_j\right). \tag{6.83}$$

By application of Eq. (5.24)—see also Section 2.3.1—we find the (time-average) power absorbed by the oscillating body:

$$P = \sum_{j=1}^{6} P_j = P_e - P_r \equiv AE(\beta) + A^*E^*(\beta) - |U|^2, \tag{6.84}$$

where we have generalised the one-mode quantities $E(\beta)$ and $U$ introduced in Section 6.2.3. Here,

$$P_e = \frac{1}{4}\sum_{j=1}^{6}(\hat{F}_{e,j}\hat{u}_j^* + \hat{F}_{e,j}^*\hat{u}_j) = \frac{1}{4}(\hat{\mathbf{F}}_e^{\mathrm{T}}\hat{\mathbf{u}}^* + \hat{\mathbf{F}}_e^{\dagger}\hat{\mathbf{u}}) = AE(\beta) + A^*E^*(\beta) \tag{6.85}$$

is the excitation power. (Note that we may, if we wish, replace $\hat{\mathbf{F}}_e^{\dagger}\hat{\mathbf{u}}$ with $\hat{\mathbf{u}}^{\mathrm{T}}\hat{\mathbf{F}}_e^*$.) Moreover,

$$P_r = \frac{1}{2}\sum_{j=1}^{6}\sum_{j'=1}^{6} R_{jj'}\hat{u}_{j'}\hat{u}_j^* = \frac{1}{2}\hat{\mathbf{u}}^{\dagger}\mathbf{R}\hat{\mathbf{u}} = \frac{1}{2}\hat{\mathbf{u}}^{\mathrm{T}}\mathbf{R}\hat{\mathbf{u}}^* \equiv |U|^2 \tag{6.86}$$

is the radiated power.

Here we have generalised to a six-mode case the two, complex, collective scalar parameters, the *collective excitation-power coefficient* $E(\beta)$ and the *collective oscillation amplitude* $U$, which are introduced in Section 6.2.3, for a single-mode case. These results may be extended, not only to a six-mode WEC body, but even to an array of WEC bodies, as shown in more detail in Section 8.1. Comparison of Eqs. (6.32) and (6.84) demonstrates that the *wave-power 'island'* (cf. Figure 6.5) is applicable even for a WEC body oscillating in up to six different modes [63].

In Eqs. (6.85) and (6.86), we have applied the matrix notation introduced in Chapter 5. When deriving Eq. (6.86), we have utilised the fact that the radiation resistance matrix $\mathbf{R}$ is real and symmetric. Moreover, we have made use of the reciprocity relationship (5.175) and also Eq. (5.176), from which it follows that

$$\tfrac{1}{4}(Z_{jj'} + Z_{j'j}^*) = \tfrac{1}{4}(Z_{jj'} + Z_{jj'}^*) = \tfrac{1}{2}R_{jj'}. \tag{6.87}$$

Assuming that all $\hat{F}_{e,j}$ and all $\hat{u}_j$ are known, the total absorbed power $P$ is given by Eqs. (6.84)–(6.86), provided that all matrix elements $R_{jj'}$ are also known. It is not necessary to know the imaginary part of the radiation impedance matrix. However, in order to determine the partition of absorbed power among the six individual oscillating modes, Eq. (6.83) shows that knowledge of the imaginary part of all $Z_{jj'}$, or, thus, of the added-mass matrix, is also required. This knowledge is also necessary in order to find from Eq. (6.81) the total wave force $\hat{F}_{t,j}$ in oscillating mode $j$.

As shown in the following, the last term in Eq. (6.84), the radiated power $P_r$, is necessarily nonnegative. Thus, it is possible for it to be represented as the modulus square of some complex quantity $U$. The argument is as follows.

In the case of no incident wave ($A = 0$), we have $F_{ej} \equiv 0$ and, hence, $P_e = 0$. If, in spite of this, the immersed body oscillates (due to external forcing), it would work as a wave generator (that means $P < 0$), except in particular cases when an immersed body may oscillate in an ideal fluid without the existence of any nonzero radiated wave in the far-field region (in these cases, $P = 0$, which is possible only if the radiation resistance matrix is singular). In this situation, it is impossible to have a positive absorbed power $P$. Otherwise, the principle of conservation of energy would be violated. Equation (6.84) then tells us that $P_r \geq 0$ or, in view of Eq. (6.86),

$$\hat{\mathbf{u}}^{\dagger} \mathbf{R} \hat{\mathbf{u}} \geq 0 \tag{6.88}$$

for all possible values of $\hat{\mathbf{u}}$. Thus, according to this inequality, the radiation resistance matrix $\mathbf{R}$ is positive semidefinite [16, p. 50], since $\mathbf{R}$ is a real symmetric matrix ($\mathbf{R}^{\dagger} = \mathbf{R}^{\mathrm{T}} = \mathbf{R} = \mathbf{R}^*$). If $\mathbf{R}$ is a non-singular matrix, it is positive definite. However, in many cases, as we have seen in Sections 5.7 and 5.8, the radiation resistance matrix is singular (its rank is less than its order.) For instance, with a general axisymmetric body, it is possible to choose nonzero complex velocity amplitudes for the surge and pitch modes in such a way that, by mutual cancellation, the resulting radiated wave vanishes. Hence, in the general case, we may only state that the matrix is positive semidefinite. Thus, because Eq. (6.88) holds, it is possible to state that the left-hand side of Eq. (6.86) is equal to the modulus square of a certain complex quantity $U$.

So far, we have no requirement on the arbitrary phase (argument) of the complex quantity $U$. We shall, however, find it convenient to let it have the same phase as $A^* E^*(\beta)$ appearing in Eq. (6.84). Then $A^* E^*(\beta)/U$ is a real positive quantity, which is, notably, independent of the complex velocity amplitude $\hat{\mathbf{u}}$. This means that, for example, Eq. (6.37) is applicable even for this six-mode WEC body.

### 6.5.1 Maximum Absorbed Power and Useful Power

Let us now consider the situation in which the incident wave has a relatively small amplitude, while there is no constraint on how to choose the complex velocity amplitudes $\hat{u}_j$ for the body's oscillating modes. We wish to determine these amplitudes so that the absorbed power $P$ becomes a maximum. To this end, we define the column vector

$$\delta = \hat{\mathbf{u}} - \mathbf{U}, \tag{6.89}$$

where $\mathbf{U}$ is a solution of the linear algebraic equation

$$\mathbf{R}\mathbf{U} = \tfrac{1}{2}\hat{\mathbf{F}}_e. \tag{6.90}$$

We now utilise the fact that $\mathbf{R}$ is a matrix which is real and symmetric. Thus, the transpose of Eq. (6.90) is $\hat{\mathbf{F}}_e^{\mathrm{T}}/2 = \mathbf{U}^{\mathrm{T}}\mathbf{R}^{\mathrm{T}} = \mathbf{U}^{\mathrm{T}}\mathbf{R}$. Inserting $\hat{\mathbf{u}} = \mathbf{U} + \delta$ into Eqs. (6.85) and (6.86), we obtain from Eq. (6.84)

$$P = P(\hat{\mathbf{u}}) \equiv \tfrac{1}{4}(\hat{\mathbf{F}}_e^T \hat{\mathbf{u}}^* + \hat{\mathbf{F}}_e^\dagger \hat{\mathbf{u}}) - \tfrac{1}{2}\hat{\mathbf{u}}^\dagger \mathbf{R}\hat{\mathbf{u}}$$

$$= \tfrac{1}{4}\hat{\mathbf{F}}_e^T \mathbf{U}^* + \tfrac{1}{4}\hat{\mathbf{F}}_e^\dagger \mathbf{U} + \tfrac{1}{4}\hat{\mathbf{F}}_e^T \boldsymbol{\delta}^* + \tfrac{1}{4}\hat{\mathbf{F}}_e^\dagger \boldsymbol{\delta} \tag{6.91}$$

$$- \tfrac{1}{2}\mathbf{U}^\dagger \mathbf{R}\mathbf{U} - \tfrac{1}{2}\boldsymbol{\delta}^\dagger \mathbf{R}\mathbf{U} - \tfrac{1}{2}\mathbf{U}^\dagger \mathbf{R}\boldsymbol{\delta} - \tfrac{1}{2}\boldsymbol{\delta}^\dagger \mathbf{R}\boldsymbol{\delta}.$$

Of these eight terms, only the first and eighth terms remain after mutual cancellation among the other terms. Since all terms are scalars, they equal their transposes. By transposing the third term, we see that it is cancelled by the sixth term after utilising Eq. (6.90). Applying the transposed conjugate of Eq. (6.90) in the seventh term, we see that it cancels the fourth term, and similarly, the second and the fifth terms cancel each other. Hence,

$$P = P(\hat{\mathbf{u}}) = P(\mathbf{U} + \boldsymbol{\delta}) = \tfrac{1}{4}\hat{\mathbf{F}}_e^T \mathbf{U}^* - \tfrac{1}{2}\boldsymbol{\delta}^\dagger \mathbf{R}\boldsymbol{\delta}. \tag{6.92}$$

So far, this is just another version of Eqs. (6.84)–(6.86). However, observing inequality (6.88), which means that $\mathbf{R}$ is positive semidefinite, we see that the maximum absorbed power is

$$P_{MAX} = P(\mathbf{U}) = \tfrac{1}{4}\hat{\mathbf{F}}_e^T \mathbf{U}^*, \tag{6.93}$$

since the last term in Eq. (6.92) can only reduce, and not increase, $P$.

If the radiation resistance matrix $\mathbf{R}$ is non-singular, its inverse $\mathbf{R}^{-1}$ exists and Eq. (6.90) has a unique solution for the optimum velocity:

$$\hat{\mathbf{u}}_{OPT} \equiv \mathbf{U} = \tfrac{1}{2}\mathbf{R}^{-1}\hat{\mathbf{F}}_e. \tag{6.94}$$

Then only $\boldsymbol{\delta} = 0$ can maximise Eq. (6.92). If $\mathbf{R}$ is singular, the algebraic equation (6.90) is indeterminate. We know that then the last term in Eq. (6.92) can vanish for some particular non-vanishing values of $\boldsymbol{\delta}$. (See, for instance, Section 5.7.2.) Adding such a $\boldsymbol{\delta}$ to $\mathbf{U}$, we obtain another optimum velocity $\boldsymbol{\delta} + \mathbf{U}$, which gives the same maximum absorbed power as given by Eq. (6.93). Thus, even though the optimum velocity $\hat{\mathbf{u}}_{OPT}$ is ambiguous when $\mathbf{R}$ is singular, the maximum absorbed power is unique. This maximum necessarily results if the velocity is given as one of the possible solutions of Eq. (6.90) (also see Problem 8.3.)

Because $P_{MAX}$ is real and scalar, we may transpose and conjugate the right-hand side of Eq. (6.93). If we also use Eqs. (6.85), (6.86) and (6.90), we may choose among several expressions for the maximum absorbed power, such as

$$P_{MAX} = P_{r,OPT} = \tfrac{1}{2}P_{e,OPT} = \tfrac{1}{2}\hat{\mathbf{u}}_{OPT}^T \mathbf{R}\hat{\mathbf{u}}_{OPT}^* = \tfrac{1}{4}\hat{\mathbf{F}}_e^T \hat{\mathbf{u}}_{OPT}^* = \tfrac{1}{4}\hat{\mathbf{u}}_{OPT}^T \hat{\mathbf{F}}_e^* = |U_0|^2. \tag{6.95}$$

If $\mathbf{R}$ is non-singular, it is also possible to use Eq. (6.94) to obtain

$$P_{MAX} = \tfrac{1}{8}\hat{\mathbf{F}}_e^T \mathbf{R}^{-1}\hat{\mathbf{F}}_e^* = \tfrac{1}{8}\hat{\mathbf{F}}_e^\dagger \mathbf{R}^{-1}\hat{\mathbf{F}}_e. \tag{6.96}$$

We now take into consideration energy loss due to friction and viscosity, by applying the simplified modelling proposed in Section 5.9. In analogy with Eq. (6.86) for the radiated power, the lost power is

$$P_{\text{loss}} = \tfrac{1}{2}\hat{\mathbf{u}}^{\mathrm{T}}\mathbf{R}_f\hat{\mathbf{u}}^*, \tag{6.97}$$

where $\mathbf{R}_f$ is the friction resistance matrix introduced in Eq. (5.340). The difference between the absorbed power and lost power is

$$P_u = P - P_{\text{loss}} = P_e - P_r - P_{\text{loss}}, \tag{6.98}$$

which we shall call useful power. Using Eqs. (6.85), (6.86) and (6.97), we get

$$P_u = \tfrac{1}{4}(\hat{\mathbf{F}}_e^{\mathrm{T}}\hat{\mathbf{u}}^* + \hat{\mathbf{F}}_e^{\dagger}\hat{\mathbf{u}}) - \tfrac{1}{2}\hat{\mathbf{u}}^{\mathrm{T}}(\mathbf{R} + \mathbf{R}_f)\hat{\mathbf{u}}^*. \tag{6.99}$$

If in Eq. (6.90) we replace $\mathbf{R}$ with $(\mathbf{R}+\mathbf{R}_f)$, we get the optimum condition for the oscillation velocities which maximise the useful power. If we use these optimum velocities in Eq. (6.95) and also replace $\mathbf{R}$ by $(\mathbf{R} + \mathbf{R}_f)$, we obtain the maximum useful power. Even though $\mathbf{R}$ is in many cases a singular matrix, we expect that $\mathbf{R} + \mathbf{R}_f$ is usually non-singular. Then we have the maximum useful power

$$P_{u,\text{MAX}} = \tfrac{1}{8}\hat{\mathbf{F}}_e^{\dagger}(\mathbf{R} + \mathbf{R}_f)^{-1}\hat{\mathbf{F}}_e^* \tag{6.100}$$

and the corresponding vector

$$\hat{\mathbf{u}} = \tfrac{1}{2}(\mathbf{R} + \mathbf{R}_f)^{-1}\hat{\mathbf{F}}_e \tag{6.101}$$

for the optimum complex velocity-amplitude components.

Due to the off-diagonal elements of the radiation resistance matrix $\mathbf{R}$, condition (6.90) or (6.101) for the optimum tells us that, for mode number $i$, it is not necessarily the best, in the multi-mode case, to have the velocity $u_i$ in phase with the excitation force $F_{e,i}$.

### 6.5.2 Axisymmetric WEC Body

Let us now consider some examples of maximum power absorption with no constraints on the amplitudes.

With heave motion as the only mode of oscillation for a single axisymmetric body, the radiation resistance matrix $\mathbf{R}$ and the excitation force vector $\hat{\mathbf{F}}_e$ are simplified to the scalars $R_{33}$ and $F_{e,3} = f_{30}A$, respectively. Then we have, according to Eqs. (6.94) and (6.96) [see also Eq. (3.45)],

$$P_{\text{MAX}} = \frac{|\hat{F}_{e,3}|^2}{8R_{33}} \quad \text{when} \quad \hat{u}_3 = \hat{u}_{3,\text{OPT}} = \frac{\hat{F}_{e,3}}{2R_{33}}. \tag{6.102}$$

Using Eqs. (4.130) and (5.302) gives

$$P_{\text{MAX}} = \frac{|f_3|^2}{8R_{33}}|A|^2 = \frac{|f_{30}|^2}{8R_{33}}|A|^2 = \frac{2\rho g^2 D(kh)}{8\omega k}|A|^2 = \frac{\rho g v_g}{2k}|A|^2 = \frac{1}{k}J, \tag{6.103}$$

where $J$ is the incident wave power transport per unit frontage of the incident wave. Thus, the maximum absorption width is

$$d_{a,\mathrm{MAX}} = \frac{P_{\mathrm{MAX}}}{J} = \frac{1}{k} = \frac{\lambda}{2\pi}. \tag{6.104}$$

This result was first derived independently by Newman [34], Budal and Falnes [70] and Evans [75].

Next let us consider the case of a single axisymmetric body oscillating in just the three translational modes: surge, sway and heave. The oscillation velocity components, the excitation force components and the radiation resistance matrix may be written as

$$\hat{\mathbf{u}} = \begin{bmatrix} \hat{u}_1 \\ \hat{u}_2 \\ \hat{u}_3 \end{bmatrix}, \quad \hat{\mathbf{F}}_e = \begin{bmatrix} \hat{F}_{e,1} \\ \hat{F}_{e,2} \\ \hat{F}_{e,3} \end{bmatrix}, \quad \mathbf{R} = \begin{bmatrix} R_{11} & 0 & 0 \\ 0 & R_{11} & 0 \\ 0 & 0 & R_{33} \end{bmatrix}, \tag{6.105}$$

where we have set $R_{22} = R_{11}$ for reasons of symmetry. The conditions for optimum (6.90) and for maximum power (6.95) give

$$\mathbf{U} = \begin{bmatrix} \hat{F}_{e,1}/2R_{11} \\ \hat{F}_{e,2}/2R_{22} \\ \hat{F}_{e,3}/2R_{33} \end{bmatrix} \tag{6.106}$$

$$P_{\mathrm{MAX}} = \frac{1}{4}\hat{\mathbf{F}}_e^{\mathrm{T}}\mathbf{U}^* = \frac{1}{8}\left( \frac{|f_1|^2}{R_{11}} + \frac{|f_2|^2}{R_{22}} + \frac{|f_3|^2}{R_{33}} \right)|A|^2. \tag{6.107}$$

From Eqs. (5.285) and (5.289), we have

$$f_1(\beta) = -f_{10}\cos\beta \tag{6.108}$$
$$f_2(\beta) = -f_{20}\sin\beta \tag{6.109}$$
$$f_3(\beta) = f_{30}, \tag{6.110}$$

and by using Eq. (5.302), we finally arrive at

$$\begin{aligned} P_{\mathrm{MAX}} &= \frac{1}{8}\frac{4\rho g^2 D(kh)}{\omega k}\left( \cos^2\beta + \sin^2\beta + \frac{1}{2} \right)|A|^2 \\ &= \frac{3}{k}\frac{\rho g^2 D(kh)}{4\omega}|A|^2 = \frac{3}{k}J. \end{aligned} \tag{6.111}$$

The maximum absorption width is

$$d_{a,\mathrm{MAX}} = \frac{P_{\mathrm{MAX}}}{J} = \frac{3}{k} = \frac{3}{2\pi}\lambda. \tag{6.112}$$

This result was first obtained by Newman [34].

For the case with oscillation in just the three modes—surge, heave and pitch—the radiation resistance matrix is singular. (It is a $3 \times 3$ matrix of rank 2.) As shown in Problem 6.3, the result is again and $d_{a,\mathrm{MAX}} = 3/k = (3/2\pi)\lambda$.

Let us now consider the more general case in which all six modes of motion are allowed. The maximum power as given by Eq. (6.111) corresponds to optimum interference of the incident wave with three radiated circular waves, which have three possible different $\theta$ variations, in accordance with Eq. (5.285). So, for instance, if surge is already involved and optimised, it is not possible to improve the optimum situation by also involving pitch, since both these modes produce radiated waves with the same $\theta$ variation. If three modes of motion corresponding to three different $\theta$ variations [see Eq. (5.285)] are already involved with their complex amplitudes chosen to yield optimum wave interference between the three radiated circular waves and the plane incident wave, then it is not possible to increase the absorbed power by increasing the number of degrees of freedom beyond three. Hence, the rank of the radiation resistance matrix for an axisymmetric system, as given by Eq. (5.302), cannot be more than three. Thus, the matrix is necessarily singular if its dimension is larger than $3 \times 3$.

## Problems

### Problem 6.1: Power Absorbed by a Heaving Buoy

An axisymmetric buoy has a cylindrical part, of height $2l = 8$ m and of diameter $2a = 6$ m, above a hemispherical lower end. The buoy is arranged to have a draught of $l + a = 7$ m in its equilibrium position. The heave amplitude $|\hat{s}|$ should be limited to $l$. Assume that this power buoy is equipped with a machinery by which the oscillatory motion can be controlled. Firstly, the heave speed is controlled to be in phase with the heave excitation force. Secondly, the damping of the oscillation should be adjusted for (optimum amplitude corresponding to) maximum absorbed power with small waves and for necessary limitation of the heave amplitude in larger waves. Apart from the amplitude limitation, we shall assume that linear theory is applicable. Moreover, we shall assume that the wave, as well as the heave oscillation, is sinusoidal. Deep water is assumed. The density of sea water is $\rho = 1020$ kg/m$^3$, and the acceleration of gravity $g = 9.81$ m/s$^2$.

(a) Using numerical results from Problem 5.12, calculate and draw graphs for the absorbed power $P$ as function of the amplitude $|A|$ of the incident wave when the wave period $T$ is 6.3 s, 9.0 s, 11.0 s and 15.5 s. For each period, it is necessary to determine the wave amplitude $|A_c|$ above which the design limit of the heave amplitude comes into play.

(b) Let us next assume that the average absorbed power has to be limited to $P_{max} = 300$ kW due to the design capacity of the installed machinery. For this reason, it is desirable (as an alternative to completely stopping the heave oscillation) to adjust the phase angle $\gamma$ between the velocity and the excitation force to a certain value ($\gamma \neq 0$) such that the absorbed power becomes $P_{max}$. For this situation of very large incident wave, calculate and draw a graph for $\gamma$ as a function of $|A|$ for each of the four mentioned wave periods.

(c) Discuss (verbally) how this will be modified if we, instead of specifying $P_{\mathrm{max}} = 300$ kW for the machinery, specify a maximum external damping force amplitude $|F_u|_{\mathrm{max}} = 2.4 \times 10^5$ N or

(d) a maximum load resistance $R_u = 1.0 \times 10^5$ N s m$^{-1}$.

## Problem 6.2: Ratio of Absorbed Power to Volume

Assume that an incident regular (sinusoidal) wave of amplitude $|A| = 1$ m is given (on deep water). Consider wave periods in the interval 5 s $< T <$ 16 s. A heaving buoy is optimally controlled for maximum absorption of wave energy (cf. Problem 6.1). The buoy is shaped as a cylinder with a hemispherical bottom, and it has a diameter of $2a$ and a total height of $a + 2l$, where $2l$ is the height of the cylindrical part. The equilibrium draught is $a + l$. Apart from the design limit $l$ for the heave amplitude, we shall assume that linear theory is valid.

(a) Derive an expression for the ratio $P/V$ between the absorbed power $P$ and the volume $V$ of the buoy, in terms of $\rho$, $g$, $a$, $l$, $|A|$, $\omega = 2\pi/T$, $|f_3|/S$ and $\omega R_{33}/S$. (See Problem 5.12 for a definition of some of these symbols.) With a fixed value for $l$, show that $P/V$ has its largest value when $a \to 0$. Draw a curve for $(P/V)_{\mathrm{max}}$ versus $T$.

(b) Using results from Problem 6.1, draw a curve for $P/V$ versus $T$, when $a = 3$ m and $l = 4$ m. In addition to the scale for $P/V$ (in kW/m$^3$), include a scale for $P$ (in kW). Also draw a curve for how $P$ would have been if there had been no limitation of the heave amplitude.

## Problem 6.3: Maximum Absorbed Power by an Axisymmetric Body

The maximum power absorbed by an oscillating body in a plane incident wave is

$$P_{\mathrm{MAX}} = \frac{1}{2}\mathbf{U}^{\mathrm{T}}\mathbf{R}\mathbf{U}^*,$$

where $\mathbf{R}$ is the radiation resistance matrix and where the optimum velocity vector $\mathbf{U}$ has to satisfy the algebraic equation

$$\mathbf{R}\mathbf{U} = \hat{\mathbf{F}}_e/2,$$

where $\hat{\mathbf{F}}_e$ is a vector composed of all (six) excitation force components of the body. Let $\lambda$ denote the wavelength and $J$ the wave power transport of the incident plane harmonic wave.

Use the reciprocity relation between the radiation resistance matrix and the excitation force vector to show that an axisymmetric body oscillating with optimum complex amplitudes absorbs a power

(a) $P_{\text{MAX}} = (3\lambda/2\pi)J$ if it oscillates in the surge and heave modes only,
(b) $P_{\text{MAX}} = (\lambda/\pi)J$ if it oscillates in the surge and pitch modes only and
(c) $P_{\text{MAX}} = (3\lambda/2\pi)J$ if it oscillates in the surge, heave and pitch modes only.

State the conditions which the optimum complex velocity amplitudes have to satisfy in each of the three cases, (a)–(c). It is assumed that linear theory is applicable; that is, the wave amplitude is sufficiently small to ensure that the oscillation amplitudes are not restricted.

## Problem 6.4: Alternative Derivation of the Wave-Power 'Island'

Let us define $U_0 = \hat{F}_e/\sqrt{8R_{jj}}$ and $U = \hat{u}_j\sqrt{R_{jj}/2}$. Then, on the basis of Eqs. (6.4)–(6.6) and (6.12), show that

$$P_{\text{MAX}} - P = |U_0 - U|^2.$$

## Problem 6.5: Maximum Absorbed Power with Optimised Amplitude

The maximum absorbed power when the amplitude but not the phase of the oscillation is optimised is, for two different situations, given by Eqs. (3.41) and (6.10). In the former case, the oscillation is restricted by the dynamic equation (3.30), whereas in the latter case, the oscillation amplitude is chosen at will (which may necessitate that a control device is part of the system).

(a) For the situation to which Eq. (3.41) pertains, determine the phase angle $\gamma$ between the velocity and the excitation force, in terms of $x \equiv (\omega m_m + \omega m_r - S/\omega)/R_r$, and express $P_{a,\text{max}}$ in terms of $\hat{F}_e$, $R_r$ and $x$.
(b) Setting $\gamma_j = \gamma$, $R_{jj} = R_r$ and $\hat{F}_{e,j} = \hat{F}_e$, apply Eq. (6.10) to express $P_{\text{max}}$ in terms of $\hat{F}_e$, $R_r$ and $x$.
(c) Then express $p \equiv P_{\text{max}}/P_{a,\text{max}}$ in terms of $x$, and show that $p \geq 1$. Finally, show that $p = 1$ for resonance ($x = 0$).

# CHAPTER SEVEN

# Oscillating Water Columns and Other Types of Wave-Energy Converters

## 7.1 Oscillating Water Column WECs

Many of the wave-energy converters which have been investigated in several countries are of the oscillating water column (OWC) type. By 'oscillating water column', we understand the water contained inside a hollow structure with a submerged opening, where the water inside the structure is communicating with the water of the open sea (see Figure 4.1). The power takeoff may be hydraulic, but pneumatic power takeoff using air turbines is more common. In the latter case, there is a dynamic air pressure above the water surface inside the OWC chamber. In such a case, the OWC may be referred to as a 'periodic surface pressure' [97] or an 'oscillating surface-pressure distribution' [98]. It is this kind of OWC which is the subject of study in the present section. As with an oscillating body (cf. Chapter 5), two kinds of interaction are considered: the radiation problem and the excitation problem. The radiation problem concerns the radiation of waves due to an oscillating dynamic air pressure above the interface. The excitation problem concerns the oscillation due to an incident wave when the dynamic air pressure is zero. Comparisons are made with wave–body interactions. Wave-energy extraction by OWCs is also discussed.

## 7.1.1 The Applied-Pressure Description for a Single OWC

In approximate theoretical studies of an OWC, one may think of the internal water surface $S_k$ as an imaginary, weightless, rigid piston which is considered as an oscillating body. Such a theory does not correctly model the hydrodynamics because the boundary condition (4.38) is not exactly satisfied. It may, however, give good approximate results for low frequencies, when the wavelength is very long compared to the (characteristic) horizontal length of the internal water surface $S_k$. Then, far from resonances of this surface, it moves approximately as a horizontal plane piston.

As opposed to this rigid-piston approximation, general and more correct theoretical results are presented here, based on the linearised hydrodynamical theory for an ideal irrotational fluid [97, 98].

The basic equations for the complex amplitude of the velocity potential are given by Eqs. (4.33)–(4.35) as

$$\nabla^2 \hat{\phi} = 0 \quad \text{in the fluid region,} \tag{7.1}$$

$$\left[\frac{\partial \hat{\phi}}{\partial n}\right]_{S_b} = 0 \quad \text{on fixed solid surfaces } S_b, \tag{7.2}$$

$$\left[-\omega^2 \hat{\phi} + g\frac{\partial \hat{\phi}}{\partial z}\right]_{S_0} = 0 \quad \text{on the free water surface } S_0, \tag{7.3}$$

$$\left[-\omega^2 \hat{\phi} + g\frac{\partial \hat{\phi}}{\partial z}\right]_{S_k} = -\frac{i\omega}{\rho}\hat{p}_k \quad \text{on the internal water surface } S_k. \tag{7.4}$$

Of these four linear equations in $\hat{\phi}$, the first three are homogeneous, whereas Eq. (7.4) is inhomogeneous. The right-hand side of Eq. (7.4) is a driving term due to the dynamic air pressure or 'air-pressure fluctuation' $p_k$. This system of linear equations may be supplemented with a radiation condition at infinite distance, as discussed in Sections 4.1, 4.3 and 4.6. In the case of an incident wave, this may serve as the driving function.

Let there be an incident wave [cf. Eqs. (4.88) and (4.98)]

$$\hat{\phi}_0 = \frac{-g}{i\omega}e(kz)\hat{\eta}_0, \tag{7.5}$$

where

$$\hat{\eta}_0 = A\exp[-ik(x\cos\beta + y\sin\beta)]. \tag{7.6}$$

It is assumed that the sea bed is horizontal at a depth $z = -h$. The angle of incidence is $\beta$ with respect to the $x$-axis. We decompose the resulting velocity potential as

$$\phi = \phi_0 + \phi_d + \phi_r, \tag{7.7}$$

where the two last terms represent a diffracted wave and a radiated wave, respectively. We require

$$\nabla^2 \begin{bmatrix} \phi_0 \\ \phi_d \\ \phi_r \end{bmatrix} = 0 \quad \text{in the fluid region,} \tag{7.8}$$

$$\left(-\omega^2 + g\frac{\partial}{\partial z}\right)\begin{bmatrix} \hat{\phi}_0 \\ \hat{\phi}_d \\ \hat{\phi}_r \end{bmatrix} = 0 \quad \text{on } S_0, \tag{7.9}$$

and

$$\frac{\partial}{\partial n}\begin{bmatrix} \phi_0 + \phi_d \\ \phi_r \end{bmatrix} = 0 \quad \text{on } S_b. \tag{7.10}$$

(Note that $\partial\phi_0/\partial n = 0$ and, hence, also $\partial\phi_d/\partial n = 0$ on the plane sea bed $z = -h$, which is a part of $S_b$.) Furthermore,

$$\left(-\omega^2 + g\frac{\partial}{\partial z}\right)(\hat{\phi}_0 + \hat{\phi}_d) = 0 \qquad \text{on } S_k \tag{7.11}$$

$$\left(-\omega^2 + g\frac{\partial}{\partial z}\right)\hat{\phi}_r = -\frac{i\omega}{\rho}\hat{p}_k \qquad \text{on } S_k. \tag{7.12}$$

With the use of Eq. (7.7), it is easy to verify that if the homogeneous Eqs. (7.8)–(7.11) and the inhomogeneous boundary condition (7.12) are all satisfied, then the required Eqs. (7.1)–(7.4) are necessarily satisfied.

Writing

$$\hat{\phi}_r = \varphi_k \hat{p}_k, \tag{7.13}$$

we find that the proportionality coefficient $\varphi_k$ must satisfy the Laplace equation, the preceding homogeneous boundary conditions for $\hat{\phi}_r$ on $S_0$ and $S_b$, and the inhomogeneous boundary condition

$$\left(-\omega^2 + g\frac{\partial}{\partial z}\right)\varphi_k = -\frac{i\omega}{\rho} \qquad \text{on } S_k. \tag{7.14}$$

The coefficient $\varphi_k$ introduced in Eq. (7.13) is analogous to the coefficient $\varphi_j$ in Eq. (5.9) or (5.35).

The volume flow produced by the oscillating internal water surface is given by

$$Q_{t,k} = \iint_{S_k} v_z \, dS = \iint_{S_k} \frac{\partial\phi}{\partial z} \, dS. \tag{7.15}$$

The SI unit for volume flow is m³/s. It is convenient to decompose the total volume flow into two terms:

$$Q_{t,k} = Q_{e,k} + Q_{r,k}, \tag{7.16}$$

where we have introduced an *excitation volume flow*

$$Q_{e,k} = \iint_{S_k} \frac{\partial}{\partial z}(\phi_0 + \phi_d) \, dS \tag{7.17}$$

and a *radiation volume flow*

$$Q_{r,k} = \iint_{S_k} \frac{\partial\phi_r}{\partial z} \, dS. \tag{7.18}$$

Since $\phi_0$ and $\phi_d$ are linear in $A$ and $\phi_r$ is linear in $p_k$, we have in terms of complex amplitudes that

$$\hat{Q}_{t,k} = \hat{Q}_{e,k} + \hat{Q}_{r,k} = q_{e,k}A - Y_{kk}\hat{p}_k, \tag{7.19}$$

where we have introduced the *excitation volume-flow coefficient*

$$q_{e,k} = \hat{Q}_{e,k}/A = \iint_{S_k} \frac{\partial}{\partial z}(\hat{\phi}_0 + \hat{\phi}_d)\frac{1}{A}\, dS \tag{7.20}$$

and the *radiation admittance*

$$Y_{kk} = -\iint_{S_k} \frac{\partial \varphi_k}{\partial z}\, dS. \tag{7.21}$$

Note that the excitation volume flow $Q_{e,k}$ is the volume flow when the air-pressure fluctuation is zero ($p_k = 0$). This is a consequence of the decompositions (7.7) and (7.16) which we have chosen (following Evans [98]). Also observe that the larger $|Y_{kk}|$ is, the larger $|Q_{r,k}|$ is admitted for a given $|\hat{p}_k|$. The term 'admittance' for the ratio between the complex amplitudes of volume flow and air-pressure fluctuation is adopted from the theory of electric circuits, where electric admittance is the inverse of electric impedance. Their real/imaginary parts are called conductance/susceptance and resistance/reactance, respectively. Thus, we may decompose the radiation admittance into real and imaginary parts, $Y_{kk} = G_{kk} + iB_{kk}$, where

$$G_{kk} = \mathrm{Re}\{Y_{kk}\} = \tfrac{1}{2}(Y_{kk} + Y_{kk}^*) \tag{7.22}$$

is the *radiation conductance* and

$$B_{kk} = \mathrm{Im}\{Y_{kk}\} \tag{7.23}$$

is the *radiation susceptance*. The SI unit for $Y_{kk}, G_{kk}$ and $B_{kk}$ is $\mathrm{m^3\,s^{-1}/Pa} = \mathrm{m^5\,s^{-1}\,N^{-1}}$. Thus, in the present context, the product of mechanical impedance (cf. Chapters 2 and 5) and (pneumatic/hydraulic) admittance is not a dimensionless quantity. Its SI unit would be $(\mathrm{N\,s\,m^{-1}})(\mathrm{m^5\,s^{-1}\,N^{-1}}) = \mathrm{m^4}$.

### 7.1.2 Absorbed Power and Radiation Conductance

Whereas the (pneumatic/hydraulic) admittance represents the ratio between volume flow and pressure fluctuation, the product of these two quantities has the dimension of power, in SI units, $(\mathrm{m^3\,s^{-1}})(\mathrm{Nm^{-2}}) = \mathrm{N\,m\,s^{-1}} = \mathrm{W}$. Proceeding in analogy with Eqs. (6.3)–(6.8), we have for the (time-average) power $P$ absorbed from the wave,

$$P = \overline{p_k(t)Q_{t,k}(t)} = \tfrac{1}{2}\mathrm{Re}\{\hat{p}_k\hat{Q}_{t,k}^*\} = \tfrac{1}{2}\mathrm{Re}\{\hat{p}_k(\hat{Q}_{e,k} - Y_{kk}\hat{p}_k)^*\}, \tag{7.24}$$

where we have used Eq. (7.19) and an analogue of Eqs. (2.75)–(2.76). As in Eq. (6.4), we write the absorbed power as

$$P = P_e - P_r. \tag{7.25}$$

The excitation power is

$$P_e = \tfrac{1}{4}\hat{p}_k\hat{Q}_{e,k}^* + \tfrac{1}{4}\hat{p}_k^*\hat{Q}_{e,k} = \tfrac{1}{2}\mathrm{Re}\{\hat{p}_k\hat{Q}_{e,k}^*\} = \tfrac{1}{2}\mathrm{Re}\{\hat{p}_k q_{e,k}^* A^*\}, \tag{7.26}$$

and the radiated power is

$$P_r = \tfrac{1}{4}\hat{p}_k \hat{Y}_{kk}^* \hat{p}_k^* + \tfrac{1}{4}\hat{p}_k^* \hat{Y}_{kk}\hat{p}_k = \tfrac{1}{2}G_{kk}|\hat{p}_k|^2, \tag{7.27}$$

where $G_{kk}$ is given by Eq. (7.22). Since $P_r \geq 0$, the radiation conductance $G_{kk}$ cannot be negative.

### 7.1.3 Reactive Power and Radiation Susceptance

Whereas the radiation conductance $G_{kk}$ is related to radiated power, the radiation susceptance $B_{kk}$ represents reactive power, and it may be related to the difference $W_k - W_p$ between kinetic energy and potential energy in the near-field region of the wave radiated as a result of the air-pressure fluctuation $p_k$. To be more explicit, it can be shown [63] that

$$\frac{1}{4\omega}B_{kk}|\hat{p}_k|^2 = W_k - W_p, \tag{7.28}$$

which is analogous to Eq. (5.190). Also compare Eqs. (2.85)–(2.88).

As the frequency goes to zero, the kinetic energy $W_k$ tends to zero, whereas the potential energy $W_p$, in accordance with Eqs. (7.69)–(7.72), reduces to

$$W_{p0} = \rho g S_k |\hat{\eta}_{k0}|^2/4 - \rho g S_k |\eta_{k0}|^2/2 = -\rho g S_k |\hat{\eta}_{k0}|^2/4. \tag{7.29}$$

Note that $\hat{p}_k \to -\rho g \hat{\eta}_k$ as $\omega \to 0$. Thus, as $\omega \to 0$,

$$B_{kk} \to -\frac{4\omega W_{p0}}{|\hat{p}_k|^2} = \frac{\omega S_k}{\rho g}. \tag{7.30}$$

Note that the effect of hydrostatic stiffness is included in the radiation susceptance $B_{kk}$. This is in contrast to the case of a heaving body, where the effect is represented by a separate term—such as the term $S_b$ in Eq. (5.346)—and not in the radiation reactance (or added mass). From the discussion in Section 5.9.1 on resonances for heaving bodies, it appears plausible that OWCs also have resonances. From physical arguments, we expect that Eq. (5.356) represents approximately the resonance frequency for an OWC contained in a surface-piercing vertical tube which is submerged to a depth $l$ and which has a diameter that is small in comparison with $l$. Since the effect of hydrostatic stiffness is included in the radiation susceptance $B_{kk}$, resonance occurs for frequencies where $B_{kk} = 0$—that is, for frequencies where the radiation admittance $Y_{kk}$ is real.

### 7.1.4 Example: Axisymmetric OWC

As an example, consider a circular vertical tube, the lower end of which just touches the water surface (Figure 7.1). This is an OWC chamber with a particularly simple geometry.

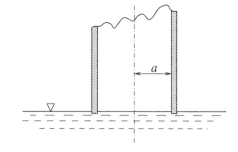

Figure 7.1: A vertical circular tube just penetrating the free water surface represents a simple form of an OWC chamber.

Since diffraction is negligible in this case, it is easy [cf. Problem 7.1 and Eq. (7.45)] to derive expressions for the excitation volume flow $Q_{e,k}$ and the radiation conductance $G_{kk}$, namely

$$\hat{Q}_{e,k} = q_{e,k}A \quad \text{with } q_{e,k} = \frac{i\omega}{k}2\pi a J_1(ka), \tag{7.31}$$

$$G_{kk} = \frac{k}{8J}|\hat{Q}_{e,k}|^2 = \frac{\omega k}{2\rho g^2 D}|q_{e,k}|^2 = \frac{2\omega}{\rho g}[\pi a J_1(ka)]^2. \tag{7.32}$$

Here $J_1$ is the first-order Bessel function of the first kind. Since

$$(2/ka)J_1(ka) = 1 + \mathcal{O}\{k^2a^2\} \quad \text{as } ka \to 0, \tag{7.33}$$

we have

$$\hat{Q}_{e,k} \approx i\omega A\pi a^2 \tag{7.34}$$

in the long-wavelength limit—that is, for $ka \ll 1$. Since $i\omega A$ is the complex amplitude of the vertical component of the fluid velocity corresponding to the incident wave at the origin, this is a result to be expected in the long-wavelength case.

We know that $G_{kk}$ can never be negative. However, $G_{kk}$ (as well as $\hat{Q}_{e,k}$) has the same zeros as $J_1(ka)$, namely $ka = 3.832, 7.016, 10.173, \ldots$.

The derivation of the radiation susceptance

$$B_{kk} = \text{Im}\{Y_{kk}\} \tag{7.35}$$

is more complicated. Results for deep water are given by Evans [98]. The graph in Figure 7.2 demonstrates that $B_{kk}$ is positive for low frequencies, whereas negative values are most typically found for high frequencies. The zero crossings in the graph show the first seven resonances. The first of them is at $ka = 1.96$ corresponding to a disc of radius of approximately three-tenths of a wavelength ($a/\lambda = ka/2\pi = 1.96/2\pi = 0.31$). The corresponding angular eigenfrequency is

$$\omega_0 = \sqrt{k_0 g} = \sqrt{1.96\, g/a} = 1.4\sqrt{g/a}. \tag{7.36}$$

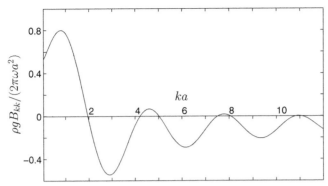

Figure 7.2: Radiation susceptance $B_{kk} = \text{Im}\{Y_{kk}\}$ for a circular OWC with a negligible submergence [98]. Here $(\rho g/2\omega\pi a^2)B_{kk} \to 1/2$ as $ka \to 0$, in agreement with Eq. (7.30).

If we associate this lowest eigenfrequency with that of a heaving, rigid, weightless, circular disc of radius $a$ and, hence, with hydrostatic stiffness $S_b = \rho g \pi a^2$, its added mass at $\omega_0$ would be

$$m_{33} = \frac{S_b}{\omega_0^2} = \frac{\rho g \pi a^2}{1.96 \, g/a} = 0.51 \, \pi a^3 \rho = 0.76 \frac{2\pi}{3} a^3 \rho. \tag{7.37}$$

This equals approximately three-quarters of the mass of water displaced by a hemisphere of radius $a$. Note that the effect of hydrostatic stiffness is included in the radiation susceptance $B_{kk}$ (as opposed to the radiation reactance of a floating body such as, for instance, the envisaged rigid disk). In the long-wavelength limit ($ka \to 0$), this hydrostatic effect is the sole contribution to $B_{kk}$. Thus, in accordance with Eq. (7.30),

$$B_{kk} \to \omega\pi a^2/(\rho g) \quad \text{as } ka \to 0. \tag{7.38}$$

Note that the considered circular tube with negligible submergence can sustain only infinitesimal amplitudes. A more practical case would require a finite submergence of the tube.

### 7.1.5 Maximum Absorbed Power

In the present section, let us make a discussion analogous to that of Section 6.2.1 for the case of an oscillating body. The excitation power, as given by Eq. (7.26), may be written as

$$P_e = \frac{1}{2}|\hat{p}_k \hat{Q}_{e,k}| \cos(\gamma_k) = \frac{1}{2}|\hat{p}_k q_{e,k} A| \cos(\gamma_k), \tag{7.39}$$

where $\gamma_k$ is the phase angle between the air-pressure fluctuation $\hat{p}_k$ and the excitation volume flow $\hat{Q}_{e,k}$. Note that, for a given incident wave and a given phase angle, $P_e$ is linear in the air-pressure amplitude $|\hat{p}_k|$, whereas the radiated power $P_r$, as given by Eq. (7.27), is quadratic in $|\hat{p}_k|$. Hence, the absorbed power

$P = P_e - P_r$ may be represented by a parabola, just as in the case of an oscillating body (cf. Figure 6.3). It is easy to show that $P$ is maximum when $\hat{p}_k$ has the optimum value

$$|\hat{p}_k|_{\text{opt}} = \frac{|\hat{Q}_{e,k}|}{2G_{kk}} \cos(\gamma_k), \tag{7.40}$$

corresponding to the maximum absorbed power

$$P = P_{\max} \equiv \frac{|\hat{Q}_{e,k}|^2}{8G_{kk}} \cos^2(\gamma_k). \tag{7.41}$$

Also note that

$$P_{r,\text{opt}} = \tfrac{1}{2} P_{e,\text{opt}} = P_{\max} \tag{7.42}$$

and that $P = 0$ for $|\hat{p}_k| = 2|\hat{p}_k|_{\text{opt}}$.

Further, if it may be accomplished (by resonance or by phase control) that the air-chamber pressure fluctuation $p_k$ is in phase with the excitation volume flow $Q_{e,k}$, then $\gamma_k = 0$, and the absorbed power takes the maximum value

$$P_{\text{MAX}} = \frac{|\hat{Q}_{e,k}|^2}{8G_{kk}} \tag{7.43}$$

when

$$\hat{p}_k = \hat{p}_{k,\text{OPT}} = \frac{\hat{Q}_{e,k}}{2G_{kk}}. \tag{7.44}$$

This maximum absorbed power must correspond to maximum destructive interference, as discussed in Section 6.1. Thus, we expect that Eqs. (6.103)–(6.104) also hold if the optimally absorbing system is an axisymmetric OWC. With this in mind, it follows from Eqs. (7.20), (7.43) and (6.103) that

$$G_{kk} = \frac{k}{8J} |\hat{Q}_{e,k}|^2 = \frac{\omega k}{2\rho g^2 D(kh)} |q_{e,k}|^2. \tag{7.45}$$

This reciprocity relation for the axisymmetric case, where $\hat{Q}_{e,k} = q_{e,k} A$ is independent of the angle $\beta$ of wave incidence, is analogous to the reciprocity relation (5.146) for a heaving axisymmetric body.

If we define complex quantities $U$ and $E(\beta)$ as

$$|U|^2 = P_r = \tfrac{1}{2} G_{kk} |p_k|^2, \tag{7.46}$$
$$E(\beta) = p_k^* q_e(\beta)/4, \tag{7.47}$$

we may see that the absorbed power $P$ of an OWC as given by Eqs. (7.25)–(7.27) may be expressed in terms of $U$ and $E(\beta)$ in the same way as Eq. (6.32). This means that Eqs. (6.34)–(6.37)—especially the simple formula (6.37), which represents the wave-power island in Figure 6.5—are also applicable to an OWC. We may see, for instance, that Eqs. (6.34) and (6.35) lead to the same expressions for $P_{\text{MAX}}$ and $\hat{p}_{k,\text{OPT}}$ as given by Eqs. (7.43) and (7.44).

### 7.1.6 Reciprocity Relations for an OWC

For the OWC, there are reciprocity relations similar to those of the oscillating body. At present, let us just state some of the relations and defer the proofs to Section 8.2.6. Relations for the three-dimensional case as well as for the two-dimensional case are given next.

In presenting the reciprocity relations, we need the far-field coefficients for radiated waves. In the three-dimensional case, this coefficient $a_k(\theta)$ is, in analogy with Eq. (5.123), given by the asymptotic expression for the radiated wave:

$$\hat{\phi}_r \sim \hat{p}_k a_k(\theta) e(kz)(kr)^{-\frac{1}{2}} e^{-ikr} \tag{7.48}$$

as $kr \to \infty$. The two-dimensional far-field coefficient $a_k^{\pm}$ is, in analogy with Eqs. (4.88) and (5.321), defined by the asymptotic expression

$$\hat{\eta}_r \sim \hat{p}_k a_k^{\pm} e^{-ik|x|}, \tag{7.49}$$

where the plus and minus signs refer to radiation in the positive and negative $x$-directions, respectively.

For the three-dimensional case, the relations analogous to Eqs. (5.135) and (5.145) are as follows. The radiation conductance $G_{kk} = \mathrm{Re}\{Y_{kk}\}$ is related to the far-field coefficient $a_k(\theta)$ by

$$G_{kk} = \frac{\omega\rho D(kh)}{2k} \int_0^{2\pi} |a_k(\theta)|^2 \, d\theta \tag{7.50}$$

and to the excitation volume flow $\hat{Q}_{e,k}(\beta)$ by

$$G_{kk} = \frac{k}{16\pi J} \int_{-\pi}^{\pi} |\hat{Q}_{e,k}(\beta)|^2 \, d\beta. \tag{7.51}$$

We may observe that Eq. (7.45) follows directly from Eq. (7.51) when $\hat{Q}_{e,k}$ is independent of $\beta$. The analogue of the Haskind relation is

$$q_{e,k}(\beta) = \hat{Q}_{e,k}(\beta)/A = -\rho g[D(kh)/k]\sqrt{2\pi} a_k(\beta \pm \pi) e^{i\pi/4}. \tag{7.52}$$

The minus sign is not a misprint, as the reader might suspect when comparing with Eq. (5.142). The minus sign is there because a positive air pressure (that is, an increase in air pressure) pushes the OWC downward instead of upward, resulting in a negative volume flow. On the other hand, a positive vertical force (that is, an upward force) on a floating body pushes it also upward, resulting in a positive body velocity.

For the case of propagation in a wave channel of width $d$ under conditions of no cross waves—that is, for $kd < \pi$ (see Problem 4.6)—the reciprocity relations for the radiation conductance per unit width are

$$G'_{kk} = \frac{\rho g^2 D(kh)}{2\omega} (|a_k^+|^2 + |a_k^-|^2) \tag{7.53}$$

$$G'_{kk} = \frac{1}{8J} \left( \left| \frac{\hat{Q}_{e,k}(0)}{d} \right|^2 + \left| \frac{\hat{Q}_{e,k}(\pi)}{d} \right|^2 \right) = \frac{1}{8J} \left( |\hat{Q}'_{e,k}(0)|^2 + |\hat{Q}'_{e,k}(\pi)|^2 \right), \qquad (7.54)$$

whereas the excitation volume flow may be expressed as

$$\hat{Q}_{e,k}(0) = A \frac{\rho g^2 D(kh) d}{\omega} a_k^- \qquad (7.55)$$

if the incident wave originates from $x = -\infty$ (that is, if $\beta = 0$) and

$$\hat{Q}_{e,k}(\pi) = A \frac{\rho g^2 D(kh) d}{\omega} a_k^+ \qquad (7.56)$$

if it originates from $x = +\infty$ (i.e., $\beta = \pi$). These relations apply, of course, to the two-dimensional case, since it was assumed that no cross wave exists in the wave channel. For the two-dimensional case, we have introduced quantities per unit width (in the $y$-direction) denoted by $\hat{Q}'_{e,k}$ for the excitation volume flow and $G'_{kk}$ for the radiation conductance. These reciprocity relations for a two-dimensional OWC are analogous to similar reciprocity relations for two-dimensional bodies. See Eqs. (5.329)–(5.333).

Let us, as an application, consider the two-dimensional situation with an incident wave (in the positive $x$-direction) in a wave channel of width $d < \pi/k$. A combination of Eqs. (7.54) and (7.43) gives, for this case,

$$P_{\text{MAX}} = \frac{Jd |\hat{Q}_{e,k}(0)|^2}{|\hat{Q}_{e,k}(0)|^2 + |\hat{Q}_{e,k}(\pi)|^2} \qquad (7.57)$$

or, alternatively,

$$P_{\text{MAX}} = \frac{Jd}{1 + |a_k^+/a_k^-|^2}, \qquad (7.58)$$

where Eqs. (7.55) and (7.56) have been used. Note that the OWC result, Eq. (7.58), is analogous to the oscillating-body result, Eq. (8.147). [To see this, one may find it helpful to use Eqs. (5.324) and (5.325) to express the Kochin functions in terms of far-field coefficients.] For an OWC which radiates symmetrically (equal waves in both directions, i.e., $a_k^+ = a_k^-$), we have $P_{\text{MAX}} = Jd/2$, which corresponds to 50% absorption of the incident wave power. On the other hand, if the OWC is unable to radiate in the positive direction ($a_k^+ = 0$), then $P_{\text{MAX}} = Jd$, which means 100% wave absorption. The case of $a_k^+ = 0$ may be realised, for instance, by an OWC spanning a wave channel at the downstream end where the absorbing beach has been removed. With an open air chamber ($\hat{p}_k = 0$), no power is absorbed ($P = 0$) and the incident wave will be totally reflected. With optimum air-chamber pressure, according to Eq. (7.44), we have 100% absorption, which means that the optimum radiated wave in the negative direction just cancels out the reflected one.

### 7.1.7 OWC with Pneumatic Power Takeoff

Next, let us discuss the practical possibility of achieving an optimum phase and optimum amplitude of the air-chamber pressure. We assume that the air chamber has a volume $V_a$ and that an air turbine is placed in a duct between the chamber and the outer atmosphere. For simplicity, we represent the turbine by a pneumatic admittance $\Lambda_t$, which we assume to be a constant at our disposal. (To a reasonable approximation, a Wells turbine [99] is linear, and its $\Lambda_t$ is real and, within certain limits, inversely proportional to the speed of rotation.) We assume that $\Lambda_t$ is independent of the air-chamber pressure $\hat{p}_k$, which requires that

$$|\hat{p}_k| \ll p_a, \tag{7.59}$$

where $p_a$ is the ambient absolute air pressure.

If we neglect air compressibility, the volume flow at the turbine equals that at the internal water surface, which means that the air pressure is given by

$$\hat{p}_k = \hat{Q}_{t,k}/\Lambda_t. \tag{7.60}$$

If air compressibility is taken into consideration, it can be shown (see Problem 7.2) that

$$\hat{p}_k = \hat{Q}_{t,k}/\Lambda, \tag{7.61}$$

where

$$\Lambda = \Lambda_t + i\omega(V_a/\kappa p_a) \tag{7.62}$$

and $\kappa$ is the exponent in the gas law of adiabatic compression ($\kappa = 1.4$ for air). The imaginary part of $\Lambda$ may be of some importance in a full-scale OWC, but it is usually negligible in downscaled laboratory model experiments.

Thus, the dynamics of the OWC determines the air-chamber pressure by the relation

$$\hat{p}_k = q_{e,k}A/(Y_{kk} + \Lambda), \tag{7.63}$$

which is obtained by combining Eqs. (7.19) and (7.61). Hence, the optimum phase condition (that $\hat{p}_k$ is in phase with the excitation volume flow $\hat{Q}_{e,k} = q_{e,k}A$ or that $\gamma_k = 0$) is fulfilled if

$$\text{Im}\{Y_{kk} + \Lambda\} = B_{kk} + \frac{\omega V_a}{\kappa p_a} = 0. \tag{7.64}$$

Compare Eqs. (7.23) and (7.62). If air compressibility is negligible, this corresponds to the hydrodynamic resonance condition $B_{kk} = 0$. Moreover, comparison of Eqs. (7.63) and (7.44) shows that the amplitude $\hat{p}_k$ is optimum if the additional condition

$$\text{Re}\{\Lambda\} = G_{kk} \equiv \text{Re}\{Y_{kk}\} \tag{7.65}$$

is satisfied. That is, we have to choose a (real) turbine admittance $\Lambda_t$ which is equal to the radiation conductance $G_{kk}$.

As a numerical example, let us consider an axisymmetric OWC, as shown in Figure 7.1, and let the diameter be $2a = 8$ m. Thus, the internal water surface is $S_k = \pi a^2 = 50\,\text{m}^2$. Further, assume that the average air-chamber volume is $V_a \approx 300$ m$^3$ and the ambient air pressure is $p_a = 10^5$ Pa. For a typical wave period $T = 2\pi/\omega = 9$ s, the (deep-water) wavelength is $\lambda = 2\pi/k = 126$ m—that is, $ka = 0.20$. Thus $J_1(ka) = 0.099$, and Eq. (7.32) gives $G_{kk} = 2.2 \times 10^{-4}$ m$^5$/(s N). From the curve in Figure 7.2, we see that $(\rho g/2\pi\omega a^2)B_{kk} \approx 0.5$—that is, $B_{kk} \approx 0.0018$ m$^5$/(s N), which is larger than $G_{kk}$ by one order of magnitude. Taking into consideration the effect of air compressibility, we have Im$\{\Lambda\} = \omega V_a/(\kappa p_a) = 0.0015$ m$^5$/(s N), which is also positive and of the same order of magnitude as $B_{kk}$. Hence, the optimum phase condition is far from being satisfied in this example. If, however, the phase could be optimised by some artificial means—for instance, by control of air valves in the system—then the optimum turbine admittance would be $\Lambda_t = G_{kk} \approx 0.0002$ m$^5$/(s N).

Finally, let us add some remarks of practical relevance. We have used linear analysis and neglected viscous effects. This is a reasonable approximation to reality only if the oscillation amplitude $\hat{s}_b$ of the water at the barrier (the inlet mouth) of the OWC does not exceed the radius of curvature $\rho_b$ at the barrier. Then the so-called Keulegan–Carpenter number $N_{KC} = \pi|\hat{v}_b|/(\omega\rho_b)$, where $v_b$ is the water velocity at the barrier, is small and is less than $\pi$. See Eq. (6.80). In practice, the circular tube of the OWC has to have a finite wall thickness and a finite submergence, in contrast to the case discussed in connection with Figure 7.1, where both these dimensions were assumed to approach zero. If the wall thickness is, say, 1 m, the radius of curvature at the lower end could be 0.5 m. Then we may expect that the present linear theory is applicable if the volume flow does not exceed $0.5\,\text{m} \times 50\,\text{m}^2 \times 2\pi/(9\,\text{s}) \approx 17\,\text{m}^3/\text{s} \sim 20\,\text{m}^3/\text{s}$, which, according to Eq. (7.43), corresponds to $P_{MAX}$ not exceeding $(20\ \text{m}^3/\text{s})^2/(8 \times 0.0002\ \text{m}^5\,\text{s}^{-1}\,\text{N}^{-1}) = 0.25 \times 10^6$ W $= 0.25$ MW. This limit for linear behaviour may be increased by increasing the radius of curvature at the lower barrier end. This may be achieved, for instance, by making the lower end of the vertical tube somewhat horn shaped.

### 7.1.8 Potential Energy of an Oscillating Water Column

The potential energy inside an OWC is associated with water being lifted against gravity and also against the air pressure. Per unit horizontal area of the mean air–water interface $S_k$ of the OWC, the potential energy relative to the sea bed (cf. Figure 4.1 and Section 4.4.1) is equal to

$$(\rho g/2)(h + z_k + \eta_k)^2 + (p_{k0} + p_k)\eta_k. \tag{7.66}$$

The increase in relation to calm water is

$$(\rho g/2)\eta_k^2 + \rho g h \eta_k + p_{\text{atm}}\eta_k + p_k\eta_k, \tag{7.67}$$

where we have used Eq. (4.13). The second and third terms have vanishing average values. Hence, the time-average potential energy per unit horizontal area is

$$E_p|_{S_k} = (\rho g/2)\overline{\eta_k^2(t)} + \overline{p_k(t)\eta_k(t)}, \tag{7.68}$$

which agrees with Eq. (4.114), applicable for the case in which the dynamic air pressure $p_k$ is zero. Note that although $p_k(t)$ does not vary with the horizontal coordinates, $\eta_k(t)$ may depend on the horizontal coordinates $(x, y)$. Thus, $E_p|_{S_k}$, which by definition is independent of time, may also depend on the horizontal coordinates $(x, y)$.

For cases in which $\eta_k$ and $p_k$ vary sinusoidally with angular frequency $\omega$, we may write $E_p|_{S_k}$ in terms of complex amplitudes:

$$E_p|_{S_k} = (\rho g/4)|\hat{\eta}_k|^2 + (1/2)\text{Re}\{\hat{p}_k\hat{\eta}_k^*\} = (\rho g\hat{\eta}_k\hat{\eta}_k^* + \hat{p}_k\hat{\eta}_k^* + \hat{p}_k^*\hat{\eta}_k)/4. \tag{7.69}$$

In the case of zero dynamic air pressure, only the first term remains, in agreement with Eq. (4.115).

Note that the first term in the preceding expressions for $E_p|_{S_k}$, if multiplied by infinitesimal horizontal area $\Delta S$, may be interpreted as the potential energy of a vertical water column of cross section $\Delta S$ and stiffness $\rho g \Delta S$. This is easily seen by comparing with Eqs. (2.84) and (2.85). For a semisubmerged column-shaped rigid body, the corresponding stiffness is called buoyancy stiffness or hydrostatic stiffness; see Section 5.9.1.

Next, we wish to express $E_p|_{S_k}$ in terms of the velocity potential. Using Eqs. (4.38) and (7.4), we find

$$\hat{\eta}_k = \frac{1}{i\omega}\left[\frac{\partial\hat{\phi}}{\partial z}\right]_{S_k}, \tag{7.70}$$

$$\hat{p}_k = -i\omega\rho\left[\hat{\phi} - \frac{g}{\omega^2}\frac{\partial\hat{\phi}}{\partial z}\right]_{S_k}. \tag{7.71}$$

Thus, we may express the time-average potential-energy surface density (which is a real quantity) in terms of the (complex) velocity potential as

$$E_p|_{S_k} = -\frac{\rho g}{4\omega^2}\left[\frac{\partial\hat{\phi}}{\partial z}\frac{\partial\hat{\phi}^*}{\partial z}\right]_{S_k} + \frac{\rho}{4}\left[\hat{\phi}\frac{\partial\hat{\phi}^*}{\partial z} + \hat{\phi}^*\frac{\partial\hat{\phi}}{\partial z}\right]_{S_k}. \tag{7.72}$$

On the interface $S_0$ between the water and open air, where zero dynamic air pressure is assumed, the simpler boundary condition (7.3) replaces (7.4). There the surface density of potential energy may be expressed in various ways as

$$E_p|_{S_0} = \frac{\rho g}{4}|\hat{\eta}|^2 = \frac{\rho g}{4\omega^2}\left[\frac{\partial\hat{\phi}}{\partial z}\frac{\partial\hat{\phi}^*}{\partial z}\right]_{S_0} = \frac{\rho\omega^2}{4g}\left[\hat{\phi}^*\hat{\phi}\right]_{S_0}$$

$$= \frac{\rho}{4}\left[\hat{\phi}\frac{\partial\hat{\phi}^*}{\partial z}\right]_{S_0} = \frac{\rho}{4}\left[\hat{\phi}^*\frac{\partial\hat{\phi}}{\partial z}\right]_{S_0}. \tag{7.73}$$

This expression was applied in Eq. (5.194). We may note that expression (7.72) for $E_p|_{S_k}$ is valid for $E_p|_{S_0}$ if $S_k$ is replaced by $S_0$. (However, the opposite procedure—to replace $S_0$ in Eq. (7.73) by $S_k$—will yield $E_p|_{S_k}$ only if $\hat{p}_k = 0$.)

## 7.2 WEC Bodies Oscillating in Unconventional Modes of Motion

There exists WEC bodies which oscillate in modes other than the six rigid-body modes we have described in Section 5.1. Examples include WEC bodies which oscillate in an incline such as the sloped IPS buoy [100], articulated WEC bodies such as the articulated raft [101, 102], WEC bodies that oscillate by expansion and contraction of their physical volumes such as the Archimedes Wave Swing (AWS) device [103] and WECs utilising flexible bodies [104–106]. The analysis of such WECs is facilitated by the use of generalised modes [107], which we shall describe next. As we shall see in a moment, the theory also encompasses the conventional rigid-body modes described earlier.

### 7.2.1 Generalised Modes of Motion

Following Newman [107], we can define a general mode of motion in terms of a 'shape function' $\vec{S}_j(\mathbf{x})$, which is a vector with Cartesian components $S_{jx}(\mathbf{x}), S_{jy}(\mathbf{x})$ and $S_{jz}(\mathbf{x})$. The displacement of an arbitrary point $\mathbf{x} = (x, y, z)$ within the body is given by

$$\vec{s}(\mathbf{x}) = \sum_j \hat{s}_j \vec{S}_j(\mathbf{x}), \tag{7.74}$$

where $\hat{s}_j$ is the complex displacement amplitude of the body in mode $j$. With this definition, the shape function $\vec{S}_j$ for conventional rigid-body translation ($j = 1, 2, 3$) is a unit vector in the corresponding direction, whereas for rigid-body rotation ($j = 4, 5, 6$) about a reference point $\mathbf{x}_0$, it is $\vec{S}_j = \vec{S}_{j-3} \times \vec{s}$, where $\vec{s}$ is the position of point $\mathbf{x}$ relative to the reference point. For example, the shape function for the surge mode ($j = 1$) is simply a unit vector in the $x$-direction— that is, $\vec{S}_1 = (1, 0, 0)^{\mathrm{T}}$—whereas the shape function for the roll mode ($j = 4$) is $\vec{S}_4 = \vec{S}_1 \times \vec{s} = (0, -s_z, s_y)^{\mathrm{T}}$. In general, however, the shape function $\vec{S}_j(\mathbf{x})$ can take any form corresponding to the general mode of motion.

The normal component of $\vec{S}_j(\mathbf{x})$ on the wet body surface $S$ is expressed as

$$n_j = \vec{S}_j \cdot \vec{n} = S_{jx}n_x + S_{jy}n_y + S_{jz}n_z, \tag{7.75}$$

where the unit normal vector $\vec{n}$ points out of the body and into the fluid domain. It is easy to show that Eqs. (5.5)–(5.6), which are applicable to conventional rigid-body modes, agree with the preceding general expression, since they can be written as

$$n_j = \begin{cases} \vec{n} \cdot \vec{S}_j & \text{for } j = 1, 2, 3 \\ (\vec{s} \times \vec{n}) \cdot \vec{S}_{j-3} = (\vec{S}_{j-3} \times \vec{s}) \cdot \vec{n} & \text{for } j = 4, 5, 6, \end{cases} \tag{7.76}$$

with $\vec{S}_j$ for $j = 1,2,3$ defined as a unit vector in the corresponding direction.

In accordance with linear theory, the generalised hydrodynamic force corresponding to mode $j$ is defined as

$$\hat{F}_j = i\omega\rho \iint_S \hat{\phi} n_j \, dS;$$ (7.77)

cf. Eqs. (5.21)–(5.22). Here, $\hat{\phi}$ is the complex amplitude of the total velocity potential, which may be expressed as the sum of the incident wave potential $\hat{\phi}_0$, diffracted wave potential $\hat{\phi}_d$ and radiated wave potential $\hat{\phi}_r$.

The contribution from the incident and diffracted wave potentials is defined as the wave excitation force $\hat{F}_{e,j}$ [cf. Eq. (5.26)]:

$$\hat{F}_{e,j} = i\omega\rho \iint_S (\hat{\phi}_0 + \hat{\phi}_d) n_j \, dS = i\omega\rho \iint_S (\hat{\phi}_0 + \hat{\phi}_d)\frac{\partial \varphi_j}{\partial n} \, dS,$$ (7.78)

where $\varphi_j$ is the unit-amplitude radiation potential, according to the definition

$$\hat{\phi}_r = \sum_j \hat{u}_j \varphi_j,$$ (7.79)

where $\hat{u}_j$ is the velocity amplitude corresponding to mode $j$. The second equality in Eq. (7.78) is the consequence of the body boundary condition $\partial\hat{\phi}_r/\partial n = \hat{u}_n = \sum_j \hat{u}_j n_j$; cf. Eq. (5.8).

The contribution from the radiation potential, as in Section 5.2.1, can be expressed in terms of the radiation impedance $Z_{jj'}$:

$$\hat{F}_{r,j} = -\sum_{j'} Z_{jj'} \hat{u}_{j'},$$ (7.80)

where

$$Z_{jj'} = -i\omega\rho \iint_S \varphi_{j'} n_j \, dS = -i\omega\rho \iint_S \varphi_{j'}\frac{\partial \varphi_j}{\partial n} \, dS,$$ (7.81)

which in turn can be expressed in terms of the added inertia $m_{jj'}$ and radiation resistance $R_{jj'}$:

$$Z_{jj'} = R_{jj'} + i\omega m_{jj'}.$$ (7.82)

The hydrostatic restoring force coefficients—that is, the generalised hydrostatic force in mode $j$ due to a unit displacement in mode $j'$—are given as [107]

$$K_{jj'} = -\rho g \iint_S n_{j'}(S_{jz} + zD_j)dS,$$ (7.83)

where $D_j$ is the divergence of the shape function—that is, $D_j = \nabla \cdot \vec{S}_j(\mathbf{x})$. For certain modes, such as the conventional rigid-body modes, $D_j = 0$, and, thus, $K_{jj'}$ reduces to

$$K_{jj'} = -\rho g \iint_S n_{j'} S_{jz} dS.$$ (7.84)

For example, the hydrostatic restoring force coefficient for rigid-body heave is $K_{33} = -\rho g \iint_S n_z dS = \rho g S_w$, where $S_w$ is the mean water-plane area of the body, in agreement with Eq. (5.343).

It should be noted that Eq. (7.83) gives the restoring force coefficients due to the hydrostatic pressure alone. The gravitational force due to the body mass also contributes to the total restoring force, and this must be considered in the dynamics of the body. For example, from physical arguments, we know that a displacement in sway of a free floating body symmetric about the $y = 0$ plane should not result in any restoring moment in roll. However, Eq. (7.83) gives $K_{42} = -\rho g \iint_S n_y s_y dS = -\rho g V$, where $V$ is the mean displaced volume of the body. This hydrostatic contribution is exactly balanced by the gravitational contribution, which is given by $m_m g$, where $m_m$ is the mass of the body. Since $m_m = \rho V$ for a free floating body, the gravitational contribution exactly balances the hydrostatic contribution, resulting in zero total roll restoring moment due to displacement in sway.

The coefficients of the generalised mass matrix are given as

$$M_{jj'} = \iiint_V \rho_m(\mathbf{x}) \mathbf{S}_j^T \mathbf{S}_{j'} dS, \tag{7.85}$$

where $\rho_m(\mathbf{x})$ is the mass density of the body, and the integral is taken over the total volume of the body (not to be confused with the displaced volume in the preceding paragraph). This expression follows from the definition of the kinetic energy of the body and from writing the velocity of any point $\mathbf{x}$ within the body as $\sum_j \hat{u}_j \vec{S}_j(\mathbf{x})$.

### 7.2.2 Absorbed Power

Notice that expressions (7.77)–(7.82) have the same form as the corresponding expressions for a rigid body oscillating in conventional modes of motion, as treated in Chapter 5. It follows that the power absorbed by WECs oscillating in general modes of motion is given by the same equations as found in Chapter 6, such as Eqs. (6.84)–(6.86) and (6.95). The wave-power island (cf. Figure 6.5) is also applicable for these WECs.

### 7.2.3 Example: Two-Body Axisymmetric System

To illustrate the theory just described, let us consider a simple example of a two-body axisymmetric system, constrained to oscillate in heave. In Sections 5.7.3 and 8.1.4, this two-body system is analysed by considering the heave motion of each body as two independent modes. Alternatively, we may treat the two-body system as one combined body and consider the heave motion of the combined body as the first mode and the heave motion of the second body relative to the combined body as the second mode. In this case, the first mode is a conventional rigid-body mode, whereas the second mode is a generalised mode.

Using the first approach—that is, considering the heave motion of each body as two independent modes—we may write the total hydrodynamic force acting on the bodies as

$$
\begin{bmatrix} \hat{F}_3 \\ \hat{F}_9 \end{bmatrix} = \begin{bmatrix} \hat{F}_{e,3} \\ \hat{F}_{e,9} \end{bmatrix} - \begin{bmatrix} Z_{33} & Z_{39} \\ Z_{93} & Z_{99} \end{bmatrix} \begin{bmatrix} \hat{u}_3 \\ \hat{u}_9 \end{bmatrix}, \tag{7.86}
$$

where the modes are numbered according to Eq. (5.154). Alternatively, with the second approach—that is, with a conventional heave mode and a generalised relative heave mode—we write

$$
\begin{bmatrix} \hat{F}_0 \\ \hat{F}_7 \end{bmatrix} = \begin{bmatrix} \hat{F}_{e,0} \\ \hat{F}_{e,7} \end{bmatrix} - \begin{bmatrix} Z_{00} & Z_{07} \\ Z_{70} & Z_{77} \end{bmatrix} \begin{bmatrix} \hat{u}_0 \\ \hat{u}_7 \end{bmatrix}, \tag{7.87}
$$

where $j = 7$ denotes the generalised mode and $j = 0$ denotes the heave mode of the combined body, to distinguish it from $j = 3$, which has been used to denote the heave mode of the first body. By definition,

$$
\hat{u}_7 = \hat{u}_9 - \hat{u}_3, \tag{7.88}
$$

$$
\hat{u}_0 = \hat{u}_3, \tag{7.89}
$$

which, according to Eq. (7.79), lead to

$$
\varphi_0 = \varphi_3 + \varphi_9, \tag{7.90}
$$

$$
\varphi_7 = \varphi_9. \tag{7.91}
$$

The shape function of the generalised mode can be written as

$$
\vec{S}_7 = \begin{cases} [0,0,1]^{\mathrm{T}} & \text{for } \mathbf{x} \text{ within body 2} \\ 0 & \text{elsewhere.} \end{cases} \tag{7.92}
$$

Thus,

$$
n_7 = \vec{S}_7 \cdot \vec{n} = \begin{cases} n_z & \text{for } \mathbf{x} \text{ on body 2} \\ 0 & \text{elsewhere.} \end{cases} \tag{7.93}
$$

It follows from definitions (7.77) and (7.78) that

$$
\hat{F}_7 = \hat{F}_9, \tag{7.94}
$$

$$
\hat{F}_0 = \hat{F}_3 + \hat{F}_9, \tag{7.95}
$$

$$
\hat{F}_{e,7} = \hat{F}_{e,9}, \tag{7.96}
$$

$$
\hat{F}_{e,0} = \hat{F}_{e,3} + \hat{F}_{e,9}. \tag{7.97}
$$

Substituting Eqs. (7.88)–(7.89) and (7.94)–(7.97) into Eqs. (7.86)–(7.87), we arrive at these identities:

$$
Z_{00} = Z_{33} + Z_{99} + 2Z_{39}, \tag{7.98}
$$

$$
Z_{07} = Z_{39} + Z_{99}, \tag{7.99}
$$

$$
Z_{77} = Z_{99}, \tag{7.100}
$$

which are in agreement with definition (7.81) for the radiation impedance if we make use of Eqs. (7.90)–(7.91). Hence, the two approaches, Eqs. (7.86) and (7.87), are equivalent.

If the two bodies are both floating and piercing the water surface as in the system shown in Figure 5.18 but without the submerged body, then the mass matrix and the restoring matrix of the system read as

$$\mathbf{M} = \begin{bmatrix} m_{m1} & 0 \\ 0 & m_{m2} \end{bmatrix}, \quad \mathbf{S}_b = \rho g \begin{bmatrix} S_{w1} & 0 \\ 0 & S_{w2} \end{bmatrix} \tag{7.101}$$

if the first approach is used, whereas if the second approach is used, they read as

$$\mathbf{M} = \begin{bmatrix} m_{m1} + m_{m2} & m_{m2} \\ m_{m2} & m_{m2} \end{bmatrix}, \quad \mathbf{S}_b = \rho g \begin{bmatrix} S_{w1} + S_{w2} & S_{w2} \\ S_{w2} & S_{w2} \end{bmatrix}. \tag{7.102}$$

Here, $m_{m1}$ and $m_{m2}$ are the masses, whereas $S_{w1}$ and $S_{w2}$ are the mean water-plane areas, of body 1 and body 2, respectively. These expressions can be obtained from Eqs. (7.85) and (7.84).

The mean power that is absorbed by the two-body axisymmetric system can be obtained from Eqs. (6.84)–(6.86). With the first approach,

$$\hat{\mathbf{F}}_e^T \hat{\mathbf{u}}^* = \hat{F}_{e,3}\hat{u}_3^* + \hat{F}_{e,9}\hat{u}_9^*, \tag{7.103}$$

whereas with the second approach,

$$\hat{\mathbf{F}}_e^T \hat{\mathbf{u}}^* = \hat{F}_{e,0}\hat{u}_0^* + \hat{F}_{e,7}\hat{u}_7^*. \tag{7.104}$$

Substituting Eqs. (7.96)–(7.97) and (7.88)–(7.89) into Eq. (7.104), we obtain the same expression as in Eq. (7.103). Thus, the excitation power $P_e$ is the same for both approaches. By making similar substitutions, we can also show that the product $\hat{\mathbf{u}}^T \mathbf{R} \hat{\mathbf{u}}^*$, and hence, the radiated power $P_r$, is the same for both approaches.

## 7.2.4 Example: Submerged Body with Movable Top

As a second example, let us consider a completely submerged body in the form of a vertical cylinder with a movable top but otherwise fixed. The AWS device [103] is a WEC that belongs to this category. The body then has only one mode, which we shall denote with the index $j = 7$. The shape function $\vec{S}_7$ is

$$\vec{S}_7(\mathbf{x}) = \begin{cases} [0,0,1]^T & \text{for } \mathbf{x} \text{ within the movable top} \\ 0 & \text{elsewhere,} \end{cases} \tag{7.105}$$

and the normal component of $\vec{S}_7$ on the wet body surface is, therefore,

$$n_7 = \vec{S}_7 \cdot \vec{n} = \begin{cases} n_z & \text{for } \mathbf{x} \text{ on the movable top} \\ 0 & \text{elsewhere.} \end{cases} \tag{7.106}$$

On the wet surface of the movable top, the unit normal vector is pointing upwards; thus, $n_z = 1$ on this surface. According to Eq. (7.84), the hydrostatic restoring coefficient is therefore negative and is given as

$$K_{77} = -\rho g S_{\text{top}},$$ (7.107)

where $S_{\text{top}}$ is the projected area of the top on the horizontal plane.

## Problems

### Problem 7.1: Circular Cylindrical OWC

The lower end of a vertical thin-walled tube of radius $a$ just penetrates the free water surface. The tube constitutes the air chamber above an OWC. Assume the fluid to be ideal, and neglect diffraction. Derive an expression for the excitation volume flux $\hat{Q}_e$ when there is an incident plane wave

$$\hat{\phi}_0 = \frac{-g}{i\omega} e(kz)\hat{\eta}_0,$$

$$\hat{\eta}_0 = Ae^{-ikx}.$$

Let the $z$-axis coincide with the axis of the tube.

Furthermore, find an expression for the radiation conductance $G_{kk}$ of the OWC. Simplify the expression for $G_{kk}$ for the case of deep water ($kh \ll 1$). [Hint: use the reciprocity relation (7.51). We may also use the integrals

$$\int_0^x t J_0(t)\, dt = x J_1(x), \qquad \int_0^{2\pi} e^{ix\cos t}\, dt = 2\pi J_0(x),$$

where $J_0(x)$ and $J_1(x)$ are Bessel functions of the first kind and of order zero and one, respectively.]

### Problem 7.2: Air Compressibility

Use the expression

$$p_a V_a^\kappa = (p_a + \Delta p_a)(V_a + \Delta V_a)^\kappa$$

for adiabatic compression to derive the last term in Eq. (7.62). Assume that $\Delta p_a$ and $\Delta V_a$ are so small that small terms of higher order may be neglected.

### Problem 7.3: Two Identical Bodies a Distance Apart

Consider a system of two identical bodies, body A and body B, say, separated at a given horizontal distance from each other. Each of the bodies is constrained to oscillate in heave only but is otherwise free to oscillate independently of the other. The motion of the system is therefore described by two independent modes. We may choose the first mode to be the heave motion of body A and the second mode to be the heave motion of body B. Let us denote these modes by $j = 3$ and $j = 9$, respectively. Alternatively, we may choose the first mode to be both bodies moving the same distance upwards and the second mode to be the bodies moving in opposite directions. Let us denote these modes by $j = 1$ and $j = 2$, respectively.

(a) Write the shape functions $\vec{S}_j(\mathbf{x})$ corresponding to mode $j = 3, 9, 1, 2$, as well as their normal components.

(b) Based on the definitions of the total hydrodynamic force, Eq. (7.77), and of the wave excitation force, Eq. (7.78), express $\hat{F}_1$ and $\hat{F}_2$ in terms of $\hat{F}_3$ and $\hat{F}_9$, and also, $\hat{F}_{e,1}$ and $\hat{F}_{e,2}$ in terms of $\hat{F}_{e,3}$ and $\hat{F}_{e,9}$.

(c) From definition (7.74), express $\hat{u}_1$ and $\hat{u}_2$ in terms of $\hat{u}_3$ and $\hat{u}_9$.

(d) Then, using definition (7.79), express $\varphi_1$ and $\varphi_2$ in terms of $\varphi_3$ and $\varphi_9$.

(e) Making use of the relationship just derived, express $Z_{11}, Z_{12}$ and $Z_{22}$ in terms of $Z_{33}, Z_{39}$ and $Z_{99}$. Also show that $Z_{12} = 0$ [Hint: make use of the fact that because the two bodies are identical, $Z_{33} = Z_{99}$].

(f) Finally, show that the expressions for the mean absorbed power obtained using the two different approaches are exactly the same.

# CHAPTER EIGHT

# Wave-Energy Converter Arrays

It is, in particular, the wavelengths of persisting wind-generated ocean waves that limit the power capacity of individual WEC units to the sub-megawatt region. Thus, a sizeable wave power plant needs to be an array of many individual WEC units. An example of such an array is shown in Figure 8.1. When a WEC unit has been successfully developed and sufficiently tested in the sea, the next development stage is to develop a future industry for mass production of WEC units. This is still far ahead. Strangely enough, however, for quite a long time, there has been a significant research interest in WEC arrays. Theoretical studies of WEC-body arrays were first made by Budal [62] and afterwards, more systematically, by Falnes [25] and Evans [61]. Budal also took into consideration absorbed-power limitation due to the finite hull volume of the array's WEC bodies [45]. WEC arrays consisting of WEC bodies and OWCs were considered by Falnes and McIver [109] and, independently, by Fernandes [110]. A more recent review is given by Falnes and Kurniawan [63]. Before proceeding to a systematic discussion, we first present a phenomenological description of the matter.

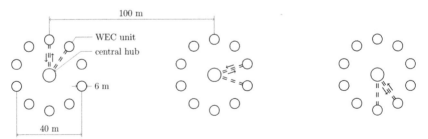

Figure 8.1: Groups of 10 WEC units, each of power capacity 200 kW, provide hydraulic power to a central hub containing gas accumulators for energy storage and a 2 MW electric generator (as proposed by Budal in 1979 [108]).

## 8.1 WEC Array Consisting of Several Bodies

The analysis of an array of WEC bodies, each oscillating in up to six degrees of freedom, is similar to the analysis of a single WEC body oscillating in several degrees of freedom, discussed in Section 6.5.

Let us consider a system of $N$ ($N \geq 1$) bodies, each of them oscillating in up to six degrees of freedom. As in Section 5.5, to each motion mode $j$ ($j = 1, 2, \ldots 6$) of body $p$ ($p = 1, 2, \ldots N$), we associate an oscillator number $i = 6(p - 1) + j$. In accordance with Eq. (5.155), the total wave force on oscillator $i$ has a complex amplitude

$$\hat{F}_{t,i} = \hat{F}_{e,i} + \hat{F}_{r,i} = \hat{F}_{e,i} - \sum_{i'=1}^{6N} Z_{ii'} \hat{u}_{i'}, \tag{8.1}$$

where $Z_{ii'}$ is an element of the radiation impedance matrix discussed in Sections 5.5.1 and 5.5.3. The only difference between this equation and Eq. (6.81), which pertains to a single WEC body oscillating in six degrees of freedom, is that the sum is here taken up to $6N$ instead of 6.

The (time-average) power absorbed by oscillator $i$ is as given by Eq. (6.82), with index $j$ replaced by $i$. Upon inserting the expression for $\hat{F}_{t,i}$ from Eq. (8.1) into Eq. (6.82), we find that Eqs. (6.82)–(6.86) are applicable to the array, provided we take the sums up to $6N$ instead of 6. Correspondingly, the velocity vector $\hat{\mathbf{u}}$ and excitation force vector $\hat{\mathbf{F}}_e$ in these equations are each a $6N$-dimensional column vector, whereas the radiation resistance matrix $\mathbf{R}$ is of dimension $6N \times 6N$. Thus, Eq. (6.95) for the maximum absorbed wave power is also applicable to the array of WEC bodies.

By extending Eqs. (6.82)–(6.86) to be applicable for an array of WEC bodies, we have also generalised the collective oscillation amplitude $U$ and the collective excitation-power coefficient $E(\beta)$, which were introduced for a single WEC body in Section 6.5 (see also Section 6.2), to be valid for an array of WEC bodies:

$$|U|^2 = \tfrac{1}{2} \hat{\mathbf{u}}^\dagger \mathbf{R} \hat{\mathbf{u}} \tag{8.2}$$

$$E(\beta) = \tfrac{1}{4} \mathbf{f}^{\mathrm{T}}(\beta) \hat{\mathbf{u}}^*, \tag{8.3}$$

according to Eqs. (6.85)–(6.86). With these more generalised definitions of $E(\beta)$ and $U$, the simple equations (6.32)–(6.37), including the wave-power-island equation

$$P_{\mathrm{MAX}}(\beta) - P = |U_0(\beta) - U|^2, \tag{8.4}$$

is applicable even for a WEC-body array. The optimum condition (6.34) leads to the condition

$$\mathbf{R} \hat{\mathbf{u}}_{\mathrm{OPT}} = \hat{\mathbf{F}}_e/2 \tag{8.5}$$

for the optimum velocity vector $\hat{\mathbf{u}}_{\mathrm{OPT}}$, as in the case of a single WEC body; cf. Eq. (6.90). Furthermore, Eq. (6.35) gives

$$P_{\mathrm{MAX}} = \tfrac{1}{4}\hat{\mathbf{F}}_e^{\mathrm{T}}\hat{\mathbf{u}}_{\mathrm{OPT}}^* = \tfrac{1}{2}\hat{\mathbf{u}}_{\mathrm{OPT}}^\dagger \mathbf{R}\hat{\mathbf{u}}_{\mathrm{OPT}}, \tag{8.6}$$

as in Eq. (6.95).

Note that $P_{\mathrm{MAX}} = |U_0(\beta)|^2$ is proportional to the wave amplitude squared and depends on the angle $\beta$ of wave incidence. Assuming that the WEC array is provided with sufficient control devices, we may consider $U$ to be an independent variable, on which the absorbed wave power $P$ depends.

We shall say a few more words about the collective oscillation amplitude $U$ and the collective excitation-power coefficient $E(\beta)$. When considering a single WEC body oscillating in one mode, we have $U$ and $E(\beta)$ expressed in terms of the radiation Kochin function; see Eqs. (6.30)–(6.31). For the array, generalised expressions can be obtained from Eqs. (8.2)–(8.3) by making use of relations (5.224) and (5.178), which yield

$$|U|^2 = \frac{\omega\rho v_p v_g}{4\pi g}\int_0^{2\pi}\hat{\mathbf{u}}^\dagger \mathbf{h}^*(\theta)\mathbf{h}^{\mathrm{T}}(\theta)\hat{\mathbf{u}}\,d\theta = \frac{\omega\rho v_p v_g}{4\pi g}\int_0^{2\pi}|H_r(\theta)|^2\,d\theta, \tag{8.7}$$

$$E(\beta) = \frac{\rho v_p v_g}{2}\mathbf{h}^{\mathrm{T}}(\beta \pm \pi)\hat{\mathbf{u}}^* = \frac{\rho v_p v_g}{2}\bar{H}_r(\beta \pm \pi). \tag{8.8}$$

Here we have also made use of Eq. (4.107) and, as in Eqs. (5.152)–(5.153), introduced the 'adjoint companion'

$$\bar{H}_r(\theta) = \mathbf{h}^{\mathrm{T}}(\theta)\hat{\mathbf{u}}^* = \hat{\mathbf{u}}^\dagger \mathbf{h}(\theta) \tag{8.9}$$

of the radiation Kochin function

$$H_r(\theta) = \mathbf{h}^{\mathrm{T}}(\theta)\hat{\mathbf{u}} = \hat{\mathbf{u}}^{\mathrm{T}}\mathbf{h}(\theta). \tag{8.10}$$

Alternative ways to arrive at the same expressions are discussed in [63]. Notice that Eqs. (8.7)–(8.8), when applied to the case of only one oscillator, lead to Eqs. (6.30)–(6.31).

Inserting expressions (8.7)–(8.8) into the optimum condition (6.34), which is applicable also for a WEC-body array, we have the condition

$$A\bar{H}_{r0}(\beta \pm \pi) = \frac{\omega}{2\pi g}\int_0^{2\pi}|H_{r0}(\theta)|^2 d\theta, \tag{8.11}$$

which the optimum radiated wave's Kochin function $H_{r0}$ and its adjoint companion $\bar{H}_{r0}$ need to satisfy. Using relation (5.152), we can alternatively express this condition as

$$AH_{r0}^*(\beta) = \frac{\omega}{2\pi g}\int_0^{2\pi}[H_d(\theta) + H_{r0}(\theta)]H_{r0}^*(\theta)\,d\theta, \tag{8.12}$$

which involves the diffracted wave's Kochin function $H_d(\theta)$.

### 8.1.1 Optimum Gain Function for WEC-Body Arrays

In Eq. (6.35), we presented several different expressions for the maximum absorbed power. We shall find it convenient to add the following expressions also:

$$P_{\text{MAX}} = \frac{P_{\text{MAX}}^2}{P_{\text{MAX}}} = \frac{(P_{e,\text{OPT}}/2)^2}{P_{r,\text{OPT}}} = \frac{|AE_0(\beta)|^2}{|U_0|^2}. \tag{8.13}$$

Inserting Eqs. (8.7)–(8.8) gives

$$P_{\text{MAX}} = \frac{|AE_0(\beta)|^2}{|U_0|^2} = \frac{\rho g v_g |A|^2}{2k} G(\beta) = \frac{J}{k} G(\beta) = J d_{a,\text{MAX}}, \tag{8.14}$$

where $J$ is the wave-energy transport or the wave-power level, as given by Eq. (4.130), and $d_a = P/J$ is the absorption width. Moreover, we have introduced the optimum gain function

$$G(\beta) = \frac{2\pi |\bar{H}_{r0}(\beta \pm \pi)|^2}{\int_0^{2\pi} |H_{r0}(\theta)|^2 d\theta}. \tag{8.15}$$

The optimum gain function for a single body oscillating in one mode is given earlier in Eq. (6.14). Observe that, with only one oscillator, we have $|\bar{H}_r(\theta)|^2 = |h_j(\theta)u_j^*|^2 = |h_j(\theta)u_j|^2 = |H_r(\theta)|^2$, and thus Eq. (8.15) reduces to Eq. (6.14). We must note, however, that in general $|\bar{H}_r(\theta)|^2 \neq |H_r(\theta)|^2$ for the multi-mode case.

Equation (8.15) tells us that in order to maximise power absorption, it is important for a WEC system to have the ability to radiate a wave in one predominant direction.

### 8.1.2 Similarity-Transformed WEC-Radiation Matrix

We have now generalised the optimum gain function for a single body oscillating in one mode, Eq. (6.14), to that for an array of bodies each oscillating in more than one mode, Eq. (8.15). When considering the one-body, one-mode case, we have also derived the directional average of the gain function, as given by Eq. (6.15), which follows immediately from Eq. (6.14). For an array of WEC bodies oscillating in multiple modes, it is less straightforward to obtain the directional average of the gain function from Eq. (8.15), and so another method needs to be employed, that is, by using similarity transformation; see, for example, [16, chapter V] and Section 2.4, where this method was briefly discussed. For this purpose, we need to diagonalise the radiation-resistance matrix **R**, which is a matrix of dimension $6N \times 6N$.

This diagonalisation procedure may preferably be performed on a slightly more general radiation-damping matrix **D**, which is complex and *Hermitian*—that is,

$$\mathbf{D}^\dagger \equiv (\mathbf{D}^{\text{T}})^* = \mathbf{D}. \tag{8.16}$$

(This radiation-damping matrix is applicable for a WEC array consisting of WEC bodies as well as OWCs). Thus, the symmetric real radiation resistance matrix $\mathbf{R}$ is a special case of such a Hermitian complex matrix $\mathbf{D}$, which we, for a general case, shall suppose to be an $M \times M$ matrix. For the particular case of an array consisting of $N$ WEC bodies and no OWCs, $M = 6N$. Then, $\mathbf{D} = \mathbf{R}$. It can be shown (see Section 8.2.3) that just as the real radiation resistance matrix $\mathbf{R}$, this more general radiation-damping matrix $\mathbf{D}$ is also positive semidefinite. Thus, $\hat{\mathbf{u}}^{\dagger}\mathbf{D}\hat{\mathbf{u}} \geq 0$ for any $M$-dimensional complex column vector $\hat{\mathbf{u}}$.

Our first step is to determine the eigenvalues $\Lambda_i$ and the corresponding normalised and mutually orthogonal eigenvectors $\mathbf{e}_i$ of the Hermitian matrix $\mathbf{D}$, for $i = 1, 2, \ldots, M - 1, M$. They have to satisfy the following system of linear homogeneous equations:

$$\mathbf{D}\mathbf{e}_i = \Lambda_i \mathbf{e}_i. \tag{8.17}$$

Because matrix $\mathbf{D}$ is Hermitian, all its eigenvalues $\Lambda_i$ are real [16, page 109], and since it is positive semidefinite, they are all nonnegative. They are the $M$ solutions of the $M$-degree equation $|\mathbf{D} - \Lambda\mathbf{I}| = 0$, where $\mathbf{I}$ is the $M \times M$ identity matrix, for which all its diagonal entries equal 1, whereas all the other entries equal 0. Further, $|\mathbf{D} - \Lambda\mathbf{I}|$ is the determinant of matrix $(\mathbf{D} - \Lambda\mathbf{I})$. We may wish to arrange the $M$ eigenvalues in a descending order—that is,

$$\Lambda_1 \geq \Lambda_2 \geq \cdots \geq \Lambda_M \geq 0. \tag{8.18}$$

Note that if $\mathbf{e}_i$ is a possible solution when $\Lambda = \Lambda_i$, then also $C_i\mathbf{e}_i$ is a solution, where $C_i$ is an arbitrary complex scalar. It is convenient to normalise the eigenvectors $\mathbf{e}_i$ in such a way that the condition

$$\mathbf{e}_i^{\dagger}\mathbf{e}_{i'} = \delta_{ii'} = \begin{cases} 1 & \text{for } i = i' \\ 0 & \text{for } i \neq i' \end{cases} \tag{8.19}$$

applies. With this choice, we may consider this particular set of eigenvectors to be a complete set of mutually orthogonal unit vectors in our $M$-dimensional complex space [16, chapter V].

Let us now define the $M \times M$ matrix

$$\mathbf{S} = \begin{bmatrix} \mathbf{e}_1 & \mathbf{e}_2 & \mathbf{e}_3 & \cdots & \mathbf{e}_M \end{bmatrix}. \tag{8.20}$$

It then follows from Eq. (8.19) that $\mathbf{S}^{\dagger}\mathbf{S} = \mathbf{I}$. Hence, $\mathbf{S}^{-1} = \mathbf{S}^{\dagger}$, or the inverse of matrix $\mathbf{S}$ is equal to its conjugate transpose.

We are now ready to carry out a similarity transformation of vectors $\hat{\mathbf{u}}$ and $\hat{\mathbf{F}}_e$. Let

$$\hat{\mathbf{u}}' = \mathbf{S}^{-1}\hat{\mathbf{u}} = \mathbf{S}^{\dagger}\hat{\mathbf{u}}, \qquad \hat{\mathbf{F}}_e' = \mathbf{S}^{-1}\hat{\mathbf{F}}_e = \mathbf{S}^{\dagger}\hat{\mathbf{F}}_e \tag{8.21}$$

and conversely, $\hat{\mathbf{u}} = \mathbf{S}\hat{\mathbf{u}}'$ and $\hat{\mathbf{F}}_e = \mathbf{S}\hat{\mathbf{F}}_e'$. Then, the collective oscillation amplitude $U$ and the collective excitation-power coefficient $E(\beta)$ satisfy the two equations

$$|U|^2 = \hat{\mathbf{u}}^\dagger \mathbf{D}\hat{\mathbf{u}}/2 = \hat{\mathbf{u}}'^\dagger \mathbf{D}'\hat{\mathbf{u}}'/2, \tag{8.22}$$

$$E(\beta) = \hat{\mathbf{u}}^\dagger \hat{\mathbf{F}}_e/(4A) = \hat{\mathbf{u}}'^\dagger \hat{\mathbf{F}}'_e/(4A), \tag{8.23}$$

where

$$\mathbf{D}' = \mathbf{S}^\dagger \mathbf{D}\mathbf{S}. \tag{8.24}$$

Using Eq. (8.17), we find that, except in its main diagonal, this similarity-transformed radiation matrix

$$\mathbf{D}' = \mathbf{S}^{-1}\mathbf{D}\mathbf{S} = \mathbf{S}^{-1}\mathbf{S}\,\mathrm{diag}\,(\Lambda_1, \Lambda_2, \cdots, \Lambda_M) = \mathrm{diag}\,(\Lambda_1, \Lambda_2, \cdots, \Lambda_M) \tag{8.25}$$

has only zero elements. The diagonal elements are given by the eigenvalues. Note that the Hermitian matrix $\mathbf{D}$, in general, may be complex. However, the similarity-transformed, diagonalised, matrix $\mathbf{D}'$ is necessarily real since all the eigenvalues are real.

For some particular systems, matrix $\mathbf{D}$ is singular. Then $|\mathbf{D}| = 0$, and thus, at least one of its eigenvalues equals zero. If

$$\Lambda_1 \geq \Lambda_2 \geq \cdots \geq \Lambda_{r_\mathbf{D}} > \Lambda_{r_\mathbf{D}+1} = \Lambda_{r_\mathbf{D}+2} = \cdots = \Lambda_M = 0, \tag{8.26}$$

where the integer $r_\mathbf{D}$ satisfies $1 \leq r_\mathbf{D} < M$, then we say that the radiation-damping matrix $\mathbf{D}$ is singular and of rank $r_\mathbf{D}$. (The matrix is, however, non-singular if $r_\mathbf{D} = M$.) An example of such systems, an axisymmetric system of three concentric bodies, is discussed in Section 5.7.2. For this system, $M = 18$ possible oscillating modes, but the rank $r_\mathbf{R}$ of the radiation-resistance matrix $\mathbf{R}$ is no more than 3.

### 8.1.3 Alternative Expression for Maximum Absorbed Power of a WEC-Body Array

By now, it is not yet obvious how the diagonalisation procedure described earlier can be useful. This will become clear in a moment.

For our WEC array consisting of $N$ immersed WEC bodies and $M$ oscillation modes, where $M \leq 6N$, the $M \times M$ radiation resistance matrix may, according to Eq. (5.145), be written as

$$\mathbf{R} = \frac{k}{16\pi J} \int_{-\pi}^{\pi} \hat{\mathbf{F}}_e(\beta)\hat{\mathbf{F}}_e^\dagger(\beta)\,d\beta. \tag{8.27}$$

In agreement with Eqs. (8.21) and (8.25), the similarity-transformed radiation-resistance matrix is

$$\mathbf{R}' = \mathbf{S}^\dagger \mathbf{R}\mathbf{S} = \frac{k}{16\pi J} \int_{-\pi}^{\pi} \mathbf{S}^\dagger \hat{\mathbf{F}}_e(\beta)\hat{\mathbf{F}}_e^\dagger(\beta)\mathbf{S}\,d\beta = \frac{k}{16\pi J} \int_{-\pi}^{\pi} \hat{\mathbf{F}}'_e(\beta)\hat{\mathbf{F}}_e'^\dagger(\beta)\,d\beta. \tag{8.28}$$

From Eqs. (8.25) and (8.28), we can conclude that the only nonzero elements of matrix $\mathbf{R}'$ are the following main diagonal elements:

$$R'_{ii} = \Lambda_i = \frac{k}{16\pi J} \int_{-\pi}^{\pi} |\hat{F}'_{e,i}(\beta)|^2 \, d\beta, \quad \text{for } i = 1, 2, \ldots, M-1, M. \tag{8.29}$$

By the use of Eqs. (8.21) and (8.24), the optimum condition (8.5) may be reformulated as

$$\mathbf{R}'\hat{\mathbf{u}}'_{\text{OPT}} = \hat{\mathbf{F}}'_e/2, \tag{8.30}$$

that is,

$$R'_{ii}\hat{u}'_{i,\text{OPT}}(\beta) = \Lambda_i\hat{u}'_{i,\text{OPT}}(\beta) = \tfrac{1}{2}\hat{F}'_{e,i}(\beta) \quad \text{for } i = 1, 2, \ldots, M, \tag{8.31}$$

and the maximum absorbed wave power, Eq. (8.6), may be rewritten as

$$P_{\text{MAX}}(\beta) = \frac{1}{2}\hat{\mathbf{u}}'^{\dagger}_{\text{OPT}}(\beta)\mathbf{R}'\hat{\mathbf{u}}'_{\text{OPT}}(\beta) = \frac{1}{2}\sum_{i=1}^{M}\Lambda_i|\hat{u}'_{i,\text{OPT}}(\beta)|^2. \tag{8.32}$$

Note that the use of Eqs. (8.22) and (8.23) in combination with Eqs. (6.34) and (6.35) would lead to the same results.

In certain cases, the radiation resistance matrix $\mathbf{R}$ may be singular and of rank $r_{\mathbf{R}}$, which is an integer that satisfies $1 \leq r_{\mathbf{R}} < M$. Then, because $\Lambda_i = 0$ for $r_{\mathbf{R}} < i \leq M$, we have [45, equation (135)]

$$P_{\text{MAX}}(\beta) = \frac{1}{2}\sum_{i=1}^{M}\Lambda_i|\hat{u}'_{i,\text{OPT}}(\beta)|^2 = \frac{1}{2}\sum_{i=1}^{r_{\mathbf{R}}}\Lambda_i|\hat{u}'_{i,\text{OPT}}(\beta)|^2 = \frac{1}{8}\sum_{i=1}^{r_{\mathbf{R}}}\frac{|\hat{F}'_{e,i}(\beta)|^2}{\Lambda_i}, \tag{8.33}$$

where we, in the last step, made use of Eq. (8.31). Observe that if the smallest positive eigenvalue $\Lambda_{r_{\mathbf{R}}}$ is very small, the required optimum velocity amplitude $|\hat{u}'_{i,\text{OPT}}(\beta)|$ may be impractically large! Note that for $r_{\mathbf{R}} < i \leq M$, we have no particular optimum requirement for $\hat{u}'_{i,\text{OPT}}(\beta)$. Then the optimum value of $\hat{u}'_{i,\text{OPT}}(\beta)$ is ambiguous, although the maximum absorbed power is unambiguous, as is made clear by Eq. (8.33).

Notice that because of similarity transformation, matrix $\mathbf{R}'$ is a diagonal matrix with the main diagonal elements being the eigenvalues of matrix $\mathbf{R}$; see Eq. (8.29). The consequence of this is that the maximum absorbed power can be expressed as a simple sum, as given by Eq. (8.33); cf. Eq. (8.6).

With this alternative expression for $P_{\text{MAX}}(\beta)$, the WEC-body array's maximum absorbed power averaged over all directions is

$$P_{\text{MAX,average}} = \frac{1}{2\pi}\int_{-\pi}^{\pi} P_{\text{MAX}}(\beta) \, d\beta = \frac{1}{16\pi}\sum_{i=1}^{r_{\mathbf{R}}}\frac{1}{\Lambda_i}\int_{-\pi}^{\pi}|\hat{F}'_{e,i}(\beta)|^2 \, d\beta. \tag{8.34}$$

If we here eliminate the integral by using Eq. (8.29), we obtain the simple relationship

$$P_{\text{MAX,average}} = \frac{1}{2\pi} \int\limits_{-\pi}^{\pi} P_{\text{MAX}}(\beta)\, d\beta = \sum_{i=1}^{r_{\mathbf{R}}} \frac{J}{k} = \frac{r_{\mathbf{R}} J}{k} = r_{\mathbf{R}} \frac{J\lambda}{2\pi}. \tag{8.35}$$

It is interesting to note that this is just a factor of $r_{\mathbf{R}}$ larger than the direction-averaged wave power which may be absorbed by one single, WEC body, oscillating in only one mode, as given by Eqs. (6.13) and (6.15). For the array, the optimum power-gain function $G(\beta)$ has a directional average that equals the rank $r_{\mathbf{R}}$ of the WEC array's radiation-resistance matrix $\mathbf{R}$.

### 8.1.4 Example: Two-Body Axisymmetric System

Let us consider two concentric axisymmetric buoys, $p = 1$ and $p = 2$, restricted to oscillate in the heave mode only. Then the relevant oscillating modes are $i = 6(p - 1) + 3$; that is, $i = 3$ and $i = 9$ [cf. Eq. (5.154)]. According to Eq. (5.302), the system's radiation resistance matrix is

$$\mathbf{R} = \begin{bmatrix} R_{33} & R_{39} \\ R_{93} & R_{99} \end{bmatrix} = \frac{\omega k}{2\rho g^2 D(kh)} \begin{bmatrix} |f_{30}|^2 & f_{30} f_{90}^* \\ f_{90} f_{30}^* & |f_{90}|^2 \end{bmatrix}. \tag{8.36}$$

This result is also in agreement with Eq. (5.316). It is easy to verify that the corresponding determinant vanishes, which means that $\mathbf{R}$ is a singular matrix. Choosing an arbitrary value $U_9$, we find that the indeterminate equation (6.90) for optimum gives

$$U_3 = \frac{f_{30} A/2 - R_{39} U_9}{R_{33}} = \frac{f_{90} A/2 - R_{99} U_9}{R_{93}}. \tag{8.37}$$

From Eq. (6.93), the maximum power is

$$P_{\text{MAX}} = \frac{(f_{30} U_3^* + f_{90} U_9^*) A}{4} = \frac{|f_{30} A|^2}{8 R_{33}} = \frac{f_{30} f_{90}^* |A|^2}{8 R_{93}} = \frac{|f_{90} A|^2}{8 R_{99}}, \tag{8.38}$$

which is independent of the choice of $U_9$. Inserting from Eqs. (8.36) and (4.130), we get

$$P_{\text{MAX}} = \frac{\rho g^2 D(kh)}{4\omega k} |A|^2 = \frac{J}{k}. \tag{8.39}$$

Thus, the maximum absorption width of this two-body axisymmetric system is the same as that of a single axisymmetric body oscillating in heave, as given by Eq. (6.104). This observation gives a clue to understanding why radiation resistance matrices may turn out to be singular.

The maximum absorbed power as given by Eqs. (6.103) and (8.39) corresponds to optimum destructive interference between the incident plane wave and the isotropically radiated circular wave generated by the heaving oscillation of one body or two bodies [in the case of Eq. (6.103) or Eq. (8.39), respectively]; see Figure 6.2. As for this optimum interference, it does not matter from which of the two bodies the circular wave originates. It is not possible to increase the maximum absorbed power by adding another isotropically radiating body to the system. This is the physical explanation for the singularity of the radiation resistance matrix given by Eq. (8.36). If only isotropically radiating heave modes are involved, the rank of the matrix cannot be more than one.

### 8.1.5 Example: Two Axisymmetric Bodies a Distance Apart

As a second example, let us consider a system of two equal axisymmetric bodies with their vertical symmetry axes located at $(x, y) = (\mp d/2, 0)$. We shall assume that they are oscillating in the heave mode only. Then the excitation-force vector is of the form $\hat{\mathbf{F}}_e = \left[ \hat{F}_{e,3} \hat{F}_{e,9} \right]^T$. Further, the radiation resistance matrix may be written as

$$\mathbf{R} = \begin{bmatrix} R_d & R_c \\ R_c & R_d \end{bmatrix}. \tag{8.40}$$

Note that the diagonal entry $R_d$ is positive, while the off-diagonal entry $R_c$ may be positive or negative, depending on the distance $d$ between the two bodies. Since waves radiated from the two distinct bodies cannot cancel each other in all directions in the far-field region—which may be proved by summing two asymptotic expressions (4.256) and (4.260)—the matrix $\mathbf{R}$ is non-singular, and hence, $R_c^2 < R_d^2$. (We assume that the body does not have such a peculiar shape that we may find $R_d$ to vanish for some particular frequency.)

The eigenvalues $\Lambda = \Lambda_{1,2}$ satisfy the algebraic equation

$$0 = |\mathbf{R} - \Lambda\mathbf{I}| = (R_d - \Lambda)^2 - R_c^2 = (\Lambda - R_d - R_c)(\Lambda - R_d + R_c)$$
$$\equiv (\Lambda - \Lambda_1)(\Lambda - \Lambda_2). \tag{8.41}$$

Thus,

$$\Lambda_1 = R_d + R_c > 0, \qquad \Lambda_2 = R_d - R_c > 0. \tag{8.42}$$

It is easy to show that, for this case, the corresponding eigenvectors $\mathbf{e}_1$ and $\mathbf{e}_2$, as defined by Eq. (8.17) and normalised according to Eq. (8.19), are given by

$$\begin{bmatrix} \mathbf{e}_1 & \mathbf{e}_2 \end{bmatrix} = \frac{1}{\sqrt{2}} \begin{bmatrix} 1 & 1 \\ 1 & -1 \end{bmatrix} \equiv \mathbf{S} = \mathbf{S}^T = \mathbf{S}^\dagger = \mathbf{S}^{-1}, \tag{8.43}$$

where we have also used definition (8.20). In accordance with Eq. (8.21), the excitation force and the optimum oscillation velocity vectors transform as

$$\hat{\mathbf{F}}_e' = \begin{bmatrix} \hat{F}_{e,3}' \\ \hat{F}_{e,9}' \end{bmatrix} = \mathbf{S}^{-1}\hat{\mathbf{F}}_e = \frac{1}{\sqrt{2}} \begin{bmatrix} \hat{F}_{e,3} + \hat{F}_{e,9} \\ \hat{F}_{e,3} - \hat{F}_{e,9} \end{bmatrix}, \tag{8.44}$$

$$\hat{\mathbf{u}}'_{\text{OPT}} = \begin{bmatrix} \hat{u}'_{3,\text{OPT}} \\ \hat{u}'_{9,\text{OPT}} \end{bmatrix} = \mathbf{S}^{-1}\hat{\mathbf{u}}_{\text{OPT}} = \frac{1}{\sqrt{2}} \begin{bmatrix} \hat{u}_{3,\text{OPT}} + \hat{u}_{9,\text{OPT}} \\ \hat{u}_{3,\text{OPT}} - \hat{u}_{9,\text{OPT}}, \end{bmatrix}. \tag{8.45}$$

For this example, where the $2 \times 2$ matrix $\mathbf{R}$ is non-singular, Eqs. (8.33)–(8.35) are applicable, with $r_{\mathbf{R}} = M = 2$. The maximum power absorbed by the two heaving bodies is

$$
\begin{aligned}
P_{\text{MAX}}(\beta) &= \frac{|\hat{F}_{e,3}(\beta) + \hat{F}_{e,9}(\beta)|^2}{16(R_d + R_c)} + \frac{|\hat{F}_{e,3}(\beta) - \hat{F}_{e,9}(\beta)|^2}{16(R_d - R_c)} \\
&= \frac{(R_d + R_c)|\hat{u}_{3,\text{OPT}} + \hat{u}_{9,\text{OPT}}|^2}{4} + \frac{(R_d - R_c)|\hat{u}_{3,\text{OPT}} - \hat{u}_{9,\text{OPT}}|^2}{4}.
\end{aligned}
\tag{8.46}
$$

The optimum complex velocity amplitudes of the two bodies satisfy

$$\hat{u}_{3,\text{OPT}} + \hat{u}_{9,\text{OPT}} = \frac{\hat{F}_{e,3}(\beta) + \hat{F}_{e,9}(\beta)}{2(R_d + R_c)}, \quad \hat{u}_{3,\text{OPT}} - \hat{u}_{9,\text{OPT}} = \frac{\hat{F}_{e,3}(\beta) - \hat{F}_{e,9}(\beta)}{2(R_d - R_c)}, \tag{8.47}$$

according to Eq. (8.31).

From a wave-body interaction point of view, it is interesting to note that the first rhs term in Eq. (8.46) and the first equation of (8.47) correspond to a suboptimum situation when the two equal heaving bodies cooperate as a source-mode (monopole) radiator—that is, when the constraint $\hat{u}_9 = \hat{u}_3$ is applied. Then the two bodies are constrained to heave with equal amplitudes and equal phases. In contrast, the last rhs term in Eq. (8.46) and the last equation of (8.47) correspond to a suboptimum situation when the two bodies are constrained to cooperate as a dipole-mode radiator—that is, when the constraint $\hat{u}_9 = -\hat{u}_3$ is applied. In general, Eqs. (8.46)–(8.47) may be considered to quantify the optimum situation for this combined monopole-dipole wave-absorbing system.

If the maximum radius of each body is sufficiently small—say, less than $1/40$ of a wavelength—it may be considered as a point absorber, for which the heave excitation force $\hat{F}_e$ is dominated by the Froude–Krylov force, and the diffraction force may be neglected. If, moreover, the centre-to-centre distance $d$ between the two bodies is large in comparison with the maximum body radius, then

$$\hat{\mathbf{F}}_e = \begin{bmatrix} \hat{F}_{e,3} \\ \hat{F}_{e,9} \end{bmatrix} \approx \hat{F}_{e0} \begin{bmatrix} \exp\{ik(d/2)\cos\beta\} \\ \exp\{-ik(d/2)\cos\beta\} \end{bmatrix}, \tag{8.48}$$

where $\hat{F}_{e0} = \rho g \pi a^2 A$. Here $a$ is the water plane radius of each body, and $A$ is the complex amplitude of the incident wave elevation at the chosen origin $(x, y) = (0, 0)$. For this point-absorber case, the entries in the radiation-resistance matrix $\mathbf{R}$ in Eq. (8.40) are approximately given by [25, equations 43–44] [see also Eq. (8.27)]

$$R_d \approx R_0 = \frac{k|\hat{F}_{e0}|^2}{8J} = \frac{k|\hat{F}_{e0}/A|^2}{4\rho g v_g}, \quad R_c \approx R_0 J_0(kd), \tag{8.49}$$

where $J_0$ denotes Bessel function of the first kind and order zero. We observe that matrix $\mathbf{R}$ is non-singular, and moreover,

$$R_d + R_c \approx R_0[1 + J_0(kd)], \qquad R_d - R_c \approx R_0[1 - J_0(kd)] \qquad (8.50)$$

are positive, since $-1 < J_0(kd) < 1$ for $kd > 0$.

Using Eqs. (8.46)–(8.50), we find

$$\begin{bmatrix} \hat{F}_{e,3} + \hat{F}_{e,9} \\ \hat{F}_{e,3} - \hat{F}_{e,9} \end{bmatrix} \approx 2\hat{F}_{e0} \begin{bmatrix} \cos\{(kd/2)\cos\beta\} \\ i\sin\{(kd/2)\cos\beta\} \end{bmatrix} \qquad (8.51)$$

and

$$\begin{aligned} P_{\text{MAX}}(\beta) &\approx \frac{|F_{e,0}|^2}{4R_0} \left\{ \frac{\cos^2\left[k(d/2)\cos\beta\right]}{1 + J_0(kd)} + \frac{\sin^2\left[k(d/2)\cos\beta\right]}{1 - J_0(kd)} \right\} \\ &= \frac{|F_{e,0}|^2}{4R_0} \frac{1 - J_0(kd)\cos(kd\cos\beta)}{1 - J_0^2(kd)}. \end{aligned} \qquad (8.52)$$

Note that, in general, this maximum absorbed power is not equally divided between the two bodies.

We may note from the point-absorber approximation (8.48) that, since $\hat{F}_{e0}/A$ is real, $f_i(\beta) = \hat{F}_{e,i}(\beta)/A = \hat{F}_{e,i}^*(\beta + \pi)/A^* = f_i^*(\beta + \pi)$ for $i = 3, 9$. Correspondingly, we then find from Eqs. (8.3), (8.8) and (8.9)–(8.10) that $\bar{H}_r(\beta) = h_3(\beta)\hat{u}_3^* + h_9(\beta)\hat{u}_9^* = h_3^*(\beta + \pi)\hat{u}_3^* + h_9^*(\beta + \pi)\hat{u}_9^* = \bar{H}_r^*(\beta + \pi)$ and, similarly, $\bar{H}_r(\beta + \pi) = \bar{H}_r^*(\beta)$. Thus, for the two heaving bodies, we have here explicitly demonstrated that in the point-absorber limit the term containing the integral on the rhs of Eq. (5.152) is, as expected, negligible because diffraction effects are in this case negligible.

### 8.1.6 Maximum Absorbed Power with Amplitude Constraints

In practice, there are certain limitations on excursion, velocity and acceleration of oscillating bodies. Thus, for all designed bodies of the system there are upper bounds to amplitudes. For sufficiently low waves, these limitations are not reached, and the optimisation without constraints in the previous subsections is applicable. For moderate and large wave heights, such limitations or amplitude constraints may be important.

Cases in which constraints come into play for only some of the bodies of a system are complicated to analyse. More complicated numerical optimisation has to be applied [43, 111]. It is simpler when the wave height is so large that no optimum oscillation amplitude is less than the bound due to the design specifications of the system [45, 62]. It is also relatively simple to analyse a case in which only one single, but global, constraint is involved [112, 113]. However, this case is not further pursued here.

## 8.2 WEC Array of Oscillating Bodies and OWCs

Let us consider a WEC array consisting of $N_B$ WEC bodies and $N_C$ OWC chambers. If an OWC chamber structure can oscillate, it belongs also to the set of oscillating bodies. The oscillating bodies may be partly or completely submerged. The equilibrium level of an OWC's internal water surface may differ from the mean level of the external water surface, provided the internal air pressure, at equilibrium, is correspondingly adjusted (see Figure 4.8).

We may assume that each WEC body is, in general, free to oscillate in all its six modes of motion. Then the WEC-array system has

$$M = 6N_B + N_C \tag{8.53}$$

independent oscillators. The states of the $6N_B$ WEC-body oscillators are characterised by velocity components $u_{ij}$, where the first subscript ($i = 1, 2, \ldots, N_B$) denotes the body number. The second subscript ($j = 1, 2, \ldots, 6$) denotes the mode of body motion. For the remaining oscillators, the states are given by dynamic pressures $p_k$ ($k = 1, 2, \ldots, N_C$) of the air in the OWC chambers. For a given frequency, the state of each oscillator is then given by a single complex amplitude ($\hat{u}_{ij}$ or $\hat{p}_k$).

Let a plane incident wave

$$\eta_0(x, y) = A e^{-ikr(\beta)} \tag{8.54}$$

be given as in Eq. (4.98). Here $A$ is the complex elevation amplitude at the origin $r(\beta) = 0$, and

$$r(\beta) = x \cos \beta + y \sin \beta, \tag{8.55}$$

where $(x, y)$ are horizontal Cartesian coordinates and $\beta$ is the angle of incidence [cf. Eq. (4.267)].

Let us first consider the case in which the oscillation amplitude is zero for each oscillator. The incident wave produces a hydrodynamic force on the wet surface $S_i$ of any immersed body $i$ and a volume flow due to the induced motion of the internal water surface $S_k$ of any OWC unit $k$. Provided the bodies are not moving ($u_{ij} = 0$ for all $i$ and $j$) and provided there is no dynamic air pressure above all OWCs' internal water surface ($p_k = 0$ for all $k$), the force and volume flow are called the *excitation force* $F_{e,ij}$ and the *excitation volume flow* $Q_{e,k}$, respectively. With the assumption of linear theory, their complex amplitudes are proportional to the incident wave amplitude $A$—that is, $\hat{F}_{e,ij}(\beta) = f_{ij}(\beta)A$ and $\hat{Q}_{e,k}(\beta) = q_k(\beta)A$ or, in column-vector notation,

$$\hat{\mathbf{F}}_e(\beta) = \mathbf{f}(\beta)A \quad \text{and} \quad \hat{\mathbf{Q}}_e(\beta) = \mathbf{q}(\beta)A. \tag{8.56}$$

Here, the complex coefficients of proportionality $f_{ij}(\beta)$ and $q_k(\beta)$ or, in column-vector notation, $\mathbf{f}(\beta)$ and $\mathbf{q}(\beta)$ are functions of $\beta$ and also of $\omega$. They are termed excitation coefficients, the *excitation force coefficient* and the *excitation volume flow coefficient*, respectively.

### 8.2.1 Wave-Interacting OWCs and Bodies

Next, consider the case in which the amplitudes of all oscillators may be different from zero; thus, $u_{ij} \neq 0$ and $p_k \neq 0$. Because of our assumption of linearity, we have proportionality between input and output. Let us introduce additional complex coefficients of proportionality ($Z_{ij,i'j'}$, $H_{ij,k}$, $Y_{k,k'}$ and $H_{k,ij}$). Moreover, we may use the principle of superposition. Then, in terms of its complex amplitude, we write the $j$ component of the total force acting on body $i$ as

$$\hat{F}_{t,ij} = f_{ij}A - \sum_{i'j'} Z_{ij,i'j'}\,\hat{u}_{i'j'} - \sum_{k} H_{ij,k}\,\hat{p}_k, \tag{8.57}$$

where the second sum runs from $k = 1$ to $k = N_C$. In the first sum, $i'$ runs from 1 to $N_B$ and $j'$ from 1 to 6. [Instead of using a single index $l = 6(i' - 1) + j'$, we denote a body oscillation by an apparent double index $i'j'$ to distinguish it from a pressure oscillator, denoted by a single index $k$.] In order to write Eq. (8.57) in vectorial form, we introduce the two matrices

$$\mathbf{Z} = \{Z_{ij,i'j'}\} \quad \text{and} \quad \mathbf{H} \equiv \mathbf{H}_{up} = \{H_{ij,k}\}. \tag{8.58}$$

Then we may write Eq. (8.57) in matrix form as

$$\hat{\mathbf{F}}_t = \mathbf{f}A - \mathbf{Z}\hat{\mathbf{u}} - \mathbf{H}_{up}\hat{\mathbf{p}} = \hat{\mathbf{F}}_e - \mathbf{Z}\hat{\mathbf{u}} - \mathbf{H}\hat{\mathbf{p}}. \tag{8.59}$$

Furthermore, the total volume flow due to the oscillation of the internal water surface $S_k$ is

$$\hat{Q}_{t,k} = q_k A - \sum_{k'} Y_{k,k'}\,\hat{p}_{k'} - \sum_{ij} H_{k,ij}\,\hat{u}_{ij}, \tag{8.60}$$

or, in matrix notation,

$$\hat{\mathbf{Q}}_t = \mathbf{q}A - \mathbf{Y}\hat{\mathbf{p}} - \mathbf{H}_{pu}\hat{\mathbf{u}} = \hat{\mathbf{Q}}_e - \mathbf{Y}\hat{\mathbf{p}} + \mathbf{H}^{\mathrm{T}}\hat{\mathbf{u}}, \tag{8.61}$$

where we utilised the fact that

$$H_{k,ij} = -H_{ij,k} \quad \text{and thus} \quad \mathbf{H}_{pu} = -\mathbf{H}^{\mathrm{T}}; \tag{8.62}$$

see Problem 8.2 and Section 8.2.5.

It can be shown (see Section 8.2.5) that matrices $\mathbf{Z}$ and $\mathbf{Y}$ are symmetric, which means that they do not change by transposition, or

$$\mathbf{Z}^{\mathrm{T}} = \mathbf{Z}, \quad \mathbf{Y}^{\mathrm{T}} = \mathbf{Y}. \tag{8.63}$$

The three matrices $\mathbf{Z}$, $\mathbf{Y}$ and $\mathbf{H}$ have dimensions $6N_B \times 6N_B$, $N_C \times N_C$, and $6N_B \times N_C$, respectively.

Equations (8.59) and (8.61) represent a linear system, which may be illustrated in a block diagram as shown in Figure 8.2. The two sets of coefficients $Z_{ij,i'j'}$ and $H_{ij,k}$ in Eq. (8.57) compose the radiation impedance matrix $\mathbf{Z}$ (cf. Section 5.5) and the *OWC-to-body coupling matrix* $\mathbf{H}$ for the oscillating

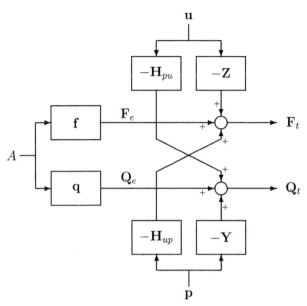

Figure 8.2: Block diagram of the system of oscillating bodies and OWCs. The incident wave, the oscillating bodies' velocities and the oscillating air-chamber pressures (represented by $A$, $\mathbf{u}$ and $\mathbf{p}$, respectively) are considered as inputs to the system. The body forces and the volume flows (represented by $\mathbf{F}_t$ and $\mathbf{Q}_t$) are considered as outputs here. The diagram illustrates Eqs. (8.59) and (8.61).

bodies. Similarly, the set of complex coefficients $Y_{k,k'}$ in Eq. (8.60) composes the radiation admittance matrix $\mathbf{Y}$ for the oscillating surface pressure distribution, as introduced first by Evans [98]. In the same equation we introduced also the body-to-OWC coupling coefficients $H_{k,ij}$ which compose the matrix $\mathbf{H}_{pu}$.

The two sets of complex coefficients $H_{ij,k}$ and $H_{k,ij}$ have the dimension of length squared for $j = 1,2,3$ and length cubed for $j = 4,5,6$. They represent hydrodynamical coupling between the WEC array's oscillating bodies and its OWCs.

Mechanical impedance, as introduced in Section 2.2.2, has the dimension of force divided by velocity. Contrary to common usage in electrical circuit theory, we have here, for notational convenience, defined an admittance $Y_{k,k'}$ which is not dimensionally inverse to impedance $Z_{ij,i'j'}$. Note that $Y'_{k,k'} \equiv Y_{k,k'}/S_k S_{k'}$ is dimensionally inverse to $Z_{ij,i'j'}$, but it is more convenient to write just $Y_{k,k'}$ instead of $S_k S_{k'} Y'_{k,k'}$ in Eq. (8.60).

Defining the three $M$ dimensional column vectors

$$\hat{\mathbf{F}}_{FQ,t} \equiv \begin{bmatrix} \hat{\mathbf{F}}_t \\ -\hat{\mathbf{Q}}_t \end{bmatrix}, \quad \hat{\mathbf{F}}_{FQ,e} \equiv \begin{bmatrix} \hat{\mathbf{F}}_e \\ -\hat{\mathbf{Q}}_e \end{bmatrix} \quad \text{and} \quad \hat{\mathbf{u}}_{FQ} \equiv \begin{bmatrix} \hat{\mathbf{u}} \\ -\hat{\mathbf{p}} \end{bmatrix}, \tag{8.64}$$

and defining the complex $M \times M$ total-radiation-impedance matrix as

$$\mathbf{Z}_{FQ} \equiv \begin{bmatrix} \mathbf{Z} & -\mathbf{H} \\ \mathbf{H}^T & \mathbf{Y} \end{bmatrix} = \begin{bmatrix} \mathbf{R} + i\mathbf{X} & -(\mathbf{C} + i\mathbf{J}) \\ (\mathbf{C} + i\mathbf{J})^T & \mathbf{G} + i\mathbf{B} \end{bmatrix}, \tag{8.65}$$

we can combine the two equations (8.59) and (8.61) into a single matrix equation

$$\hat{\mathbf{F}}_{FQ,t} = \hat{\mathbf{F}}_{FQ,e} + \hat{\mathbf{F}}_{FQ,r} = \hat{\mathbf{F}}_{FQ,e} - \mathbf{Z}_{FQ}\hat{\mathbf{u}}_{FQ}, \tag{8.66}$$

which specialises to Eq. (6.81) for the one-body case ($N_B = 1$ and $N_C = 0$).

For the purpose of further discussion to follow, we have also, in Eq. (8.65), split complex matrix entries into real and imaginary parts: $\mathbf{Z} = \mathbf{R}+i\mathbf{X}$, $\mathbf{Y} = \mathbf{G}+i\mathbf{B}$ and $\mathbf{H} = \mathbf{C} + i\mathbf{J}$, where the radiation resistance matrix $\mathbf{R}$, the radiation reactance matrix $\mathbf{X}$, the radiation conductance matrix $\mathbf{G}$, the radiation susceptance matrix $\mathbf{B}$, as well as the matrices $\mathbf{C}$ and $\mathbf{J}$, are real, whereas matrices $\mathbf{Z}$, $\mathbf{Y}$ and $\mathbf{H}$ are complex. All these nine matrices are frequency dependent.

### 8.2.2 Active Power and Reactive Power

For a simple oscillating system, we considered, in Section 2.3.1, active power and reactive power, which are related to produced or consumed power and to stored power, respectively. We also defined a 'complex power' $\mathcal{P} = R|\hat{u}|^2/2 + iX|\hat{u}|^2/2$, where the real and imaginary parts are the active power and the reactive power, respectively; see Eq. (2.90). Here, $R$ and $X$ denote resistance and reactance, respectively.

Furthermore, in Section 3.4, we introduced the concepts of radiation resistance, radiation reactance and radiation impedance, which have been a matter of discussion in most of the subsequent chapters. For instance, in Section 5.5.4, we discussed how the radiation reactance, and thus the 'added mass', is related to the—usually positive—difference between time-average kinetic and potential energies in the near-field region of oscillating bodies. As we shall discuss in some detail later, the three real matrices $\mathbf{R}$, $\mathbf{G}$ and $\mathbf{J}$ are associated with active power, while the three remaining ones, matrices $\mathbf{X}$, $\mathbf{B}$ and $\mathbf{C}$, are associated with reactive power.

Let us split the total radiation matrix $\mathbf{Z}_{FQ}$ as given by Eq. (8.65), not into real and imaginary parts, but into active and reactive parts, as follows:

$$\mathbf{Z}_{FQ} = \mathbf{Z}_{FQ,\text{active}} + \mathbf{Z}_{FQ,\text{reactive}}, \tag{8.67}$$

where

$$\mathbf{Z}_{FQ,\text{active}} = \begin{bmatrix} \mathbf{R} & -i\mathbf{J} \\ i\mathbf{J}^{\mathrm{T}} & \mathbf{G} \end{bmatrix} \equiv \mathbf{D}, \tag{8.68}$$

$$\mathbf{Z}_{FQ,\text{reactive}} = i\begin{bmatrix} \mathbf{X} & i\mathbf{C} \\ -i\mathbf{C}^{\mathrm{T}} & \mathbf{B} \end{bmatrix}. \tag{8.69}$$

According to Eqs. (8.65)–(8.69), the total radiation force is

$$\hat{\mathbf{F}}_{FQ,r} = -\mathbf{Z}_{FQ}\,\hat{\mathbf{u}}_{FQ} = -\mathbf{D}\hat{\mathbf{u}}_{FQ} + \mathbf{Z}_{FQ,\text{reactive}}\,\hat{\mathbf{u}}_{FQ}. \tag{8.70}$$

Thus, we have here, for the present WEC array, extended Eqs. (3.32)–(3.33), where we defined the radiation force $F_{r,i}$ and split it into active and reactive components. Observe that both matrices $\mathbf{D} = \mathbf{Z}_{FQ,\text{active}}$ and $(\mathbf{Z}_{FQ,\text{reactive}}/i)$ are not real and symmetric, but complex and Hermitian, since $\mathbf{D}^\dagger = \mathbf{D}$ and $(\mathbf{Z}_{FQ,\text{reactive}}/i) = (\mathbf{Z}_{FQ,\text{reactive}}/i)^\dagger$, due to relations (8.62)–(8.63). Matrices $\mathbf{D}$ and $(\mathbf{Z}_{FQ,\text{reactive}}/i)$ are real and symmetric if the WEC array contains no OWCs or no oscillating bodies, that is, in cases where $\mathbf{Z}_{FQ} = \mathbf{Z} = \mathbf{R}+i\mathbf{X}$ or $\mathbf{Z}_{FQ} = \mathbf{Y} = \mathbf{G}+i\mathbf{B}$, respectively.

As discussed in Section 8.1.2, all eigenvalues of Hermitian matrices are real [16, page 109]. Moreover, as we shall show later, matrix $\mathbf{D}$ is positive semidefinite; see Eq. (8.75). Therefore, the eigenvalues of $\mathbf{D}$ are necessarily nonnegative. The matrix $(\mathbf{Z}_{FQ,\text{reactive}}/i)$ may, however, have negative as well as positive eigenvalues.

Since matrix $\mathbf{D}$ is Hermitian, it is possible to diagonalise it as in Eq. (8.25). It follows that, for any $M$-dimensional complex column vector $\hat{\mathbf{u}}_{FQ}$, the scalar matrix product $\hat{\mathbf{u}}_{FQ}^\dagger \mathbf{D} \hat{\mathbf{u}}_{FQ} = \hat{\mathbf{u}}_{FQ}^{\prime\dagger} \mathbf{D}' \hat{\mathbf{u}}_{FQ}'$ is real. By the same argument, $\hat{\mathbf{u}}_{FQ}^\dagger \mathbf{Z}_{FQ,\text{reactive}} \hat{\mathbf{u}}_{FQ}$ is purely imaginary.

If we premultiply Eq. (8.70) by $-\hat{\mathbf{u}}_{FQ}^\dagger/2$, we obtain the 'complex radiated power'

$$\mathcal{P}_r = -\hat{\mathbf{u}}_{FQ}^\dagger \hat{\mathbf{F}}_{FQ,r}/2 = \hat{\mathbf{u}}_{FQ}^\dagger \mathbf{D} \hat{\mathbf{u}}_{FQ}/2 + \hat{\mathbf{u}}_{FQ}^\dagger \mathbf{Z}_{FQ,\text{reactive}} \hat{\mathbf{u}}_{FQ}/2, \tag{8.71}$$

where the last term, the reactive-power term $\hat{\mathbf{u}}_{FQ}^\dagger \mathbf{Z}_{FQ,\text{reactive}} \hat{\mathbf{u}}_{FQ}/2$, is purely imaginary, whereas the first term, the radiated-power term $\hat{\mathbf{u}}_{FQ}^\dagger \mathbf{D} \hat{\mathbf{u}}_{FQ}/2 \equiv P_r$, is real and nonnegative—see inequality (8.75). Moreover, if we premultiply by $\hat{\mathbf{u}}_{FQ}^\dagger/2$ the excitation vector $\hat{\mathbf{F}}_{FQ,e}$, which was introduced in Eq. (8.64), we get the 'complex excitation power'

$$\mathcal{P}_e = \hat{\mathbf{u}}_{FQ}^\dagger \hat{\mathbf{F}}_{FQ,e}/2. \tag{8.72}$$

Here the imaginary part, Im$\{\mathcal{P}_e\}$, represents reactive power; see Eq. (5.218).

### 8.2.3 Wave Power Converted by the WEC Array

For any oscillation vector $\hat{\mathbf{u}}_{FQ} = \begin{bmatrix} \hat{\mathbf{u}} & -\hat{\mathbf{p}} \end{bmatrix}^{\mathrm{T}}$, the time-average wave power absorbed by the array is $P = \text{Re}\{\mathcal{P}\} = \text{Re}\{\mathcal{P}_e\} - \text{Re}\{\mathcal{P}_r\} = P_e - P_r$, where the excitation power $P_e$ and the radiated power $P_r$ are given by

$$P_e = \text{Re}\{\mathcal{P}_e\} = \frac{\hat{\mathbf{u}}_{FQ}^\dagger \hat{\mathbf{F}}_{FQ,e} + \hat{\mathbf{F}}_{FQ,e}^\dagger \hat{\mathbf{u}}_{FQ}}{4} \quad \text{and} \quad P_r = \text{Re}\{\mathcal{P}_r\} = \frac{\hat{\mathbf{u}}_{FQ}^\dagger \mathbf{D} \hat{\mathbf{u}}_{FQ}}{2}; \tag{8.73}$$

cf. Eqs. (6.85)–(6.86) and (7.26)–(7.27). We may express this in the form of Eq. (6.32) or (6.84), provided we define the collective excitation-power coefficient $E(\beta)$ and the collective oscillation amplitude $U$ by

$$E(\beta) = \frac{\hat{\mathbf{u}}_{FQ}^\dagger \hat{\mathbf{F}}_{FQ,e}}{4A} \quad \text{and} \quad |U|^2 = UU^* = \frac{\hat{\mathbf{u}}_{FQ}^\dagger \mathbf{D} \hat{\mathbf{u}}_{FQ}}{2}, \tag{8.74}$$

where we still choose the otherwise arbitrary phase angle of $U$ such as to make $A^*E^*(\beta)/U$ a real and positive quantity.

For a case with no incident wave, thus $\hat{\mathbf{F}}_{FQ,e} = \mathbf{0}$ (which means that $P_e = 0$), energy conservation requires that the absorbed wave power $P = P_e - P_r = -P_r = -\hat{\mathbf{u}}_{FQ}^\dagger \mathbf{D}\hat{\mathbf{u}}_{FQ}/2$ cannot be positive. Hence, for all possible finite oscillation-state vectors $\hat{\mathbf{u}}_{FQ}$, we have

$$\hat{\mathbf{u}}_{FQ}^\dagger \mathbf{D}\hat{\mathbf{u}}_{FQ} \geq 0. \tag{8.75}$$

Thus, in general, the radiation damping matrix $\mathbf{D}$ is positive semidefinite. It is singular in cases when its determinant vanishes, $|\mathbf{D}| = 0$. Otherwise, it is positive definite, $\hat{\mathbf{u}}_{FQ}^\dagger \mathbf{D}\hat{\mathbf{u}}_{FQ} > 0$. This is a generalisation of inequality (6.88).

The maximum wave power absorbed by the array may be written in various ways, such as

$$P_{\text{MAX}} = \frac{P_{e,\text{OPT}}}{2} \equiv \frac{\hat{\mathbf{F}}_{FQ,e}^\dagger \hat{\mathbf{u}}_{FQ,0}}{4} = \frac{\hat{\mathbf{u}}_{FQ,0}^\dagger \hat{\mathbf{F}}_{FQ,e}}{4} = P_{r,\text{OPT}} \equiv \frac{\hat{\mathbf{u}}_{FQ,0}^\dagger \mathbf{D}\hat{\mathbf{u}}_{FQ,0}}{2}, \tag{8.76}$$

where $\hat{\mathbf{u}}_{FQ,0} \equiv \hat{\mathbf{u}}_{FQ,\text{OPT}}(\beta)$ is an optimum value of the oscillation-state vector $\hat{\mathbf{u}}_{FQ}$ that has to satisfy the optimum condition

$$\mathbf{D}\hat{\mathbf{u}}_{FQ,0}(\beta) = \hat{\mathbf{F}}_{FQ,e}(\beta)/2. \tag{8.77}$$

Assuming that the WEC array is equipped with sufficient control devices, we may here consider the complex-velocity column vector $\hat{\mathbf{u}}_{FQ}$ to be an independent variable, whereas its optimum value $\hat{\mathbf{u}}_{FQ,0}(\beta)$ is proportional to the incident wave amplitude and also dependent on the angle of wave incidence $\beta$.

By manipulating Eqs. (8.16), (8.73), (8.76) and (8.77), we can show that

$$P_{\text{MAX}}(\beta) - P = \tfrac{1}{2}[\hat{\mathbf{u}}_{FQ} - \hat{\mathbf{u}}_{FQ,0}(\beta)]^\dagger \mathbf{D}[\hat{\mathbf{u}}_{FQ} - \hat{\mathbf{u}}_{FQ,0}(\beta)]. \tag{8.78}$$

For a fixed value of the absorbed wave power $P$, where $P < P_{\text{MAX}}$, Eq. (8.78) represents an 'ellipsoid' in the complex $M$-dimensional $\hat{\mathbf{u}}_{FQ}$ space, $\mathbb{C}^M$—but reduced to an $r_\mathbf{D}$-dimensional $\hat{\mathbf{u}}_{FQ}$ space, $\mathbb{C}^{r_\mathbf{D}}$, in cases where the radiation damping matrix $\mathbf{D}$ is singular and of rank $r_\mathbf{D} < M$. The centre of the 'ellipsoid' is at the optimum point $\hat{\mathbf{u}}_{FQ} = \hat{\mathbf{u}}_{FQ,0}$. The elliptical semiaxes are $\sqrt{2(P_{\text{MAX}} - P)/\Lambda_i}$ for $i = 1, 2, \ldots, r_\mathbf{D}$, where $\Lambda_i$ is any member of the positive definite (nonzero) eigenvalues of matrix $\mathbf{D}$—cf. Eq. (8.17). The 'ellipsoid' that corresponds to $P = 0$ runs through, for example, points $\hat{\mathbf{u}}_{FQ} = \mathbf{0}$ and $\hat{\mathbf{u}}_{FQ} = 2\hat{\mathbf{u}}_{FQ,0}$. The degenerate 'ellipsoid' that corresponds to $P = P_{\text{MAX}}$ is just one point, which represents the (unconstrained) optimum situation. Choosing smaller $P$ increases the size of the 'ellipsoid'. If $M = 1$, then the 'ellipsoid' simplifies to a circle in the complex $\hat{u}_1$ plane; cf. Problem 4.16. Considering how the absorbed power $P$ varies with $\hat{\mathbf{u}}_{FQ}$, we may think of relationship (8.78) as a 'paraboloid' in the complex $M$-dimensional $\hat{\mathbf{u}}_{FQ}$ space, $\mathbb{C}^M$. The top point of this 'paraboloid' corresponds to the optimum, $(\hat{\mathbf{u}}_{FQ,0}, P_{\text{MAX}})$. Here, $M$ should be replaced by $r_\mathbf{D}$ if the radiation matrix $\mathbf{D}$ is singular.

The simple equation (6.37), which for a fixed absorbed-power value $P$ represents a circle in the complex $U$ plane, can be shown to be equivalent to Eq. (8.78) above, which represents an 'ellipsoid' in the complex $\hat{u}_{FQ}$ space, by making use of Eqs. (6.36), (8.74) and (8.77). Starting from (6.37), we have

$$P_{\text{MAX}} - P = |U_0(\beta) - U|^2 = U_0 U_0^* - U_0 U^* - U_0^* U + UU^*$$

$$= \frac{1}{2}\hat{u}_{FQ,0}^\dagger \mathbf{D}\hat{u}_{FQ,0} - AE - A^* E^* + \frac{1}{2}\hat{u}_{FQ}^\dagger \mathbf{D}\hat{u}_{FQ}$$

$$= \frac{1}{2}\hat{u}_{FQ,0}^\dagger \mathbf{D}\hat{u}_{FQ,0} - \frac{1}{4}\hat{u}_{FQ}^\dagger \hat{\mathbf{F}}_{FQ,e} - \frac{1}{4}\hat{\mathbf{F}}_{FQ,e}^\dagger \hat{u}_{FQ} + \frac{1}{2}\hat{u}_{FQ}^\dagger \mathbf{D}\hat{u}_{FQ}$$

$$= \frac{1}{2}\hat{u}_{FQ,0}^\dagger \mathbf{D}\hat{u}_{FQ,0} - \frac{1}{2}\hat{u}_{FQ}^\dagger \mathbf{D}\hat{u}_{FQ,0} - \frac{1}{2}\hat{u}_{FQ,0}^\dagger \mathbf{D}\hat{u}_{FQ} + \frac{1}{2}\hat{u}_{FQ}^\dagger \mathbf{D}\hat{u}_{FQ}$$

$$= \frac{1}{2}[\hat{u}_{FQ} - \hat{u}_{FQ,0}(\beta)]^\dagger \mathbf{D}[\hat{u}_{FQ} - \hat{u}_{FQ,0}(\beta)],$$
(8.79)

noting that the collective excitation-power coefficient $E(\beta)$ is a scalar and, thus, equals its own transpose, and recalling the Hermitian property (8.16) of the radiation-damping matrix $\mathbf{D}$.

Proof (8.79) also serves to demonstrate that, with the generalisations (8.74), Eqs. (6.32)–(6.37) are valid not only for a single, one-mode oscillating body but even for an array consisting of several WEC units—oscillating bodies as well as OWCs. In particular, Eq. (8.4), as illustrated in Figure 6.5, is applicable even to the general case of wave energy absorption by an array of oscillating bodies as well as OWCs.

The involved physical quantities, $\hat{\mathbf{F}}_{FQ,e}$ and $\hat{u}_{FQ}$, pertain to the wave-interacting surfaces of the WEC array. Thus, if inverse Fourier transformation is applied to, for example, Eqs. (6.32) and (8.74), these equations also may be applied to analyse the WEC array's wave-power absorption even in the case of non-sinusoidal time variation. (However, this is not possible for expressions where the excitation forces and volume flows are expressed in terms of Kochin functions or other far-field coefficients.)

In terms of similarity-transformed excitation amplitudes $\hat{F}'_{FQ,e,i}(\beta)$ and corresponding optimum oscillation amplitudes $\hat{u}'_{FQ,i0}(\beta)$, the maximum absorbed power may be written as

$$P_{\text{MAX}} = \sum_{i=1}^{M} \frac{|\hat{F}'_{FQ,e,i}(\beta)|^2}{8\Lambda_i} = \frac{1}{2}\sum_{i=1}^{M} \Lambda_i |\hat{u}'_{FQ,i0}(\beta)|^2,$$
(8.80)

corresponding to the optimum condition

$$\Lambda_i \hat{u}'_{FQ,i0}(\beta) = \tfrac{1}{2}\hat{F}'_{FQ,e,i}(\beta).$$
(8.81)

Likewise, Eq. (8.78) may be simplified to

$$2(P_{\text{MAX}} - P) = [\hat{u}'_{FQ} - \hat{u}'_{FQ,0}(\beta)]^\dagger \mathbf{D}'[\hat{u}'_{FQ} - \hat{u}'_{FQ,0}(\beta)]$$

$$= \sum_{i=1}^{M} \Lambda_i |\hat{u}'_{FQ,i} - \hat{u}'_{FQ,i0}(\beta)|^2.$$
(8.82)

### 8.2.4 Example: WEC Body Containing an OWC

For simplicity, let us consider a WEC body that contains one OWC and oscillates in only one body mode. It could, for example, correspond to a backward-bent duct buoy (BBDB) device [114] constrained to oscillate in pitch only or an axisymmetric wave-powered navigation buoy [115] constrained to oscillate in heave only. Both devices were invented in Japan by Yoshio Masuda. During the early 1980s, sea tests of a heaving body containing an OWC were carried out also in Norway [116].

In this simple case, the two $M$-dimensional column vectors $\hat{\mathbf{u}}_{FQ}$ and $\hat{\mathbf{F}}_{FQ,e}$, introduced by Eq. (8.64), are simplified to two-dimensional column vectors,

$$\hat{\mathbf{u}}_{FQ} = \begin{bmatrix} \hat{u} \\ -\hat{p} \end{bmatrix} \quad \text{and} \quad \hat{\mathbf{F}}_{FQ,e}(\beta) = \begin{bmatrix} \hat{F}_e(\beta) \\ -\hat{Q}_e(\beta) \end{bmatrix}. \tag{8.83}$$

Correspondingly, the radiation-damping matrix introduced by Eq. (8.68) is, in this case,

$$\mathbf{D} = \begin{bmatrix} R & -\mathrm{i}J \\ \mathrm{i}J & G \end{bmatrix}, \tag{8.84}$$

which is a Hermitian matrix of dimension $2 \times 2$.

The two eigenvalues $\Lambda_1$ and $\Lambda_2$ of this radiation damping matrix (8.84) are solutions of the second-degree algebraic equation $|\mathbf{D} - \Lambda\mathbf{I}| = \Lambda^2 - (R + G)\Lambda + RG - J^2 = 0$. Thus, $\Lambda_1$ and $\Lambda_2$ are given by

$$\Lambda_i = \left[ R + G - (-1)^i \sqrt{(R+G)^2 - 4(RG - J^2)} \right] / 2 \quad \text{for } i = 1, 2. \tag{8.85}$$

The corresponding two eigenvectors, which satisfy Eqs. (8.17) and (8.19), are

$$\mathbf{e}_i = C_i \begin{bmatrix} \mathrm{i}J \\ R - \Lambda_i \end{bmatrix}, \quad \text{where} \quad C_i = 1 / \sqrt{(R - \Lambda_i)^2 + J^2}. \tag{8.86}$$

If the system we are dealing with is a heaving axisymmetric body containing an axisymmetric OWC, we have

$$J^2 = RG, \tag{8.87}$$

and thus, from Eq. (8.85), we see that $\Lambda_1 = R + G$ and $\Lambda_2 = 0$, which means that in this case, matrix $\mathbf{D}$ is singular and of rank $r_{\mathbf{D}} = 1$. Therefore, there is only one term in the sum of the rhs of Eq. (8.82):

$$2(P_{\text{MAX}} - P) = (R + G)|\hat{u}'_{FQ,1} - \hat{u}'_{FQ,10}(\beta)|^2, \tag{8.88}$$

which represents a circle in the complex $\hat{u}'_{FQ,1}$ plane. The centre of the circle is at $\hat{u}'_{FQ,1} = \hat{u}'_{FQ,10}(\beta)$ and the radius is $\sqrt{2(P_{\text{MAX}} - P)/(R + G)}$. Because of the singularity of the radiation damping matrix, the similarity-transformed variable $\hat{u}'_{FQ,2}$ is irrelevant and may have any arbitrary value without influencing the maximum absorbed power.

It is furthermore possible to show from the optimum condition (8.77) and Eq. (8.87) that $i G \hat{F}_e = J \hat{Q}_e$ in this case, which leads to

$$P_{\mathrm{MAX}} = \frac{|\hat{F}_e|^2}{8R} = \frac{i \hat{F}_e \hat{Q}_e^*}{8J} = \frac{|\hat{Q}_e|^2}{8G} \tag{8.89}$$

upon application of Eq. (8.76). Notice that because of the singularity of the radiation damping matrix, the maximum power that can be absorbed by the system is equal to that of a single heaving body without any OWC [cf. Eq. (6.12)] or of a single OWC contained in a fixed body [cf. Eq. (7.43)].

The physical reason for the singularity is that both modes, the heaving-body mode and the OWC mode, can radiate only isotropic outgoing waves. To obtain the maximum absorbed wave power, the optimum isotropically radiated wave may be realised by any optimum combined wave radiation from the axisymmetric OWC and the heaving axisymmetric body. The transformed oscillation $\hat{u}'_{FQ,2}$ corresponds to a situation where the heave mode and the OWC mode cancel each other's radiated waves in the far-field region.

### 8.2.5 Reciprocity Relations for Radiation Parameters

Our task in the following is to prove reciprocity relations (8.62)–(8.63), which underlie the results presented so far in this section. Note that these relations are important not only for arrays of many units of WEC bodies and OWCs but also for a single WEC unit in the form of an OWC housed in a floating structure, as we have just discussed. As in Section 5.5.2, we decompose the velocity potential into incident, diffracted and radiated components. The radiated wave potential can be further decomposed as

$$\hat{\phi}_r = \sum_{ij} \varphi_{ij} \hat{u}_{ij} + \sum_k \varphi_k \hat{p}_k = \boldsymbol{\varphi}_u^{\mathrm{T}} \hat{\mathbf{u}} + \boldsymbol{\varphi}_p^{\mathrm{T}} \hat{\mathbf{p}}, \tag{8.90}$$

where $\boldsymbol{\varphi}_u$ is a column vector composed of all the complex potential coefficients $\varphi_{ij}$, which depend on $x, y, z$ and $\omega$. Similarly, $\boldsymbol{\varphi}_p$ is the column vector composed of all the coefficients $\varphi_k$.

The $j$ component of the total force on body $i$ due to the hydrodynamic pressure $\hat{p} = -i\omega \rho \hat{\phi}$ is obtained by integration,

$$\hat{F}_{t,ij} = - \iint_{S_i} \hat{p} n_{ij} \, dS = i\omega \rho \iint_{S_i} n_{ij} \hat{\phi} \, dS, \tag{8.91}$$

in accordance with Eq. (5.22). The total volume flow through the mean water surface $S_k$ is

$$\hat{Q}_{t,k} = \iint_{S_k} \hat{v}_z \, dS = \iint_{S_k} \frac{\partial \hat{\phi}}{\partial z} \, dS. \tag{8.92}$$

Using decompositions (5.160) and (8.90), we easily see that $\hat{F}_{t,ij}$ is as given by Eq. (8.57) with

$$Z_{ij,i'j'} = -i\omega\rho \iint_{S_i} n_{ij}\varphi_{i'j'}\, dS, \tag{8.93}$$

$$H_{ij,k} = -i\omega\rho \iint_{S_i} n_{ij}\varphi_k\, dS, \tag{8.94}$$

and $\hat{Q}_{t,k}$ is as given by Eq. (8.60) with

$$Y_{k,k'} = -\iint_{S_k} \frac{\partial}{\partial z}\varphi_{k'}\, dS, \tag{8.95}$$

$$H_{k,ij} = -\iint_{S_k} \frac{\partial}{\partial z}\varphi_{ij}\, dS. \tag{8.96}$$

Because of the boundary condition on the wet surfaces $S_i, i = 1, 2, \ldots, N_B$,

$$\frac{\partial \varphi_{i'j}}{\partial n} = n_{ij}\delta_{ii'} = \begin{cases} n_{ij}, & \text{for } i' = i \\ 0, & \text{for } i' \neq i, \end{cases} \tag{8.97}$$

we may rewrite expression (8.93) for the radiation impedance matrix as

$$Z_{ij,i'j'} = -i\omega\rho \iint_{S_i} \frac{\partial \varphi_{ij}}{\partial n}\varphi_{i'j'}\, dS \tag{8.98}$$

and extend the region of integration from $S_i$ to include also the wet surfaces of all the other bodies. Further, from the boundary condition for $\varphi_{ij}$ on $S_k$,

$$\left(\frac{\partial}{\partial z} - \frac{\omega^2}{g}\right)\varphi_{ij} = 0, \tag{8.99}$$

we observe that

$$0 = -i\omega\rho \iint_{S_k} \left(\frac{\partial}{\partial n} + \frac{\omega^2}{g}\right)\varphi_{ij}\,\varphi_{i'j'}\, dS, \tag{8.100}$$

since $\partial/\partial n = -\partial/\partial z$ on $S_k$. We are now in the position to extend the region of integration in Eq. (8.98) to the totality of wave-generating surfaces

$$S = \sum_{i=1}^{N_B} S_i + \sum_{k=1}^{N_C} S_k. \tag{8.101}$$

By summation, we obtain, after reverting to matrix notation,

$$\mathbf{Z} = -i\omega\rho \iint_S \frac{\partial\boldsymbol{\varphi}_u}{\partial n}\boldsymbol{\varphi}_u^{\mathrm{T}}\, dS - \sum_k i\omega\rho \iint_{S_k} \frac{\omega^2}{g}\boldsymbol{\varphi}_u\boldsymbol{\varphi}_u^{\mathrm{T}}\, dS. \tag{8.102}$$

Alternatively, since $\partial\varphi_{ij}/\partial n$ is real on $S_i$, we may replace $\partial\varphi_{ij}/\partial n$ by $\partial\varphi_{ij}^*/\partial n$ in Eq. (8.98). We may also replace $\varphi_{ij}$ by $\varphi_{ij}^*$ in Eq. (8.100). These give

$$\mathbf{Z} = -i\omega\rho \iint_S \frac{\partial\boldsymbol{\varphi}_u^*}{\partial n}\boldsymbol{\varphi}_u^{\mathrm{T}}\, dS - \sum_k i\omega\rho \iint_{S_k} \frac{\omega^2}{g}\boldsymbol{\varphi}_u^*\boldsymbol{\varphi}_u^{\mathrm{T}}\, dS. \tag{8.103}$$

Note that these expressions for the radiation impedance matrix $\mathbf{Z}$ are extensions of Eq. (5.170)–(5.171) to the case where wave-generating OWCs are also included in the system.

Subtracting from Eq. (8.102) its own transpose, we find that the last term (the sum) does not contribute because of cancellation, and hence, $\mathbf{Z} - \mathbf{Z}^{\mathrm{T}}$ vanishes due to Eqs. (5.127)–(5.128). As a consequence, $\mathbf{Z}^{\mathrm{T}} = \mathbf{Z}$, which means that $\mathbf{Z}$ is a symmetric matrix, just as in the case of no OWCs; see Section 5.5.3.

Next, let us consider the radiation admittance matrix $\mathbf{Y}$, as given in component version by Eq. (8.95). The boundary condition for $\varphi_{k'}$ on $S_k$,

$$\left(\frac{\partial}{\partial z} - \frac{\omega^2}{g}\right)\varphi_{k'} = -\frac{i\omega}{\rho g}\delta_{kk'} = \begin{cases} i\omega/\rho g & \text{for } k' = k \\ 0 & \text{for } k' \neq k, \end{cases} \tag{8.104}$$

may be written as $\delta_{kk'} = -[i\omega\rho + (\rho g/i\omega)\partial/\partial z]\varphi_{k'}$. Inserting this into Eq. (8.95) gives

$$\begin{aligned} Y_{k,k'} &= \iint_{S_k} \frac{\partial \varphi_{k'}}{\partial z}\left(i\omega\rho + \frac{\rho g}{i\omega}\frac{\partial}{\partial z}\right)\varphi_k \, dS \\ &= \sum_{k''} \iint_{S_{k''}} \frac{\partial \varphi_{k'}}{\partial z}\left(i\omega\rho + \frac{\rho g}{i\omega}\frac{\partial}{\partial z}\right)\varphi_k \, dS. \end{aligned} \tag{8.105}$$

Since $\partial/\partial n = -\partial/\partial z$ on $S_{k''}$ and $\partial\varphi_{k'}/\partial n = 0$ on $S_i$, we have

$$\begin{aligned} Y_{k,k'} &= \iint_{S} \frac{\partial \varphi_{k'}}{\partial n}\left(-i\omega\rho + \frac{\rho g}{i\omega}\frac{\partial}{\partial n}\right)\varphi_k \, dS \\ &= -i\omega\rho \iint_{S} \varphi_k \frac{\partial \varphi_{k'}}{\partial n} \, dS + \frac{\rho g}{i\omega} \iint_{S} \frac{\partial \varphi_{k'}}{\partial n}\frac{\partial \varphi_k}{\partial n} \, dS. \end{aligned} \tag{8.106}$$

Since $\varphi_k$ and $\varphi_{k'}$ satisfy the same radiation condition, $Y_{k,k'} - Y_{k',k}$ vanishes according to Eqs. (4.230) and (4.239). Hence, $\mathbf{Y}^{\mathrm{T}} = \mathbf{Y}$, which means that the radiation admittance matrix is symmetric. Note that because, for real $\omega$, we have

$$1 = -\left(i\omega\rho + \frac{\rho g}{i\omega}\frac{\partial}{\partial z}\right)\varphi_k = \left(i\omega\rho + \frac{\rho g}{i\omega}\frac{\partial}{\partial z}\right)\varphi_k^* \quad \text{on } S_k, \tag{8.107}$$

we also have

$$Y_{k,k'} = i\omega\rho \iint_{S} \varphi_k^* \frac{\partial \varphi_{k'}}{\partial n} \, dS - \frac{\rho g}{i\omega} \iint_{S} \frac{\partial \varphi_{k'}}{\partial n}\frac{\partial \varphi_k^*}{\partial n} \, dS. \tag{8.108}$$

Next, let us present a proof of the reciprocity relation (8.62). As a first step, we use the inhomogeneous boundary condition (8.97) for $\varphi_{ij}$ in Eq. (8.94), giving

$$H_{ij,k} = -i\omega\rho \sum_{i'} \iint_{S_{i'}} \varphi_k \frac{\partial \varphi_{ij}}{\partial n} \, dS. \tag{8.109}$$

In view of boundary condition (8.99) and noting that $\partial/\partial n = -\partial/\partial z$ on $S_{k'}$, we can rewrite this as

$$H_{ij,k} = -i\omega\rho \iint_S \varphi_k \frac{\partial\varphi_{ij}}{\partial n}\, dS - \frac{i\omega^3\rho}{g} \sum_{k'} \iint_{S_{k'}} \varphi_k\varphi_{ij}\, dS. \qquad (8.110)$$

The integration surface $S$ is defined by Eq. (8.101).

Moreover, using the inhomogeneous boundary condition (8.104) for $\varphi_k$ in Eq. (8.96), we find

$$H_{k,ij} = \frac{\rho g}{i\omega} \sum_{k'} \iint_{S_{k'}} \frac{\partial\varphi_{ij}}{\partial z}\left(\frac{\partial\varphi_k}{\partial z} - \frac{\omega^2}{g}\varphi_k\right)\, dS. \qquad (8.111)$$

Further, using the homogeneous boundary condition (8.99) for $\varphi_{ij}$ and also noting that $\partial/\partial z = -\partial/\partial n$ on $S_{k'}$ and $\partial\varphi_k/\partial n = 0$ on $S_i$, we obtain

$$H_{k,ij} = i\omega\rho \iint_S \varphi_{ij} \frac{\partial\varphi_k}{\partial n}\, dS + \frac{i\omega^3\rho}{g} \sum_{k'} \iint_{S_{k'}} \varphi_{ij}\varphi_k\, dS. \qquad (8.112)$$

Adding this to Eq. (8.110), we find that $H_{ij,k} + H_{k,ij} = 0$ because of Eqs. (4.230) and (4.239). Thus, we have $H_{ij,k} = -H_{k,ij}$, in agreement with statement (8.62).

## 8.2.6 Reciprocity Relations for the Radiation Damping Matrix

The radiation damping matrix $\mathbf{D}$ is expressed in terms of matrices $\mathbf{R}$, $\mathbf{G}$ and $\mathbf{J}$ according to definition (8.68). Here we shall derive reciprocity relations which express these matrices in terms of far-field coefficients, Kochin functions or excitation parameters.

First, let us express $\mathbf{R}$, $\mathbf{G}$ and $\mathbf{J}$ in terms of the surface integral (4.230). From definition (8.65), adding Eq. (8.103) to its complex conjugate—or its conjugate transpose, because $\mathbf{Z}$ is symmetric—gives the radiation resistance matrix $\mathbf{R}$. Using also definition (5.127) gives

$$\mathbf{R} = \mathrm{Re}\{\mathbf{Z}\} = \frac{1}{2}(\mathbf{Z} + \mathbf{Z}^\dagger) = \frac{i\omega\rho}{2}\mathbf{I}(\boldsymbol{\varphi}_u^*, \boldsymbol{\varphi}_u^\mathrm{T}) = -\frac{i\omega\rho}{2}\mathbf{I}(\boldsymbol{\varphi}_u, \boldsymbol{\varphi}_u^\dagger). \qquad (8.113)$$

Because the last term (the sum) in Eq. (8.103) is cancelled out, Eq. (8.113) differs from Eq. (5.176) only by the subscript $u$.

Similarly, writing Eq. (8.108) in matrix notation and adding it to its conjugate transpose gives the radiation conductance matrix $\mathbf{G}$. Using also definition (5.127), we have

$$\mathbf{G} = \mathrm{Re}\{\mathbf{Y}\} = \frac{1}{2}(\mathbf{Y} + \mathbf{Y}^\dagger) = \frac{i\omega\rho}{2}\mathbf{I}(\boldsymbol{\varphi}_p^*, \boldsymbol{\varphi}_p^\mathrm{T}) = -\frac{i\omega\rho}{2}\mathbf{I}(\boldsymbol{\varphi}_p, \boldsymbol{\varphi}_p^\dagger). \qquad (8.114)$$

We have here assumed that $\omega$ is real, and we have observed the fact that $\mathbf{G}$ is a symmetric real matrix.

For real $\omega$, the right-hand side of Eq. (8.104) is purely imaginary. Hence, we may in the integrand of Eq. (8.111) replace $\varphi_k$ with $-\varphi_k^*$. The same replacement may be made in Eq. (8.112). Further, since the right-hand side of Eq. (8.97) is real, we may replace $\varphi_{ij}$ with $\varphi_{ij}^*$ in Eq. (8.109) and, hence, also in Eq. (8.110). From definition (8.65), we have

$$J_{ij,k} = \text{Im}\{H_{ij,k}\} = \frac{1}{2i}(H_{ij,k} - H_{ij,k}^*) = \frac{1}{2i}(H_{ij,k} + H_{k,ij}^*). \tag{8.115}$$

Taking the sum of Eq. (8.110) with $\varphi_{ij}$ replaced by $\varphi_{ij}^*$ and of the complex conjugate of Eq. (8.112) with $\varphi_k$ replaced by $-\varphi_k^*$, we obtain

$$J_{ij,k} = -\frac{\omega\rho}{2}I(\varphi_k, \varphi_{ij}^*) = \frac{\omega\rho}{2}I(\varphi_{ij}^*, \varphi_k), \tag{8.116}$$

where we have used Eqs. (4.230) and (4.234). In matrix notation,

$$\mathbf{J} = \frac{\omega\rho}{2}I(\boldsymbol{\varphi}_u^*, \boldsymbol{\varphi}_p^{\mathrm{T}}) = \frac{\omega\rho}{2}I(\boldsymbol{\varphi}_u, \boldsymbol{\varphi}_p^{\dagger}). \tag{8.117}$$

By using the general relation (4.244)—as well as its complex conjugate—for waves $\psi_{i,j}$ satisfying the radiation condition, we find that Eqs. (8.113), (8.114) and (8.117) give

$$\mathbf{R} = \frac{\omega\rho D(kh)}{2k}\int_0^{2\pi} \mathbf{a}_u(\theta)\mathbf{a}_u^{\dagger}(\theta)\,d\theta = \frac{\omega\rho D(kh)}{2k}\int_0^{2\pi} \mathbf{a}_u^*(\theta)\mathbf{a}_u^{\mathrm{T}}(\theta)\,d\theta, \tag{8.118}$$

$$\mathbf{G} = \frac{\omega\rho D(kh)}{2k}\int_0^{2\pi} \mathbf{a}_p(\theta)\mathbf{a}_p^{\dagger}(\theta)\,d\theta = \frac{\omega\rho D(kh)}{2k}\int_0^{2\pi} \mathbf{a}_p^*(\theta)\mathbf{a}_p^{\mathrm{T}}(\theta)\,d\theta, \tag{8.119}$$

$$i\mathbf{J} = \frac{\omega\rho D(kh)}{2k}\int_0^{2\pi} \mathbf{a}_u^*(\theta)\mathbf{a}_p^{\mathrm{T}}(\theta)\,d\theta = -\frac{\omega\rho D(kh)}{2k}\int_0^{2\pi} \mathbf{a}_u(\theta)\mathbf{a}_p^{\dagger}(\theta)\,d\theta, \tag{8.120}$$

respectively. Observe that we have assumed $\omega$ and $k$ to be real and that $\mathbf{R}$, $\mathbf{G}$ and $\mathbf{J}$, by definition, are real. Further, by transposing and conjugating, we get

$$i\mathbf{J}^{\mathrm{T}} = \frac{\omega\rho D(kh)}{2k}\int_0^{2\pi} \mathbf{a}_p(\theta)\mathbf{a}_u^{\dagger}(\theta)\,d\theta = -\frac{\omega\rho D(kh)}{2k}\int_0^{2\pi} \mathbf{a}_p^*(\theta)\mathbf{a}_u^{\mathrm{T}}(\theta)\,d\theta. \tag{8.121}$$

By using the preceding expressions in the definition (8.65) of the radiation damping matrix, we have, similarly,

$$\mathbf{D} = \begin{bmatrix} \mathbf{R} & -i\mathbf{J} \\ i\mathbf{J}^{\mathrm{T}} & \mathbf{G} \end{bmatrix} = \frac{\omega\rho D(kh)}{2k}\int_0^{2\pi} \mathbf{a}(\theta)\,\mathbf{a}^{\dagger}(\theta)\,d\theta, \tag{8.122}$$

where $\mathbf{a}^T(\theta) \equiv [\mathbf{a}_u^T(\theta), \mathbf{a}_p^T(\theta)]$. We have now expressed the radiation parameters in terms of the far-field coefficients $a_{ij}(\theta)$ and $a_k(\theta)$, which make up the vectors $\mathbf{a}_u(\theta)$ and $\mathbf{a}_p(\theta)$, respectively.

To express the radiation parameters in terms of the excitation parameters, we proceed as follows. The excitation force and the excitation volume flow are defined as

$$\hat{F}_{e,ij} = f_{ij}A = i\omega\rho \iint_{S_i} n_{ij}(\hat{\phi}_0 + \hat{\phi}_d)\, dS, \tag{8.123}$$

$$\hat{Q}_{e,k} = q_k A = \iint_{S_k} \frac{\partial}{\partial z}(\hat{\phi}_0 + \hat{\phi}_d)\, dS \tag{8.124}$$

[cf. Eqs. (8.91) and (8.92)]. Making use of the boundary conditions on $S_i$ and $S_k$ and extending the region of integration as we did in Section 8.2.5, we can write

$$\hat{F}_{e,ij} = i\omega\rho I(\hat{\phi}_0, \varphi_{ij}), \tag{8.125}$$

$$\hat{Q}_{e,k} = -i\omega\rho I(\hat{\phi}_0, \varphi_k), \tag{8.126}$$

which is an extension of Haskind's formula (5.222). Using relation (5.140), we may further write

$$\hat{F}_{e,ij}(\beta) = \frac{\rho g D(kh)}{k} h_{ij}(\beta \pm \pi)A, \tag{8.127}$$

$$\hat{Q}_{e,k}(\beta) = -\frac{\rho g D(kh)}{k} h_k(\beta \pm \pi)A, \tag{8.128}$$

or, in vector notation,

$$\hat{\mathbf{F}}_{FQ,e}(\beta) \equiv \begin{bmatrix} \hat{\mathbf{F}}_e(\beta) \\ -\hat{\mathbf{Q}}_e(\beta) \end{bmatrix} = \frac{\rho g D(kh)}{k} A \begin{bmatrix} \mathbf{h}_u(\beta \pm \pi) \\ \mathbf{h}_p(\beta \pm \pi) \end{bmatrix} \equiv \frac{\rho g D(kh)}{k} A\mathbf{h}(\beta \pm \pi). \tag{8.129}$$

Hence,

$$\mathbf{D} = \frac{\omega\rho D(kh)}{2k} \int_0^{2\pi} \mathbf{a}(\theta)\, \mathbf{a}^\dagger(\theta)\, d\theta = \frac{\omega\rho D(kh)}{4\pi k} \int_0^{2\pi} \mathbf{h}(\theta)\, \mathbf{h}^\dagger(\theta)\, d\theta$$

$$= \frac{k}{16\pi J} \int_{-\pi}^{\pi} \hat{\mathbf{F}}_{FQ,e}(\beta)\, \hat{\mathbf{F}}_{FQ,e}^\dagger(\beta)\, d\beta, \tag{8.130}$$

where we have used Eqs. (4.272) and (4.130). Here $J$ is the wave energy transport.

For the case with only oscillating bodies and no OWCs, Eq. (8.130) specialises to Eqs. (5.135) and (5.145), and $\mathbf{D} = \mathbf{R}$ is then a real symmetrical matrix. Also, Eqs. (7.50) and (7.51) are special cases of Eq. (8.130) when the system contains just one OWC.

All diagonal elements of $\mathbf{D}$ are real, and they are also nonnegative. In particular,

$$R_{ij,ij} = \frac{\omega \rho D(kh)}{2k} \int_0^{2\pi} |a_{ij}(\theta)|^2 \, d\theta = \frac{k}{16\pi J} \int_{-\pi}^{\pi} |\hat{F}_{e,ij}(\beta)|^2 \, d\beta \geq 0, \qquad (8.131)$$

$$G_{kk} = \frac{\omega \rho D(kh)}{2k} \int_0^{2\pi} |a_k(\theta)|^2 \, d\theta = \frac{k}{16\pi J} \int_{-\pi}^{\pi} |\hat{Q}_{e,k}(\beta)|^2 \, d\beta \geq 0. \qquad (8.132)$$

Hence, the sum of all diagonal elements—that is, the so-called trace of the radiation damping matrix $\mathbf{D}$—is necessarily real and nonnegative. The trace of the matrix equals also the sum of all eigenvalues of the matrix (cf., for example, [16, p. 86]). This is consistent with inequality (8.75), which means that the Hermitian matrix $\mathbf{D}$ is positive semidefinite. Such a matrix have only real nonnegative eigenvalues (cf., for example, [16, p. 109]).

## 8.3 Two-Dimensional WEC Body

A short descriptive discussion and explanation of wave-energy conversion is given in Section 6.1, with reference to Figure 6.1. This may serve to explain how a WEC terminator, which is a two-dimensional WEC unit, may absorb energy from a plane incident wave by radiating a wave that has certain amplitude and phase relationships relative to the incident wave. A famous example of such a two-dimensional WEC unit is the Salter Duck device, well-known from a 1974 publication [73]. It is a nonsymmetrical WEC unit operating in the pitch mode. Another well-known example is the Bristol Cylinder proposed by Evans [75]. It oscillates in two modes, surge and heave, with equally large amplitudes and with phases differing by a quarter wave period. Consequently, the submerged horizontal cylinder, including its axis, moves uniformly along a circle in the $xz$-plane.

The early theory for a two-dimensional WEC body was developed independently by Newman [34], Evans [75] and Mei [117]. Alternatively, this situation may be considered as a special case of an array of equal prism-body WEC units aligned closely together along the $y$-direction. Then theoretical results for infinite WEC-body arrays may be applicable for analysing a two-dimensional WEC-body system [45].

### 8.3.1 Maximum Power Absorbed by 2-D WEC Body

Two-dimensional parameters were considered in Sections 4.7–4.8 and particularly in Section 5.8. Denoting by a prime ($'$), we introduced per-unit width values for some quantities (for example, forces $F'$, excitation force coefficients $f'_e$, radiation-resistance matrix $\mathbf{R}'$, Kochin-function coefficients $h'$, power $P'$) but not for some other quantities (for example, velocities $u$, far-field coefficients $a^{\pm}$). In the following, we discuss the theoretical maximum power absorbed by a two-dimensional WEC-body unit. For this case (see Section 5.8), Eqs. (6.95) for maximum power and Eq. (6.90) for optimum oscillation are modified into

$$P'_{MAX} = \tfrac{1}{4}\hat{\mathbf{F}}_e'^{T}\hat{\mathbf{u}}_{OPT}^* = \tfrac{1}{2}\hat{\mathbf{u}}_{OPT}^T\mathbf{R}'\hat{\mathbf{u}}_{OPT}^*, \tag{8.133}$$

where the optimum complex velocity amplitude $\hat{\mathbf{u}}_{OPT}$ has to satisfy the algebraic equation

$$\mathbf{R}'\hat{\mathbf{u}}_{OPT} = \tfrac{1}{2}\hat{\mathbf{F}}_e'. \tag{8.134}$$

We now express the excitation force and the radiation resistance (both per unit width) in terms of the Kochin functions by using Eqs. (5.329)–(5.333). We shall assume that the given incident wave is propagating in the positive $x$-direction ($\beta = 0$). Then the maximum absorbed power per unit width is rewritten as

$$P'_{MAX} = \frac{1}{4}\hat{\mathbf{F}}_e'^{T}\hat{\mathbf{u}}_{OPT}^* = \frac{1}{4}\hat{\mathbf{F}}_e'^{\dagger}\hat{\mathbf{u}}_{OPT} = \frac{\rho g D(kh)}{4k}A^*\mathbf{h}'^{\dagger}(\pi)\hat{\mathbf{u}}_{OPT}, \tag{8.135}$$

where the optimum complex amplitudes $\hat{\mathbf{u}}_{OPT}$ are given by

$$[\mathbf{h}'(0)\mathbf{h}'^{\dagger}(0) + \mathbf{h}'(\pi)\mathbf{h}'^{\dagger}(\pi)]\hat{\mathbf{u}}_{OPT} = \frac{gk}{\omega}\mathbf{h}'(\pi)A. \tag{8.136}$$

Let us introduce the relative absorbed power

$$\epsilon = \frac{P'}{J} = \frac{4\omega P'}{\rho g^2 D(kh)|A|^2} \tag{8.137}$$

and the relative optimum oscillation amplitude

$$\zeta = \frac{\omega}{gk}\frac{\hat{\mathbf{u}}_{OPT}}{A}. \tag{8.138}$$

Note that $\epsilon$ is just the fraction of the incident wave power transport (4.130) being absorbed. Its maximum value

$$\epsilon_{MAX} = \frac{4\omega}{\rho g^2 D(kh)|A|^2}\frac{\rho g D(kh)}{4k}A^*\mathbf{h}'^{\dagger}(\pi)\frac{gk}{\omega}A\,\zeta = \mathbf{h}'^{\dagger}(\pi)\,\zeta \tag{8.139}$$

is obtained when the vector of relative amplitudes $\zeta$ satisfies the equation

$$[\mathbf{h}'(0)\mathbf{h}'^{\dagger}(0) + \mathbf{h}'(\pi)\mathbf{h}'^{\dagger}(\pi)]\,\zeta = \mathbf{h}'(\pi). \tag{8.140}$$

Note that the expression within the brackets here is a normalised version of the radiation resistance matrix, which is real and, hence, is equal to its own complex conjugate. [See the last expression of Eq. (5.329).] The optimum condition (8.140) may be written as

$$\mathbf{h}'(\pi)[\mathbf{h}'^{\dagger}(\pi)\zeta - 1] + \mathbf{h}'(0)\mathbf{h}'^{\dagger}(0)\zeta = 0. \tag{8.141}$$

A matrix of the type $\mathbf{h}\mathbf{h}^\dagger$ is of rank 1 (or of rank zero in the trivial case when $\mathbf{h}^\dagger\mathbf{h} = 0$) [16, p. 239]. If the vectors $\mathbf{h}'(0)$ and $\mathbf{h}'(\pi)$ are linearly independent, the radiation resistance matrix is of rank 2. Otherwise, its rank is at most equal to 1. In general, the radiation resistance of an oscillating two-dimensional body is a $3 \times 3$ matrix. It follows that this matrix is necessarily singular, since its rank is 2 or less.

Let us assume that the two vectors $\mathbf{h}'(\pi)$ and $\mathbf{h}'(0)$ are linearly independent. Then the condition (8.141) for optimum cannot be satisfied unless

$$\mathbf{h}'^\dagger(\pi)\zeta - 1 = 0 \tag{8.142}$$
$$\mathbf{h}'^\dagger(0)\zeta = 0. \tag{8.143}$$

Thus, we have two scalar equations which the components of the optimum amplitude vector $\zeta$ have to satisfy. Hence, the system of equations is indeterminate if $\zeta$ has more than two components—that is, for a case where the radiation resistance matrix is singular, as stated previously.

Combining Eqs. (8.139) and (8.142), we obtain the unambiguous value

$$\epsilon_{MAX} = \mathbf{h}'^\dagger(\pi)\zeta = 1 \tag{8.144}$$

for the maximum relative absorbed wave power. This means that 100% of the incident wave energy is absorbed by the oscillating body. It can be shown [118, equations (53)–(59)] that when the conditions (8.142) and (8.143) are satisfied, then radiated waves cancel waves diffracted towards the left ($x \to -\infty$) and the sum of the incident wave and waves diffracted towards the right ($x \to +\infty$).

A necessary, but not sufficient, condition for the vectors $\mathbf{h}'(\pi)$ and $\mathbf{h}'(0)$ to be linearly independent is that they are of order two or more. That is, at least two modes of oscillation have to be involved.

If the body oscillates in one mode only, the aforementioned vectors simplify to scalars. Hence, Eqs. (8.139) and (8.140) in this case simplify to

$$\epsilon_{MAX} = h'^*(\pi)\zeta \tag{8.145}$$
$$(|h'(0)|^2 + |h'(\pi)|^2)\zeta = h'(\pi). \tag{8.146}$$

This means that not more than a fraction

$$\epsilon_{MAX} = \frac{1}{1 + |h'(0)/h'(\pi)|^2} \tag{8.147}$$

of the incident wave energy can be absorbed by the body. If the oscillating body is able to radiate a wave only in the negative direction, then $h'(0) = 0$, and hence, 100% absorption is possible. If equally large waves are radiated in opposite directions, we have $|h'(0)| = |h'(\pi)|$, which means that not more than 50% of the incident wave energy can be absorbed.

It is necessary that the body has a nonsymmetric radiation ability if it is to absorb more than 50% of the incident wave energy. An example of such a nonsymmetric body is the Salter Duck, intended to oscillate in the pitch mode. Already in 1974, Salter reported [73, 81] measured absorbed power corresponding to more than 80% of the incident wave power.

Next let us consider an oscillating body which (in its time-average position) is symmetric with respect to the $x = 0$ plane. It is obvious that the wave radiated by the heave motion is symmetric, while the waves radiated by the surge motion and by the pitch motion are antisymmetric. In terms of the Kochin functions, this means that

$$h_3'(\pi) = h_3'(0) \tag{8.148}$$

$$h_1'(\pi) = -h_1'(0), \quad h_5'(\pi) = -h_5'(0). \tag{8.149}$$

It may be remarked that if the body is a horizontal circular cylinder, and if the pitch rotation is about the cylinder axis, then $|h_5'| = 0$. Further, if the cylinder is completely submerged on deep water, we have

$$h_1'(0) = -ih_3'(0). \tag{8.150}$$

This means that if the cylinder axis is oscillating with equal amplitudes in heave and surge with phases differing by $\pi/2$—that is, if the centre of the cylinder is describing a circle in the $xz$-plane—then the waves generated by the cylinder motion travel away from the cylinder along the free surface, but in one direction only. This was first shown in a theoretical work by Ogilvie [119]. On this theoretical basis, Evans proposed the so-called Bristol cylinder WEC body [75].

Returning now to the case of a general symmetric body oscillating in the three modes, heave ($j = 3$), surge ($j = 1$) and pitch ($j = 5$), the vectors $\mathbf{h}'(\pi)$ and $\mathbf{h}'(0)$ are linearly independent; cf. Eqs. (8.148)–(8.149). If only two modes are involved and if the two modes are heave and surge or heave and pitch, the vectors are linearly independent. Thus, in all these cases, 100% absorption is possible if the optimum oscillation can be ideally fulfilled.

However, if surge and pitch are the only two modes involved in the body's oscillation, the vectors $\mathbf{h}'(\pi)$ and $\mathbf{h}'(0)$ are linearly dependent, i.e., $\mathbf{h}'(\pi) = -\mathbf{h}'(0)$; cf. (8.149). In this case, the optimum condition gives

$$[\mathbf{h}'(\pi) - \mathbf{h}'(0)]^\dagger \boldsymbol{\zeta} = 1 \tag{8.151}$$

and, hence,

$$\epsilon_{\mathrm{MAX}} = \mathbf{h}'^\dagger(\pi)\boldsymbol{\zeta} = \tfrac{1}{2}, \tag{8.152}$$

which means that not more than 50% power absorption is possible in this case.

We noted earlier that the $3 \times 3$ matrix for the radiation resistance is singular. There is a good reason for this. It is possible to absorb 100% of the incident wave power by optimum oscillation in two modes, one symmetric mode (heave) and one antisymmetric mode (surge or pitch). Hence, it is not possible to absorb more wave power by including a third mode. With all three modes involved in the optimisation problem, $\mathbf{R}$ is singular, and hence, the system of equations for determining the optimum values of $U_1$, $U_3$ and $U_5$ is indeterminate. The optimum complex amplitude $U_3$ is determined. If $U_1$ is arbitrarily chosen, then $U_5$ is determined, and vice versa. In this way, the antisymmetric wave, resulting from the combined surge-and-pitch oscillation, is optimum in the far-field region. If both the symmetric wave and the antisymmetric wave are optimum, all incident wave energy is absorbed by the oscillating body. (See also Problem 8.1.)

## Problems

### Problem 8.1: Maximum Absorbed Power by Symmetric 2-D Body

Show that if a plane wave is perpendicularly incident on a two-dimensional body which has a vertical symmetry plane parallel to the wave front, then, at optimum oscillation, the body absorbs

(a) half of the incident wave power if it oscillates in the surge and pitch modes only, and
(b) all incident wave power if it oscillates in surge and heave only or if it oscillates in surge, heave and pitch.

State in each case the conditions which the optimum complex velocity amplitudes have to satisfy.

### Problem 8.2: Linear Electric 2-Port

Consider an electric '2-port' (also termed '4-pole'). The input voltage, input current, output voltage and output current have complex amplitudes $U_1$, $I_1$, $U_2$ and $I_2$, respectively. Assuming that the electric circuit is linear, we may express the voltages in terms of the currents as

$$\begin{bmatrix} U_1 \\ U_2 \end{bmatrix} = \mathbf{Z} \begin{bmatrix} I_1 \\ I_2 \end{bmatrix}, \quad \text{where} \quad \mathbf{Z} = \begin{bmatrix} Z_{11} & Z_{12} \\ Z_{21} & Z_{22} \end{bmatrix}$$

is the impedance matrix of the 2-port.

By solving linear algebraic equation, we find it possible to write this relation in some other ways, such as

$$\begin{bmatrix} I_1 \\ I_2 \end{bmatrix} = \mathbf{Y} \begin{bmatrix} U_1 \\ U_2 \end{bmatrix}, \quad \begin{bmatrix} U_2 \\ I_2 \end{bmatrix} = \mathbf{K} \begin{bmatrix} U_1 \\ I_1 \end{bmatrix}, \quad \text{or} \quad \begin{bmatrix} U_1 \\ I_2 \end{bmatrix} = \mathbf{H} \begin{bmatrix} I_1 \\ U_2 \end{bmatrix},$$

where $\mathbf{Y}$ is the 'admittance' matrix, $\mathbf{K}$ the 'chain' matrix and $\mathbf{H}$ the 'hybrid' matrix. Express the four matrix elements of $\mathbf{Y}$ and of $\mathbf{H}$ in terms of the four matrix elements of $\mathbf{Z}$.

If the impedance matrix is symmetric—that is, if $Z_{21} = Z_{12}$—then the linear electric circuit is said to be 'reciprocal'. What is the corresponding condition on the hybrid matrix?

One possible electrical analogy of a mechanical system is to consider force and velocity as analogous to voltage and current, respectively. For a mechanical system consisting of one OWC and a single one-mode oscillating body, the preceding hybrid matrix may be considered as the electrical analogue of the radiation matrix in Eq. (8.65). Discuss this analogy.

## Problem 8.3: Maximum Absorbed Power

For a system of OWCs and oscillating bodies, the absorbed power may be written as [cf. Eq. (8.73)]

$$P = P(\hat{\mathbf{u}}_{FQ}) = \tfrac{1}{4}(\hat{\mathbf{F}}_{FQ,e}^{T}\hat{\mathbf{u}}_{FQ}^{*} + \hat{\mathbf{F}}_{FQ,e}^{\dagger}\hat{\mathbf{u}}_{FQ}) - \tfrac{1}{2}\hat{\mathbf{u}}_{FQ}^{\dagger}\mathbf{D}\hat{\mathbf{u}}_{FQ}. \tag{1}$$

(a) Assuming that the radiation damping matrix $\mathbf{D}$ is non-singular (that is, $\mathbf{D}^{-1}$ exists), prove that

$$P_{\text{MAX}} = \tfrac{1}{8}\hat{\mathbf{F}}_{FQ,e}^{\dagger}\mathbf{D}^{-1}\hat{\mathbf{F}}_{FQ,e},$$

and derive an explicit expression for the optimum oscillation vector $\hat{\mathbf{u}}_{FQ,0}$ in terms of the excitation vector $\hat{\mathbf{F}}_{FQ,e}$. Observe that $\mathbf{D}$ is a complex, Hermitian and positive semidefinite matrix. [Hint: introduce the vector $\hat{\boldsymbol{\delta}} = \hat{\mathbf{u}}_{FQ} - (\mathbf{D}^{*})^{-1}\hat{\mathbf{F}}_{FQ,e}/2$, and consider the nonnegative quantity $\hat{\boldsymbol{\delta}}^{T}\mathbf{D}\hat{\boldsymbol{\delta}}/2$ which would have been the radiated power if $\hat{\boldsymbol{\delta}}$ had been the complex velocity amplitude vector.]

(b) On the basis of Eq. (1), show that the maximum absorbed power is as given by Eq. (8.76) where the optimum oscillation corresponds to $\mathbf{U}$ satisfying Eq. (8.77). [Hint: make the necessary generalisation of the derivation of Eq. (6.92) from Eq. (6.91).]

(c) If the radiation damping matrix $\mathbf{D}$ is singular, Eq. (8.77) has an infinity of possible solutions for $\hat{\mathbf{u}}_{FQ,0}$. Assume that $\mathbf{U}_1$ and $\mathbf{U}_2$ are two different possible solutions. Show that $P(\mathbf{U}_1) = P(\mathbf{U}_2)$. Thus, in spite of the indeterminateness of Eq. (8.77), the maximum absorbed power $P_{\text{MAX}}$ is unambiguous. [Hint: use Eqs. (8.16), (8.76) and (8.77).]

# Bibliography

1. Chiang C. Mei, Michael Stiassnie and Dick K.-P. Yue. *Theory and Applications of Ocean Surface Waves*. World Scientific, 2005.

2. O. M. Faltinsen. *Sea Loads on Ships and Offshore Structures*. Cambridge University Press, 1990.

3. Turgut Sarpkaya. *Wave Forces on Offshore Structures*. Cambridge University Press, 2010.

4. S. K. Chakrabarti. *Hydrodynamics of Offshore Structures*. WIT Press, 1987.

5. S. H. Salter. World progress in wave energy – 1988. *International Journal of Ambient Energy*, 10(1):3–24, 1989.

6. A. D. Carmichael and J. Falnes. State of the art in wave power recovery. In Richard J. Seymour, editor, *Ocean Energy Recovery*, chapter 8, pages 182–212. American Society of Civil Engineers, 1992.

7. John Brooke. *Wave Energy Conversion*, volume 6. Elsevier, 2003.

8. J. Cruz, editor. *Ocean Wave Energy: Current Status and Future Perspectives*. Green Energy and Technology. Springer-Verlag, 2008.

9. A. F. de O. Falcão. Wave energy utilization: A review of the technologies. *Renewable and Sustainable Energy Reviews*, 14:899–918, 2010.

10. Umesh A. Korde and John Ringwood. *Hydrodynamic Control of Wave Energy Devices*. Cambridge University Press, 2016.

11. Aurélien Babarit. *Ocean Wave Energy Conversion: Resource, Technologies and Performance*. ISTE, 2017.

12. H. W. Bode. *Network Analysis and Feedback Amplifier Design*. Van Nostrand, 1945.

13. B. Friedland. *Control System Design: An Introduction to State-Space Methods*. McGraw-Hill, 1986.

14. C. Moler and C. Van Loan. Nineteen dubious ways to compute the exponential of a matrix. *Society for Industrial and Applied Mathematics Review*, 20:801–836, 1978.

15. P. Hr. Petkov, N. D. Christov and M. M. Konstantinov. *Computational Methods for Linear Control Systems*. Prentice Hall, 1991.

16. M. C. Pease. *Methods of Matrix Algebra*. Academic Press, 1965.

17. A. Papoulis. *The Fourier Integral and Its Applications*. McGraw-Hill, 1962.

18. R. N. Bracewell. *The Fourier Transform and Its Applications*. McGraw-Hill, 1986.

19. H. A. Kramers. La diffusion de la lumière par les atomes. In *Atti del Congresso Internazionale dei Fisici*, volume II, pages 545–557, Como, settembre 1927. Nicola Zanichelli, Bologna, 1928.

20. R. de L. Kronig. On the theory of dispersion of X-rays. *Journal of the Optical Society of America*, 12(6):547–557, 1926.

21. L. E. Kinsler and A. R. Frey. *Fundamentals of Acoustics*, 3rd edition. Wiley, 1982.

22. K. H. Panofsky and M. Phillips. *Classical Electricity and Magnetism*. Addison-Wesley, 1955.

23. P. McIver and D. V. Evans. The occurrence of negative added mass in free-surface problems involving submerged oscillating bodies. *Journal of Engineering Mathematics*, 18:7–22, 1984.

24. J. N. Miles. Resonant response of harbours: An equivalent circuit analysis. *Journal of Fluid Mechanics*, 46:241–265, 1971.

25. J. Falnes. Radiation impedance matrix and optimum power absorption for interacting oscillators in surface waves. *Applied Ocean Research*, 2(2):75–80, 1980.

26. E. Meyer and E. G. Neumann. *Physikalische und technische Akustik*, pages 180–182. Vierweg, 1967.

27. E. Titchmarsh. *Eigenfunction Expansion Associated with Second-Order Differential Equations*. Oxford University Press, 1946.

28. J. N. Newman. *Marine Hydrodynamics*, 40th anniversary edition. MIT Press, 2017.

29. M. S. Longuet-Higgins. The mean forces exerted by waves on floating or submerged bodies, with application to sand bars and wave power machines. *Proceedings of the Royal Society of London A*, 352(1671):463–480, 1977.

30. Yoshimi Goda. *Random Seas and Design of Maritime Structures*, 3rd edition. World Scientific, 2010.

31. M. J. Tucker and E. G. Pitt. *Waves in Ocean Engineering*, 1st edition. Elsevier, 2001.

32. M. Abramowitz and I. A. Stegun. *Handbook of Mathematical Functions*. Dover Publications, 1965.

33. J. V. Wehausen and E. V. Laitone. Surface waves. In S. Flügge, editor, *Encyclopedia of Physics*, volume IX, pages 446–778. Springer-Verlag, 1960.

34. J. N. Newman. The interaction of stationary vessels with regular waves. In *Proceedings of the 11th Symposium on Naval Hydrodynamics*, pages 491–501, London, 1976.

35. B. King. *Time-Domain Analysis of Wave Exciting Forces on Ships and Bodies*. PhD thesis, Department of Naval Architecture and Marine Engineering, University of Michigan, 1987.

36. F. T. Korsmeyer. The time domain diffraction problem. In *The Sixth International Workshop on Water Waves and Floating Bodies*, Woods Hole, MA, 1991.

37. J. Falnes. On non-causal impulse response functions related to propagating water waves. *Applied Ocean Research*, 17(6):379–389, 1995.

38. A. Erdélyi, editor. *Tables of Integral Transforms*. McGraw-Hill, 1954.

39. S. Naito and S. Nakamura. Wave energy absorption in irregular waves by feed-forward control system. In D. V. Evans and A. F. de O. Falcão, editors, *Hydrodynamics of Ocean Wave-Energy Utilization*, pages 169–280. Springer-Verlag, 1986. IUTAM Symposium, Lisbon 1985.

40. E. L. Morris, H. K. Zienkiewich, M. M. A. Pourzanjani, J. O. Flower and M. R. Belmont. Techniques for sea state prediction. In *Second International Conference on Manoeuvring and Control of Marine Craft*, pages 547–569, Southampton, UK, 1992.

41. S. E. Sand. *Three-Dimensional Deterministic Structure of Ocean Waves*. Technical Report 24, Institute of Hydrodynamics and Hydraulic Engineering (ISVA), Technical University of Denmark (DTH), 1979.

42. M. J. L. Greenhow. The hydrodynamic interactions of spherical wave-power devices in surface waves. In B. Count, editor, *Power from Sea Waves*, pages 287–343. Academic Press, 1980.

43. Å. Kyllingstad. *Approximate Analysis Concerning Wave-Power Absorption by Hydrodynamically Interacting Buoys*. PhD thesis, Institutt for eksperimentalfysikk, NTH, Trondheim, Norway, 1982.

44. M. A. Srokosz. Some relations for bodies in a canal, with application to wave-power absorption. *Journal of Fluid Mechanics*, 99:145–162, 1980.

45. J. Falnes and K. Budal. Wave-power absorption by parallel rows of interacting oscillating bodies. *Applied Ocean Research*, 4(4):194–207, 1982.

46. T. H. Havelock. Waves due to a floating sphere making periodic heaving oscillations. *Proceedings of the Royal Society of London A*, 231(1184):1–7, 1955.

47. A. Hulme. The wave forces acting on a floating hemisphere undergoing forced periodic oscillations. *Journal of Fluid Mechanics*, 121:443–463, 1982.

48. Håvard Eidsmoen. Hydrodynamic parameters for a two-body axisymmetric system. *Applied Ocean Research*, 17(2):103–115, 1995.

49. R. Eatock Taylor and E. R. Jeffreys. Variability of hydrodynamic load predictions for a tension leg platform. *Ocean Engineering*, 13(5):449–490, 1986.

50. *WAMIT User Manual*. www.wamit.com.

51. F. T. Korsmeyer, C. H. Lee, J. N. Newman and P. D. Sclavounos. The analysis of wave effects on tension-leg platforms. In *Proceedings of the Seventh International Conference on Offshore Mechanics and Arctic Engineering*, volume 2, pages 1–14, Houston, TX, 1988.

52. W. E. Cummins. The impulse response function and ship motions. *Schiffstechnik*, 9:101–109, 1962.

53. J. Kotik and V. Mangulis. On the Kramers-Kronig relations for ship motions. *International Shipbuilding Progress*, 9(97):361–368, 1962.

54. Martin Greenhow. A note on the high-frequency limits of a floating body. *Journal of Ship Research*, 28:226–228, 1984.

55. B. M. Count and E. R. Jefferys Wave power, the primary interface. In *Proceedings of the 13th Symposium on Naval Hydrodynamics*, pages 1–10, Tokyo, 1980.

56. L. J. Tick. Differential equations with frequency-dependent coefficients. *Journal of Ship Research*, 3(3):45–46, 1959.

57. S. Goldman. *Transformation Calculus and Electric Transients*. Constable and Co., London, 1949.

58. S. P. Timoshenko and J. M. Gere. *Mechanics of Materials*. Van Nostrand, 1973.

59. M. D. Haskind. The exciting forces and wetting of ships (in Russian). *Izvestyia Akademii Nauk SSSR, OtdelenieTekhnicheskikh Nauk*, 7:65–79, 1957.

60. J. N. Newman. The exciting forces on fixed bodies in waves. *Journal of Ship Research*, 6(3):10–17, 1962.

61. D. V. Evans. Some analytic results for two and three dimensional wave-energy absorbers. In B. Count, editor, *Power from Sea Waves*, pages 213–249. Academic Press, 1980.

62. K. Budal. Theory of absorption of wave power by a system of interacting bodies. *Journal of Ship Research*, 21:248–253, 1977.

63. J. Falnes and A. Kurniawan. Fundamental formulae for wave-energy conversion. *Royal Society Open Science*, 2(3):140305, 2015.

64. Å. Kyllingstad. A low-scattering approximation for the hydrodynamic interactions of small wave-power devices. *Applied Ocean Research*, 6:132–139, 1984.

65. T. Arzel, T. Bjarte-Larsson and J. Falnes. Hydrodynamic parameters for a floating WEC force-reacting against a submerged body. In *Proceedings of the Fourth European Wave Energy Conference*, Aalborg, Denmark, December 2000.

66. K. Budal. Floating structure with heave motion reduced by force compensation. In *Proceedings of the Fourth International Offshore Mechanics and Arctic Engineering Symposium*, pages 92–101, Dallas, TX, February 1985.

67. J. Falnes. Wave-energy conversion through relative motion between two single-mode oscillating bodies. *Journal of Offshore Mechanics and Arctic Engineering*, 121:32–38, 1999.

68. M. McCormick. *Ocean Wave Energy Conversion*. Wiley, 1981.

69. J. V. Wehausen. Causality and the radiation condition. *Journal of Engineering Mathematics*, 26:153–158, 1992.

70. K. Budal and J. Falnes. A resonant point absorber of ocean waves. *Nature*, 256:478–479, 1975. With Corrigendum in Vol. 257, p. 626.

71. R. Clare, D. V. Evans and T. L. Shaw. Harnessing sea wave energy by a submerged cylinder device. *Proceedings of the Institution of Civil Engineers*, 73(3):565–585, 1982.

72. D. V. Evans, D. C. Jeffrey, S. H. Salter and J. R. M. Taylor. Submerged cylinder wave energy device: Theory and experiment. *Applied Ocean Research*, 1(1):3–12, 1979.

73. S. H. Salter. Wave power. *Nature*, 249:720–724, 1974.

74. K. Budal and J. Falnes. Optimum operation of improved wave-power converter. *Marine Science Communications*, 3(2):133–150, 1977.

75. D. V. Evans. A theory for wave-power absorption by oscillating bodies. *Journal of Fluid Mechanics*, 77:1–25, 1976.

76. B. M. Count. Wave power: A problem searching for a solution. In B. M. Count, editor, *Power from Sea Waves*, pages 11–27. Academic Press, 1980.

77. J. Falnes and J. Hals. Heaving buoys, point absorbers and arrays. *Philosophical Transactions of the Royal Society A*, 370(1959):246–277, 2012.

78. J. Falnes and K. Budal. Wave-power absorption by point absorbers. *Norwegian Maritime Research*, 6(4):2–11, 1978.

79. S. H. Salter. Power conversion systems for ducks. In *Proceedings of International Conference on Future Energy Concepts*, pages 100–108, London, January 1979.

80. P. Nebel. Maximizing the efficiency of wave-energy plants using complex-conjugate control. *Journal of Systems and Control Engineering*, 206(4):225–236, 1992.

81. S. H. Salter, D. C. Jeffery and J. R. M. Taylor. The architecture of nodding duck wave power generators. *The Naval Architect*, pages 21–24, 1, 1976.

82. J. H. Milgram. Active water-wave absorbers. *Journal of Fluid Mechanics*, 43:845–859, 1970.

83. K. Budal and J. Falnes. Interacting point absorbers with controlled motion. In B. Count, editor, *Power from Sea Waves*, pages 381–399. Academic Press, 1980.

84. K. Budal, J. Falnes, T. Hals, L. C. Iversen and T. Onshus. Model experiment with a phase controlled point absorber. In *Proceedings of Second International Symposium on Wave and Tidal Energy*, pages 191–206, Cambridge, UK, 23–25 September 1981.

85. K. Budal, J. Falnes, L. C. Iversen, P. M. Lillebekken, G. Oltedal, T. Hals, T. Onshus and A. S. Høy. The Norwegian wave-power buoy project. In H. Berge, editor, *Proceedings of the Second International Symposium on Wave Energy Utilization*, pages 323–344. Tapir, Trondheim, Norway, 1982.

86. J. N. B. A. Perdigão and A. J. N. A. Sarmento. A phase control strategy for OWC devices in irregular seas. In J. Grue, editor, *The Fourth International Workshop on Water Waves and Floating Bodies*, pages 205–209, Deptartment of Mathematics, University of Oslo, 1989.

87. J. N. B. A. Perdigão. *Reactive-Control Strategies for an Oscillating-Water-Column Device*. PhD thesis, Universidade Técnica de Lisboa, Instituto Superior Técnico, 1998.

88. A. Clément and C. Maisondieu. Comparison of time-domain control laws for a piston wave absorber. In *European Wave Energy Symposium*, pages 117–122, Edinburgh, Scotland, 1993.

89. G. Chatry, A. Clément and T. Gouraud. Self-adaptive control of a piston wave absorber. In *Proceedings of the Eighth International Offshore and Polar Engineering Conference*, volume 1, pages 127–133, Montréal, Canada, May 1998.

90. J. Falnes. Small is beautiful: How to make wave energy economic. In *European Wave Energy Symposium*, pages 367–372, Edinburgh, Scotland, 1993.

91. Rod Rainey. Private communication, 2003.

92. N. Y. Sergiienko, B. S. Cazzolato, B. Ding, P. Hardy and M. Arjomandi. Performance comparison of the floating and fully submerged quasi-point absorber wave energy converters. *Renewable Energy*, 108:425–437, 2017.

93. Jørgen Hals Todalshaug. Practical limits to the power that can be captured from ocean waves by oscillating bodies. *International Journal of Marine Energy*, 3:e70–e81, 2013.

94. Ronald Shaw. *Wave energy: A Design Challenge*. Ellis Horwood Ltd., 1982.

95. K. Budal, J. Falnes, A. Kyllingstad and G. Oltedal. Experiments with point absorbers. In *Proceedings of the First Symposium on Wave Energy Utilization*, pages 253–282. Chalmers University of Technology, Gothenburg, Sweden, 1979.

96. G. H. Keulegan and L. H. Carpenter. Forces on cylinders and plates in an oscillating fluid. *Journal of Research of the National Bureau of Standards*, 60(5):423–440, 1958.

97. A. F. de O. Falcão and A. J. N. A. Sarmento. Wave generation by a periodic surface pressure and its application in wave-energy extraction. In *15th International Congress of Theoretical and Applied Mechanics*, Toronto, 1980.

98. D. V. Evans. Wave-power absorption by systems of oscillating surface pressure distributions. *Journal of Fluid Mechanics*, 114:481–499, 1982.

99. R. G. Alcorn, W. C. Beattie and R. Douglas. Turbine modelling and analysis using data obtained from the Islay wave-power plant. In *Proceedings of the Ninth International Offshore and Polar Engineering Conference*, Brest, 1999.

100. Chia-Po Lin. *Experimental Studies of the Hydrodynamic Characteristics of a Sloped Wave Energy Device*. PhD thesis, University of Edinburgh, 1999.

101. P. Haren and C. C. Mei. Wave power extraction by a train of rafts: Hydrodynamic theory and optimum design. *Applied Ocean Research*, 1(3):147–157, 1979.

102. I. F. Noad and R. Porter. Modelling an articulated raft wave energy converter. *Renewable Energy*, 114:1146–1159, 2017.

103. M. G. de Sousa Prado, F. Gardner, M. Damen and H. Polinder. Modelling and test results of the Archimedes Wave Swing. *Proceedings of the Institution of Mechanical Engineers, Part A: Journal of Power and Energy*, 220(8):855–868, 2006.

104. M. J. French. The search for low cost wave energy and the flexible bag device. In *Proceedings of the First Symposium of Wave Energy Utilization*, pages 364–377, Gothenburg, Sweden, 1979.

105. N. W. Bellamy. Development of the SEA Clam wave energy converter. In *Proceedings of the Second International Symposium on Wave Energy Utilization*, pages 175–190, Trondheim, Norway, 1982.

106. A. Kurniawan, J. R. Chaplin, D. M. Greaves and M. Hann. Wave energy absorption by a floating air bag. *Journal of Fluid Mechanics*, 812:294–320, 2017.

107. J. N. Newman. Wave effects on deformable bodies. *Applied Ocean Research*, 16:47–59, 1994.

108. J. Falnes. *Budal's 1978 Design of a Point Absorber with Hydraulic Machinery for Control and Power Take-Off*. Technical report, Institutt for fysikk, NTH, Trondheim, 1993.

109. J. Falnes and P. McIver. Surface wave interactions with systems of oscillating bodies and pressure distributions. *Applied Ocean Research*, 7:225–234, 1985.

110. A. C. Fernandes. Reciprocity relations for the analysis of floating pneumatic bodies with application to wave power absorption. In *Proceedings of the Fourth International Offshore Mechanics and Arctic Engineering Symposium*, volume 1, pages 725–730, Dallas, Texas, 1985.

111. G. P. Thomas and D. V. Evans. Arrays of three-dimensional wave-energy absorbers. *Journal of Fluid Mechanics*, 108:67–88, 1981.

112. D. V. Evans. Maximum wave-power absorption under motion constraints. *Applied Ocean Research*, 3(4):200–203, 1981.

113. D. J. Pizer. Maximum wave-power absorption of point absorbers under motion constraints. *Applied Ocean Research*, 15:227–234, 1993.

114. Y. Masuda, H. Kimura, X. Liang, X. Gao, R. M. Mogensen and T. Anderson. Regarding BBDB wave power generating plant. In G. Elliot and K. Diamantaras, editors, *Proceedings of the Second European Wave Power Conference*, pages 69–76, Lisbon, Portugal, 1995.

115. António F. de O. Falcão and João C. C. Henriques. Oscillating-water-column wave energy converters and air turbines: A review. *Renewable Energy*, 85:1391–1424, 2016.

116. J. Falnes and P. M. Lillebekken. Budal's latching-controlled-buoy type wave-power plant. In A. Lewis and G. Thomas, editors, *Proceedings of the Fifth European Wave Energy Conference*, pages 233–244, Cork, Ireland, 2003.

117. C. C. Mei. Power extraction from water waves. *Journal of Ship Research*, 20:63–66, 1976.

118. J. Falnes. Wave-power absorption by an array of attenuators oscillating with unconstrained amplitudes. *Applied Ocean Research*, 6:16–22, 1984.

119. T. F. Ogilvie. First- and second-order forces on a cylinder submerged under a free surface. *Journal of Fluid Mechanics*, 16:451–472, 1963.

# Index

The index contains **bold**, <u>underlined</u> and *italic* page numbers. **Bold** numbers refer to places where an idea or a quantity is explained or defined. References to Problems are <u>underlined</u>, while references to Figures are typed in *italic*.